Welcome to *My Math* – your very own math book! You can write in it – in fact, you are encouraged to write, draw, circle, explain, and color as you explore the exciting world of mathematics. Let's get started. Grab a pencil and finish each sentence.

My name is ____.

My favorite color is ____________________________________.

My favorite hobby or sport is ____________________________.

My favorite TV program or video game is

____.

My favorite class is ____________________________________.

McGraw Hill Education

mhmymath.com

STEM McGraw-Hill is committed to providing instructional materials in Science, Technology, Engineering, and Mathematics (STEM) that give all students a solid foundation, one that prepares them for college and careers in the 21st century.

Send all inquiries to:
McGraw-Hill Education
8787 Orion Place
Columbus, OH 43240

ISBN: 978-0-07-905766-2 (***Volume 2***)
MHID: 0-07-905766-7

Printed in the United States of America.

7 8 9 10 11 QSX 23 22 21 20 19

Grade 5 • Volume 2

Authors:
Carter • Cuevas • Day • Malloy
Altieri • Balka • Gonsalves • Grace • Krulik • Molix-Bailey
Moseley • Mowry • Myren • Price • Reynosa • Santa Cruz
Silbey • Vielhaber

GO digital ▶ connectED.mcgraw-hill.com

▶ Log In

1 Go to **connectED.mcgraw-hill.com**.

2 Log in using your username and password.

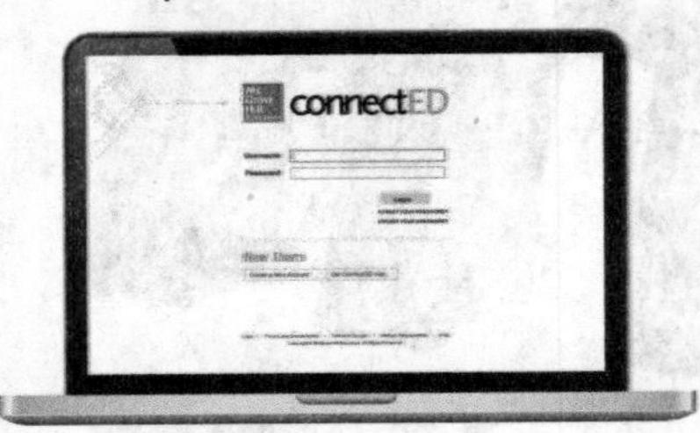

3 Click on the Student Edition icon to open the Student Center.

▶ Go to the Student Center

4 Click on Menu, then click on the **Resources** tab to see all of your online resources arranged by chapter and lesson.

5 Click on the **eToolkit** in the Lesson Resources section to open a library of eTools and virtual manipulatives.

6 Look here to find any assignments or messages from your teacher.

7 Click on the **eBook** to open your online Student Edition.

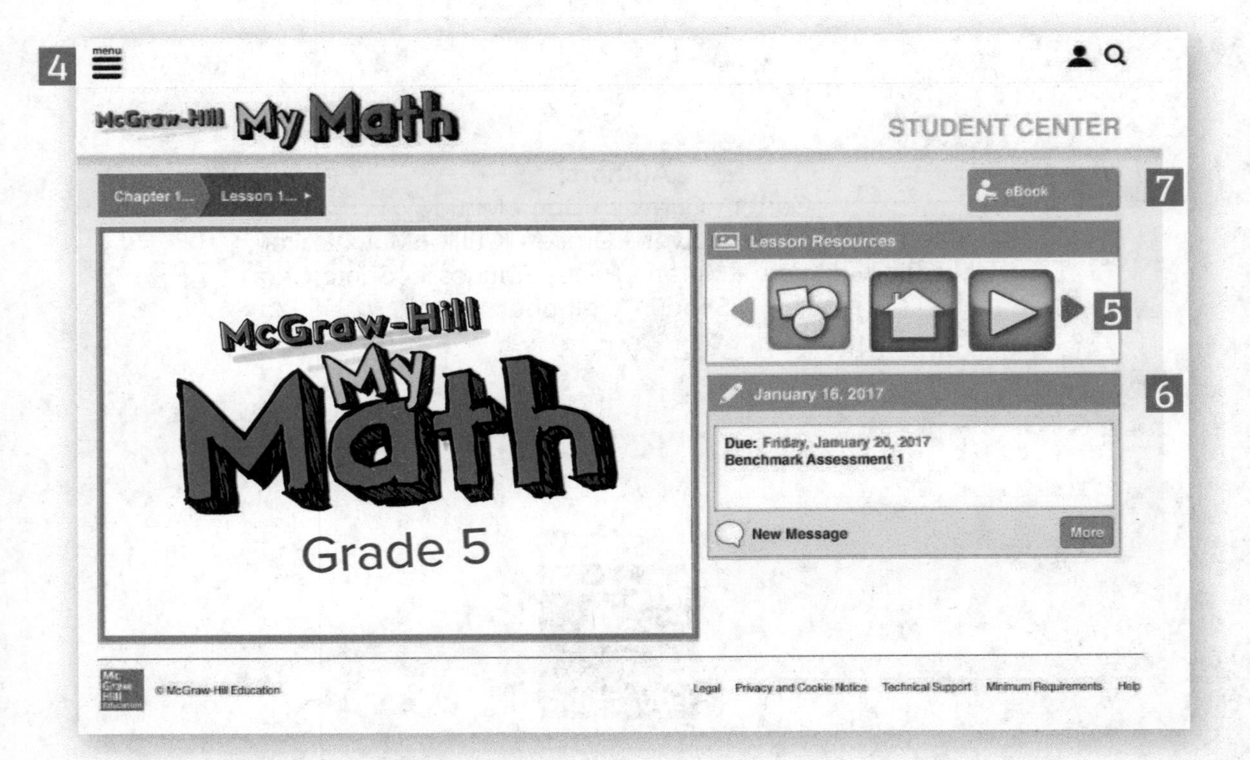

Username: ____________________

Password: ____________________

Explore the eBook!

8 Click the **speaker icon** at the top of the eBook pages to hear the page read aloud to you.

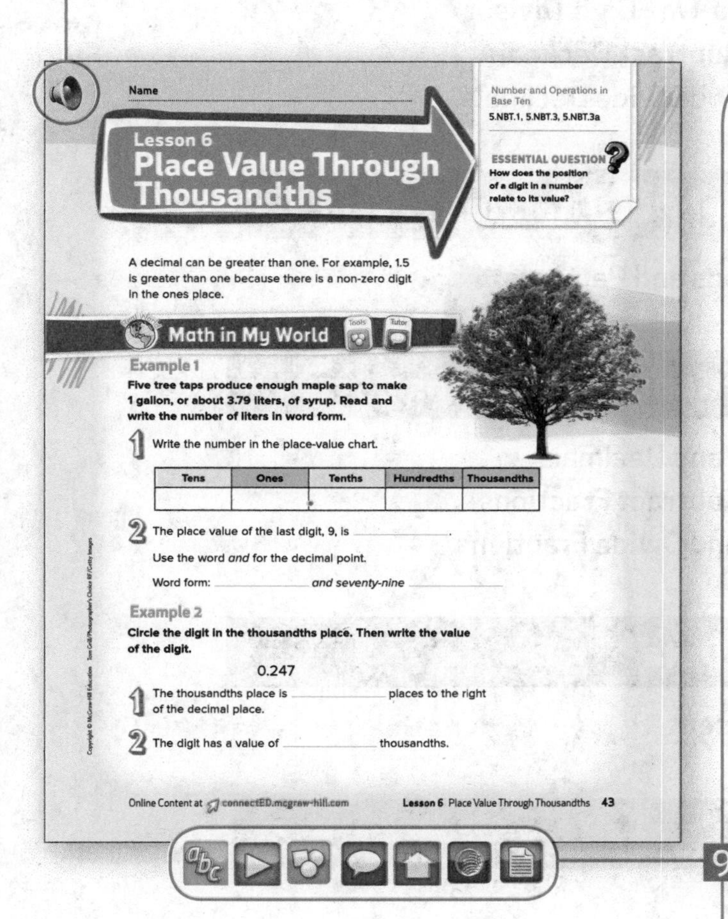
Name

Number and Operations in Base Ten
5.NBT.1, 5.NBT.3, 5.NBT.3a

Lesson 6
Place Value Through Thousandths

ESSENTIAL QUESTION
How does the position of a digit in a number relate to its value?

A decimal can be greater than one. For example, 1.5 is greater than one because there is a non-zero digit in the ones place.

Math in My World

Example 1

Five tree taps produce enough maple sap to make 1 gallon, or about 3.79 liters, of syrup. Read and write the number of liters in word form.

1 Write the number in the place-value chart.

Tens	Ones	Tenths	Hundredths	Thousandths

2 The place value of the last digit, 9, is ________.

Use the word *and* for the decimal point.

Word form: ________ *and seventy-nine* ________

Example 2

Circle the digit in the thousandths place. Then write the value of the digit.

0.247

1 The thousandths place is ________ places to the right of the decimal place.

2 The digit has a value of ________ thousandths.

Online Content at connectED.mcgraw-hill.com

Lesson 6 Place Value Through Thousandths 43

9

More resources can be found by clicking the icons at the bottom of the eBook pages.

Practice and review your Vocabulary.

Animations and videos allow you to explore mathematical topics.

Explore concepts with eTools and virtual manipulatives.

Personal Tutors are online virtual teachers that walk you through the steps of the lesson problems.

eHelp helps you complete your homework.

Explore these fun digital activities to practice what you learned in the classroom.

Worksheets are PDFs for Math at Home, Problem of the Day, and Fluency Practice.

Contents in Brief

Organized by Domain

Processes & Practices Woven Throughout

Number and Operations in Base Ten

ESSENTIAL QUESTION
How does the position of a digit in a number relate to its value?

Chapter

Place Value

Getting Started

Lessons and Homework

Wrap Up

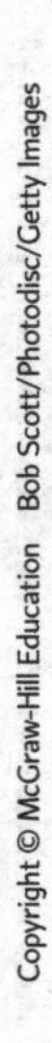

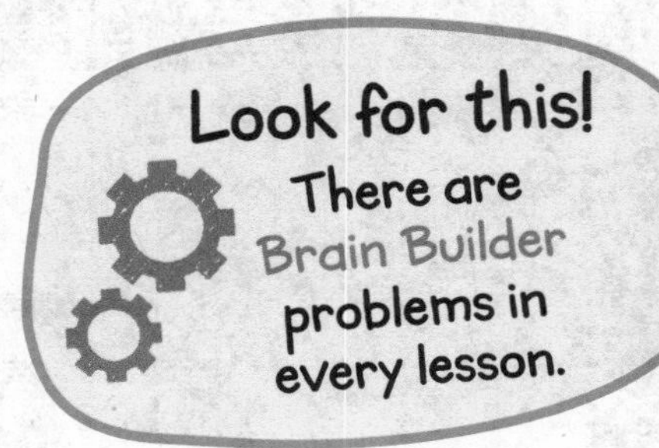

Number and Operations in Base Ten

ESSENTIAL QUESTION
What strategies can be used to multiply whole numbers?

Chapter

Multiply Whole Numbers

Getting Started

Lessons and Homework

Wrap Up

Chapter

3 Divide by a One-Digit Divisor

Number and Operations in Base Ten

ESSENTIAL QUESTION

What strategies can be used to divide whole numbers?

Getting Started

Lessons and Homework

Wrap Up

Look for this! eHelp
Click online and you can get more help while doing your homework.

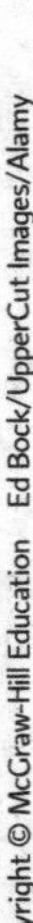

Number and Operations in Base Ten

ESSENTIAL QUESTION

What strategies can I use to divide by a two-digit divisor?

Chapter 4

Divide by a Two-Digit Divisor

Getting Started

Lessons and Homework

Wrap Up

connectED.mcgraw-hill.com

Chapter

Add and Subtract Decimals

Number and Operations in Base Ten

ESSENTIAL QUESTION

How can I use place value and properties to add and subtract decimals?

Getting Started

Lessons and Homework

Wrap Up

Number and Operations in Base Ten

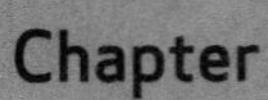

ESSENTIAL QUESTION
How is multiplying and dividing decimals similar to multiplying and dividing whole numbers?

Chapter 6 Multiply and Divide Decimals

Getting Started

Lessons and Homework

Wrap Up

Chapter

Expressions and Patterns

Operations and Algebraic Thinking

ESSENTIAL QUESTION

How are patterns used to solve problems?

Getting Started

Lessons and Homework

Wrap Up

Look for this!

Click online and you can watch a teacher solving problems.

ESSENTIAL QUESTION
How are factors and multiples helpful in solving problems?

Chapter 8 Fractions and Decimals

Getting Started

Lessons and Homework

Wrap Up

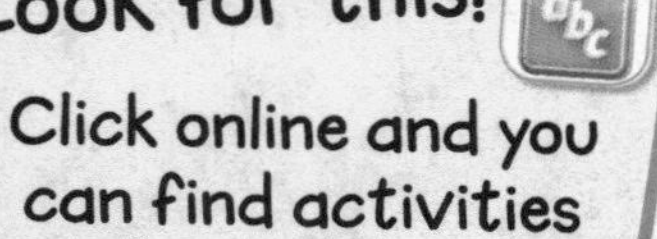

Click online and you can find activities to help build your vocabulary.

Chapter

9 Add and Subtract Fractions

Number and Operations — Fractions

ESSENTIAL QUESTION

How can equivalent fractions help me add and subtract fractions?

Getting Started

Lessons and Homework

Wrap Up

ESSENTIAL QUESTION
What strategies can be used to multiply and divide fractions?

Chapter 10 Multiply and Divide Fractions

Getting Started

Lessons and Homework

Wrap Up

Chapter

11 Measurement

Measurement and Data

ESSENTIAL QUESTION

How can I use measurement conversions to solve real-world problems?

Getting Started

Lessons and Homework

Wrap Up

ESSENTIAL QUESTION
How does geometry help me solve problems in everyday life?

Chapter 12 Geometry

Getting Started

Lessons and Homework

Wrap Up

Chapter

7 Expressions and Patterns

ESSENTIAL QUESTION

How are patterns used to solve problems?

Fun with My Friends

Watch

Watch a video!

Name

Brain Builders

MY Chapter Project

Recycling Rules

1. Today you will research the process of recycling and estimating how much your school can recycle in one week. Your teacher will put you in a group and assign your group with a recyclable: glass, metal, plastic, or paper.

2. Your teacher will give you a poster board. Divide the poster in half by drawing a line from top to bottom. Then draw lines to split the right column into three horizontal sections.

3. Use the Internet to research your recyclable and what happens to it when it is recycled. Write down this process on the left side of your poster.

4. Estimate how many pounds of your specific recyclable the average student will recycle in one week. Use that number to make a rule that shows what happens when more than one student recycles the same number of pounds, and write that rule on the top box on the right side of your poster. Your variable should be the number of students.

5. Use your rule to determine how much of your assigned recyclable your whole class can recycle, and how much your whole school can recycle. Write the answers in the other two boxes on the right side of your poster. Get permission from your principal to hang these posters in your hallway, and to place recycling bins around your school.

Name

Find each missing number.

1. 5 + ________ = 7

2. ________ − 4 = 9

3. 27 − ________ = 18

4. Devry needs to read 36 pages by Friday. He read 10 on Tuesday and 15 on Wednesday. How many pages does he need to read on Thursday in order to read 36 pages by Friday?

____

5. Quinton struck out 12 batters in his first baseball game, 5 batters in the second game, and 8 batters in the third game. How many batters did he strike out in all three games?

____

Identify each pattern.

6. 2, 5, 8, 11 . . .

____

7. 4, 9, 14, 19 . . .

____

8. 27, 21, 15, 9 . . .

____

9. Antoinette runs each day. What is the rule for the pattern shown in her running log?

____

Running Log					
Day	1	2	3	4	5
Distance (mi)	2	4	6	8	10

How Did I Do?

Shade the boxes to show the problems you answered correctly.

1	2	3	4	5	6	7	8	9

Name

MY Math Words

Vocab

Review Vocabulary

perpendicular

Making Connections

Use the review vocabulary to complete each section of the bubble map.

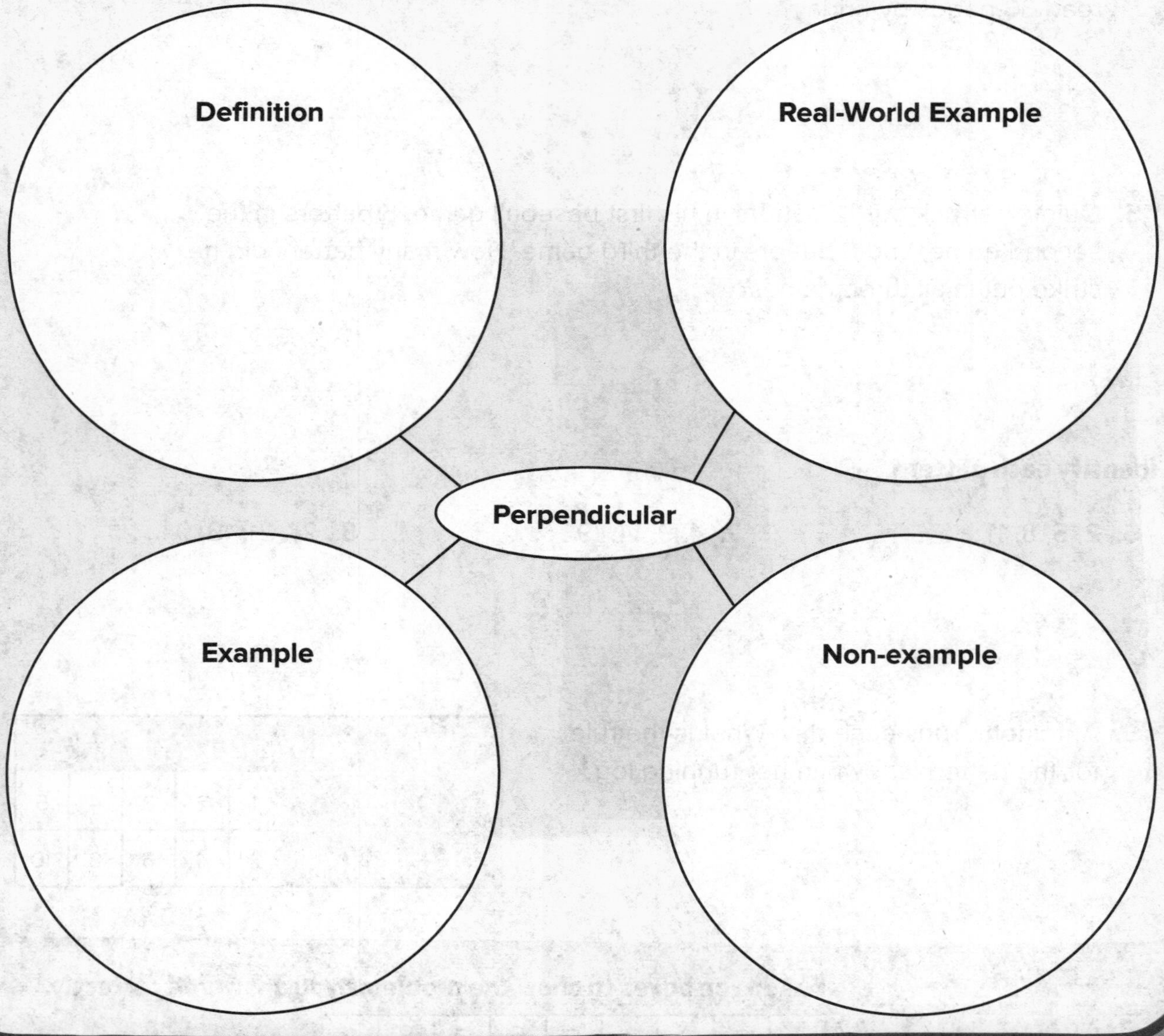

Definition

Real-World Example

Perpendicular

Example

Non-example

MY Vocabulary Cards

Processes & Practices

Lesson 7-8

coordinate plane

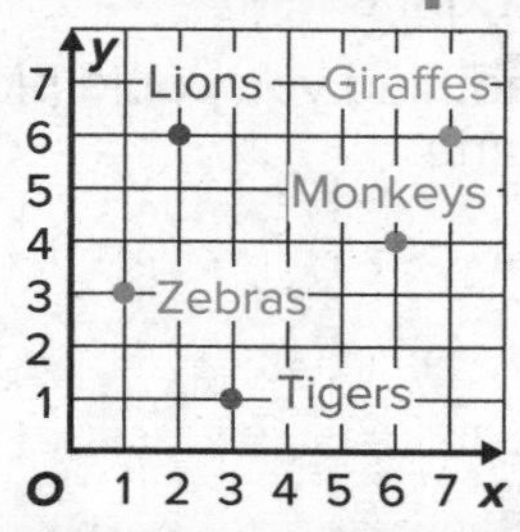

Lesson 7-1

evaluate

$5 + 7 = 12$

$15 - 10 = 5$

Lesson 7-1

numerical expression

$25 \div 5 - 3$

$15 + 7 - 8$

Lesson 7-8

ordered pair

(4, 3)

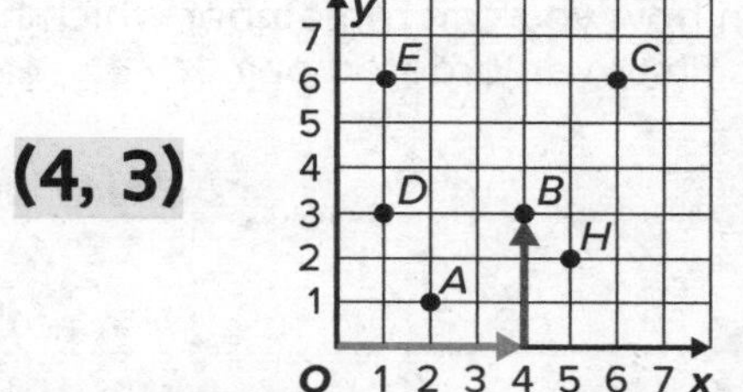

Lesson 7-2

order of operations

$[20 + 2 \times (7 - 5)] - 8$

$[20 + 2 \times 2] - 8$

$[20 + 4] - 8$

$24 - 8 = 16$

Lesson 7-8

origin

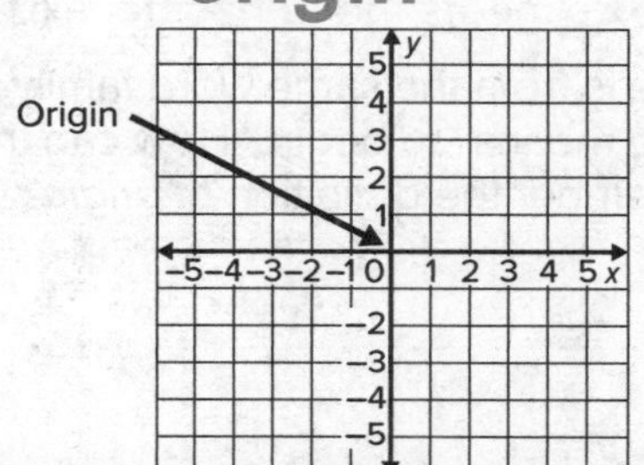

Lesson 7-6

sequence

2, 4, 6, 8, . . .

Lesson 7-6

term

2, 4, 6, 8, . . .

Ideas for Use

- Group 2 or 3 common words. Add a word that is unrelated to the group. Then work with a friend to name the unrelated word.
- Design a crossword puzzle. Use the definition for each word as the clues.

To find the value of a numerical expression by completing each operation.

When evaluating $5 + 3 + 6$, does the order in which you add matter? Explain.

A plane that is formed when two number lines intersect at a right angle.

Describe something you could plot on a coordinate plane.

A pair of numbers that is used to name a point on a coordinate plane.

Explain how you can remember which number comes first in an ordered pair.

A combination of numbers and at least one operation, such as $9 - 4$.

What is the difference between a numerical expression and an equation?

The point on a coordinate plane where the vertical axis meets the horizontal axis.

Originate is from the same word family as *origin*. *Originate* means "to begin." How can this help you remember the definition of *origin*?

The order in which operations on numbers should be done: parentheses, exponents, multiply and divide, add and subtract.

Write a sentence you could use to help you remember the correct order of operations.

A number in a pattern or sequence.

Term is a multiple-meaning word. Write a sentence using another meaning of *term*.

A list of numbers that follows a specific pattern.

Write a real-world example that is an example of a *sequence*.

MY Vocabulary Cards

Processes & Practices

Lesson 7-8

x-coordinate

(4, 3)

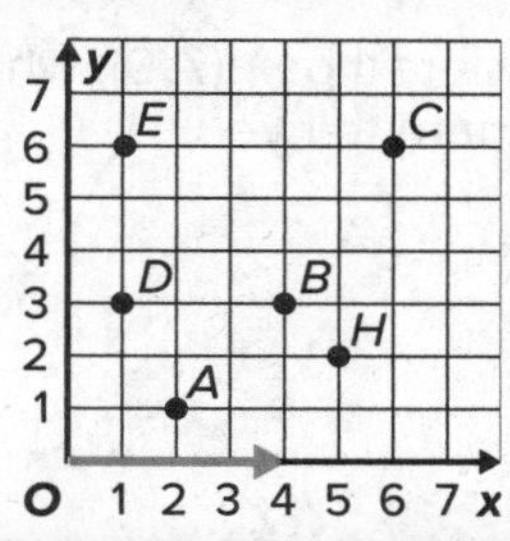

Lesson 7-8

y-coordinate

(4, 3)

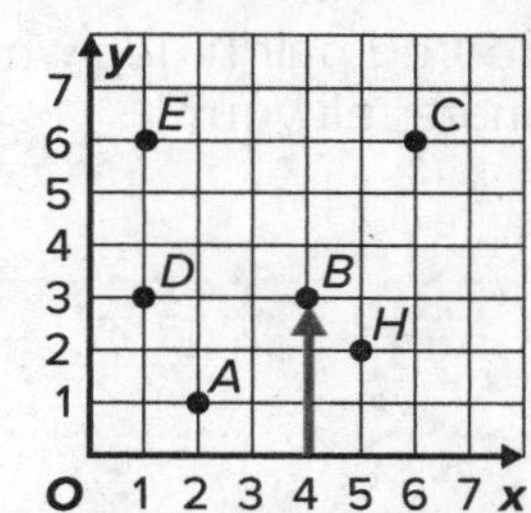

Ideas for Use

- Write the names of lessons on the front of blank cards. Write a few study tips on the back of each card.
- Use the blank cards to write your own vocabulary cards.

The second part of an ordered pair that shows how far away from the *x*-axis the point is.

In the ordered pair (1, 12), what does the *y*-coordinate tell you?

The first part of an ordered pair that shows how far away from the *y*-axis the point is.

In the ordered pair (7, 9), what does the *x*-coordinate tell you?

MY Foldable

FOLDABLES® Follow the steps on the back to make your Foldable.

Parentheses

( )

Exponents

2^2

2nd

Multiply and Divide

× ÷

3rd

Add and Subtract

+ −

4th

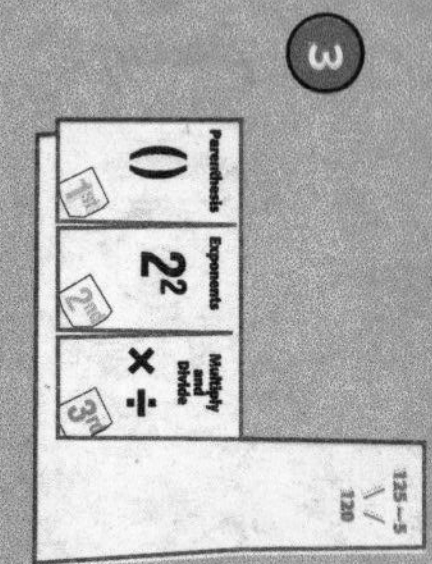

$(15 \div 3) \times 5^2 - 5$

$5 \times 5^2 - 5$

$5 \times 5^2 - 5$

$5 \times 25 - 5$

$5 \times 25 - 5$

$125 - 5$

$125 - 5$

120

Name

Lesson 1
Hands On
Numerical Expressions

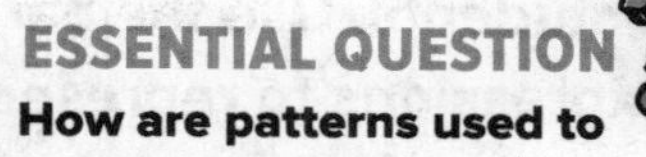

ESSENTIAL QUESTION
How are patterns used to solve problems?

A **numerical expression**, such as 8 + 7, is a combination of numbers and at least one operation. You can find the value, or **evaluate**, the numerical expression by completing each operation.

Draw It

Gregory and his family went hiking over the weekend. On Saturday, they hiked 5 miles and on Sunday, they hiked 5 miles. Use the bar diagram to write and evaluate two numerical expressions to represent the total number of miles hiked.

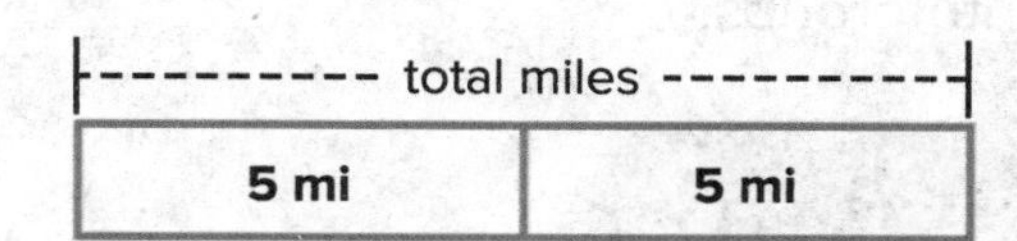

1 Use the bar diagram to write an addition expression.

5 + ______

Evaluate the expression.

5 + ______ = ______

2 Use the bar diagram to write a multiplication expression.

______ × 5

Evaluate the expression.

______ × 5 = ______

So, they hiked a total of ______ miles.

Try It

Mrs. Yearling has two groups of 5 students and two groups of 4 students. Use the bar diagram to write and evaluate two numerical expressions to represent the total number of students.

total students			
5 students	**5 students**	**4 students**	**4 students**

Use the bar diagram to write an expression using only addition.

5 + ______ + 4 + ______

Evaluate the expression.

5 + ______ + 4 + ______ = ______

Use the bar diagram to write an expression using multiplication and addition.

(______ × 5) + (______ × 4)

Evaluate the expression.

(______ × 5) + (______ × 4) = ______

Helpful Hint

Parentheses tell you which numbers to group together. Perform operations inside parentheses first.

So, there are ______ students that are divided into groups.

Talk About It

1. Evaluate the addition expressions to find the sum. Does the order in which the expression is written change the sum? Explain.

addition expression	addition expression
7 + 7 + 5 + 5	7 + 5 + 7 + 5

____

____

2. **Processes & Practices 4** **Model Math** Suppose Mrs. Yearling also had another group of 4 students. Write two new numerical expressions to represent the total number of students.

Expression 1	**Expression 2**
5 + ______ + 4 + 4 + ______	(2 × ______) + (______ × 4)

Name

Practice It

3. Caleb's music class is divided into 5 groups of 4 students for a project. Use the bar diagram to write and evaluate two numerical expressions to represent the total number of students in his music class.

total students

4 students	4 students	4 students	4 students	4 students

Write and evaluate an expression using only addition.

____

Write and evaluate an expression using multiplication.

____

So, there are ______ students in his music class.

4. Processes & Practices 7 **Identify Structure** Bailey's soccer team had snacks after the game that included 12 granola bars, 12 mini muffins, and 14 bananas. Use the bar diagram to write and evaluate two numerical expressions to represent the total number of snacks after the soccer game.

total snacks

12 granola bars	12 mini muffins	14 bananas

Write and evaluate an expression using only addition.

____

Write and evaluate an expression using multiplication and addition.

____

So, there are ______ total snacks after the game.

Apply It

5. Wasah went bird watching and spotted 6 robins, 5 sparrows, 6 cardinals, and 6 doves. Use the bar diagram to write and evaluate two numerical expressions to represent the total number of birds Wasah spotted.

total number of birds

6 robins	5 sparrows	6 cardinals	6 doves

Write and evaluate an expression using only addition.

____

Write and evaluate an expression using multiplication and addition.

____

So, Wasah spotted a total of ______ birds.

6. **Processes & Practices 3** **Which One Doesn't Belong?** The bar diagram below can be represented by three of the four expressions underneath it. Find the value of each expression and circle the one that does not represent the bar diagram.

5	4	5	4	5	4

$5 + 4 + 5 + 4 + 5 + 4$	$(3 \times 5) + (3 \times 4)$	$3 + 5 + 3 + 4$	$3 \times (5 + 4)$
______	______	______	______

Write About It

7. How can bar diagrams be used to model numerical expressions?

____

____

Name ....

Lesson 1

Hands On: Numerical Expressions

Homework Helper

Need help? connectED.mcgraw-hill.com

Paul has a fruit basket that includes 3 oranges, 4 apples, 3 bananas, and 2 grapefruits. Use the bar diagram to write and evaluate two numerical expressions to represent the total number of pieces of fruit in the basket.

total number of fruit

3 oranges	4 apples	3 bananas	2 grapefruits

1 Use the bar diagram to write an expression using only addition.

$3 + 4 + 3 + 2$

Evaluate the expression.

$3 + 4 + 3 + 2 = 12$

2 Use the bar diagram to write an expression using multiplication and addition.

$(2 \times 3) + 4 + 2$

Evaluate the expression.

$(2 \times 3) + 4 + 2 = 12$

So, there are 12 total pieces of fruit in the basket.

Practice

1. Refer to the Homework Helper. Suppose Paul went to the store and bought 4 peaches to add to the basket. Write two new numerical expressions to represent the total number of fruit in the basket.

Expression 1

$3 +$ ______ $+ 3 + 2 +$ ______

Expression 2

$(2 \times$ ______ $) + ($ ______ $\times 4) + 2$

Real World Problem Solving

2. Deborah's playlist on her MP3 player had a variety of songs on it. She had 21 country songs, 18 alternative songs, 16 pop songs, and 18 rock songs. Use the bar diagram to write and evaluate two numerical expressions to represent the total number of songs on Deborah's MP3 player.

total songs			
21 country songs	**18 alternative songs**	**16 pop songs**	**18 rock songs**

Write and evaluate an expression using only addition.

____

Write and evaluate an expression using multiplication and addition.

____

So, there are ______ total songs on Deborah's MP3 player.

3. Processes & Practices 7 **Identify Structure** Mrs. Conrad's art class was painting murals. She had 4 bottles of red paint, 5 bottles of green paint, 4 bottles of yellow paint, 6 bottles of blue paint, and 5 bottles of orange paint. Use the bar diagram to write and evaluate two numerical expressions to represent the total number of bottles of paint.

total number of bottles				
4 red	**5 green**	**4 yellow**	**6 blue**	**5 orange**

Write and evaluate an expression using only addition.

____

Write and evaluate an expression using multiplication and addition.

____

So, Mrs. Conrad has a total of ______ bottles of paint.

Name ....

Lesson 2
Order of Operations

ESSENTIAL QUESTION
How are patterns used to solve problems?

Math in My World

Example 1

The table shows the number of Calories burned in one minute for two different activities. Nathan swims for 4 minutes and then runs for 8 minutes. How many Calories has Nathan burned in all?

Activity	Calories Burned per Minute
Swimming	12
Running	10

Evaluate the expression $12 \times 4 + 10 \times 8$.

Write the expression. $12 \times 4 + 10 \times 8$

Multiply 12 by 4. ______ $+ 10 \times 8$

Multiply 10 by 8. ______ $+$ ______

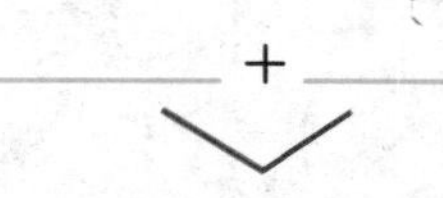

Add. ______

So, Nathan has burned ______ Calories.

The **order of operations** is a set of rules to follow when more than one operation is used in an expression.

Key Concept Order of Operations

1. Perform operations in parentheses.
2. Find the value of exponents.
3. Multiply and divide in order from left to right.
4. Add and subtract in order from left to right.

Parentheses include brackets [ ] as well as braces { }. Perform operations inside parentheses first, then perform operations inside brackets, and finally, perform operations inside braces.

Example 2

Evaluate 20 − {4 + [4 + (10 ÷ 2)]}.

Write the expression. ______ − {4 + [______ + (10 ÷ ______)]}

Divide 10 by 2. 20 − {4 + [4 + ______]}

Add. 20 − {4 + ______}

Add. 20 − ______

Subtract. ______

parentheses 1st

brackets 2nd

braces 3rd

So, 20 − {4 + [4 + (10 ÷ 2)]} = ______.

Guided Practice

1. Evaluate {28 + [(2 × 4^2) ÷ 8]}.

Write the expression. {______ + [(2 × 4^2) ÷ ______]}

Find 4^2. {28 + [(2 × ______) ÷ 8]}

Multiply. {28 + [______ ÷ 8]}

Divide. {28 + ______}

Add. ______

parentheses 1st

brackets 2nd

braces 3rd

So, {28 + [(2 × 4^2) ÷ 8]} = ______.

Explain why it is important to follow the order of operations when evaluating 15 + 3 × 4.

Name

Independent Practice

Evaluate each expression.

2. $5 \times (92 - 18) =$ ______

3. $12 + (4^2) - 11 =$ ______

4. $(15 - 5) \times [(9 \times 3) + 3] =$ ______

5. $58 - 6 \times 7 =$ ______

6. $55 - [(5^2 \times 3) - 5^2] =$ ______

7. $7 \times 10 + 3 \times 30 =$ ______

8. $2^2 + \{[1 \times (5 - 2)] \times 3\} =$ ______

9. $\{2 \times [4 - (6 \div 2)]\} \times 3 =$ ______

Algebra Find each unknown.

10. $3^3 + 3 \times 5 = k$

$k =$ ______

11. $12 - [(3^2 \times 4) - 30] = b$

$b =$ ______

Problem Solving

12. Three students are on the same team for a relay race. They finish the race in 54.3 seconds. The runners' times are shown in the table. Evaluate 54.3 − (18.8 + 17.7) to find the time of the third runner. Record your answer in the table.

Relay Times	
Runner	**Time (seconds)**
1	18.8
2	17.7
3	

13. You can find the temperature in degrees Celsius by using the expression 5 × (°F − 32) ÷ 9. If the temperature of a cup of hot chocolate is 104°F, what is the temperature of the cup of hot chocolate in degrees Celsius?

Brain Builders

14. Ryan and Maggie evenly split the cost of a $12 pizza. They also have a coupon for $2 off. Write and evaluate a numerical expression to find the cost each person will pay.

15. Processes &Practices 1 **Plan Your Solution** Show two different ways you can use parenthesis to rewrite the expression 4 + 2 × 3 − 1 so that the value is different. Explain why each expression has a different value.

16. **Building on the Essential Question** When and why does order matter?

Name ....

Lesson 2

Order of Operations

Homework Helper

Need help? connectED.mcgraw-hill.com

Evaluate $\{5^3 \div [1 \times (10 - 5)]\} - 20$.

Write the expression.	$\{5^3 \div [1 \times (10 - 5)]\} - 20$	parentheses 1st
Subtract 5 from 10.	$\{5^3 \div [1 \times 5]\} - 20$	brackets 2nd
Multiply.	$\{5^3 \div 5\} - 20$	
Find 5^3.	$\{125 \div 5\} - 20$	braces 3rd
Divide.	$25 - 20$	
Subtract.	5	

So, $\{5^3 \div [1 \times (10 - 5)]\} - 20 = 5$.

Practice

1. Evaluate $64 \div [4 \times (27 - 5^2)]$.

Write the expression.	____ $\div [4 \times$ ____ $(- 5^2)]$	
Find 5^2.	$64 \div [4 \times (27 -$ ____ $)]$	parentheses 1st
Subtract.	$64 \div [4 \times$ ____ $]$	brackets 2nd
Multiply.	$64 \div$ ____	
Divide.	____	

So, $64 \div [4 \times (27 - 5^2)] =$ ____.

Problem Solving

2. Processes & Practices 4 **Model Math** Kishauna rode her bike for 35 minutes each on Monday, Wednesday, and Saturday and 55 minutes each on Tuesday and Thursday. Write an expression that shows the total amount of time she spent riding her bike. Then evaluate the expression.

____

Brain Builders

3. Place parentheses and brackets in the expression below so that it equals 18.
$2^3 \times 4 \div 2 + 2$

____

4. Kylie and her three friends equally divided the cost to rent a movie for $4 and order sandwiches for $15. They also have a coupon for $3 off the sandwiches. Write and evaluate a numerical expression to find the cost each person will pay.

____

Vocabulary Check

5. Fill in each blank with the correct word to complete the sentence. The rules of the order of operations tell you to multiply and divide in order from ________ to ________.

6. **Test Practice** Keiko's class collected money to donate to charity. When Keiko counted the money, there were 140 five-dollar bills, and 255 one-dollar bills. What expression could he use to find out how much money was collected?

Ⓐ (140 × $5) + (255 × $1)

Ⓑ (140 × $1) + (255 × $5)

Ⓒ (140 + $5 × 255 + $1)

Ⓓ 140 + $5 + 255 + $1

Name ....

Lesson 3
Write Numerical Expressions

ESSENTIAL QUESTION
How are patterns used to solve problems?

Math in My World

Example 1

Terrell went to dinner with his friends and ordered 3 tacos. Each taco costs $2 and he has a coupon for a dollar off his purchase. The total cost in dollars of Terrell's purchase is represented by the phrase multiply three by two, then subtract one. Write the total cost as a numerical expression.

Write the phrase in parts.

Part 1 multiply three by ________

Part 2 then subtract ________

Write each part as a numerical expression.

Part 1 multiply three by two ——→ ____________

Part 2 then subtract one ——→ ____________

3 Combine the numerical expressions to represent the total cost in dollars. Add parentheses if needed.

____

Example 2

A ticket to a baseball game costs $25 and popcorn is $8. Three friends bought tickets and popcorn. The expressions below give the cost for one friend and for three friends. Compare the two expressions without evaluating them.

One Friend	Three Friends
25 + 8	(25 + 8) × 3

Both expressions contain the same addition expression. Write the addition expression. ________

For three friends, the addition expression is multiplied by ______.

So, the second expression is ______ times as large as the first expression.

Guided Practice

1. Write the phrase *add 7 and 11, then divide by 2* as a numerical expression.

   Write the phrase in parts.

   Part 1 ____________________

   Part 2 ____________________

   Write each part as a numerical expression.

   Part 1 add 7 and 11 ⟶ ________

   Part 2 then divide by 2 ⟶ ________

   Combine the numerical expressions. Add parentheses if needed.

   ____

Talk MATH

Write a real-world problem that could be represented by a numerical expression.

Name ....

Independent Practice

Write each phrase as a numerical expression.

2. divide 15 by 3, then add 13 ______________

3. subtract 4 from 20, then divide by 2 ______________

4. add 9 and 4, then multiply by 2 ______________

Processes & Practices 1 **Make Sense of Problems** Compare each pair of numerical expressions without evaluating them.

5. **Expression 1** $(7 \times 4) \div 2$ **Expression 2** 7×4

Both expressions contain the same multiplication expression.

Write the multiplication expression. __________

In Expression 1, the product is divided by ______.

So, Expression 1 is ______ as large as Expression 2.

6. **Expression 1** $2 + 5 + 8$ **Expression 2** $4 \times (2 + 5 + 8)$

Both expressions contain the same addition expression.

Write the addition expression. ______________

In Expression 2, the addition expression is multiplied by ______.

So, Expression 2 is ______ times as large as Expression 1.

Problem Solving

7. Robin wants to find the area of the triangle shown. To find the area of a triangle, multiply the base times the height and then divide by 2. The base and height of the triangle are shown. Represent the area of the triangle with a numerical expression.

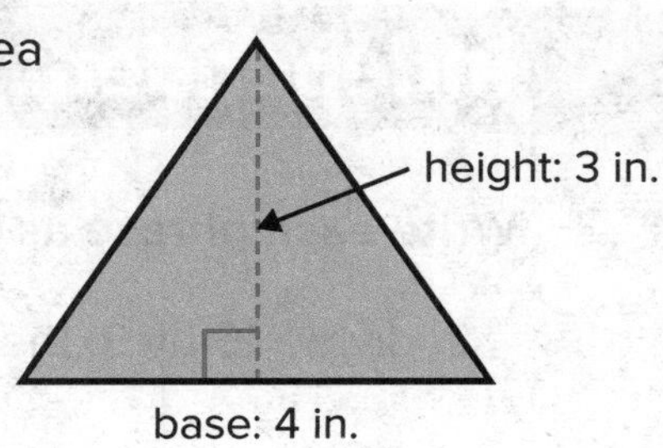

____

8. Deirdre doubled her savings account balance of \$100. Then she withdrew \$30 to buy some new clothes. Represent this situation with a numerical expression.

____

Brain Builders

9. **Processes & Practices 2** **Use Number Sense** Use a real-world model to illustrate why the numerical expression 3 less than 16 is written as $16 - 3$ and not $3 - 16$.

____

____

____

10. **Processes & Practices 7** **Identify Structure** Circle the numerical expression that is four times as large as $52 - 9$.

$52 - (9 \times 4)$	$(52 - 9) + 4$	$(52 - 9) \times 4$	$(52 - 9) \div 4$

Write the expression that is four less than 52 minus 9.

____

11. **Building on the Essential Question** Give 2 numerical expressions that can be compared without calculating the values. Explain.

____

____

____

Name ____________________

Lesson 3

Write Numerical Expressions

Homework Helper

Need help? connectED.mcgraw-hill.com

Admission to a county fair is $10 for adults and $6 for children. The total cost in dollars of admission for 1 adult ticket and 4 children's tickets is represented by the phrase four multiplied by six, then add ten. Write the total cost of admission as a numerical expression.

Write the phrase in parts.

Part 1 four multiplied by six

Part 2 then add ten

Write each part as a numerical expression.

Part 1 four multiplied by six ⟶ 4×6

Part 2 then add ten ⟶ $+ 10$

Combine the numerical expressions to represent the total cost in dollars. Add parentheses if needed.

$4 \times 6 + 10$

Practice

1. Compare the two numerical expressions without evaluating them.

Expression 1	**Expression 2**
$8 - 3$	$(8 - 3) \times 4$

Both expressions contain the same subtraction expression.

Write the subtraction expression. ________

In Expression 2, the difference is multiplied by ________.

So, Expression 2 is ________ times as large as Expression 1.

Problem Solving

2. Jeffrey purchased and downloaded 12 songs on Monday. He purchased an additional 3 songs on Tuesday. The cost to download each song is \$2. Write a numerical expression to represent this situation.

____

Brain Builders

3. Mora bought 3 bags of apples for her class. One full bag has 8 apples, and each apple weighs 6 ounces. Write two different numerical expressions you could use to represent this situation.

____

4. Processes & Practices 2 **Use Number Sense** Jane wants to find the area of the trapezoid. To find the area of a trapezoid, add the two bases, multiply by the height, then divide by 2. The bases and height of the trapezoid are shown. Represent the area of the trapezoid with a numerical expression.

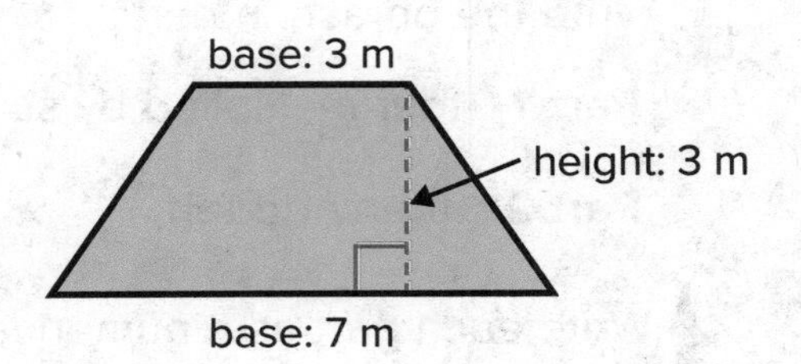

____

Draw and label a trapezoid that has the same area but different measurements. Represent the area of this trapezoid with a numerical expression.

____

____

Vocabulary Check

5. Fill in the blank with the correct term or number to complete the sentence.

A ______________ expression like $(3 + 5) \times (4 - 1)$ is a combination of numbers and at least one operation.

6. **Test Practice** Denzel and three friends go to the movies. Each person buys a movie ticket for \$8, a snack for \$4, and a drink for \$2. Which numerical expression represents the total cost of the trip to the movies for Denzel and his friends?

Ⓐ 4 + (\$8 × \$4 × \$2)

Ⓑ 4 × (\$8 + \$4 + \$2)

Ⓒ (4 × \$8) + (\$4 × \$2)

Ⓓ (4 × \$8 + \$4) + (4 × \$4 + \$2)

Name ______________________

Lesson 4
Problem-Solving Investigation
STRATEGY: Work Backward

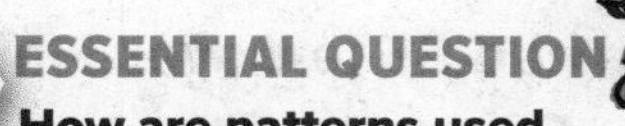

ESSENTIAL QUESTION
How are patterns used to solve problems?

Learn the Strategy

The Nature Club raised $125 to buy and install bird houses at a wildlife site. Each house costs $5. It costs $75 to rent a bus so the members can travel to the site. How many boxes can the club buy?

1 Understand

What facts do you know?

________ is available to buy and install the nesting boxes.

Each box costs ________ and the bus rental costs ________.

What do you need to find?

How many ________ can the club buy?

2 Plan

I can work backward to solve the problem.

3 Solve

Subtract the cost of the bus. Then divide by the cost for each box.

$125 − $75 = ________ ________ ÷ $5 = ________

So, ________ boxes can be bought.

4 Check

Is my answer reasonable? Explain.

Multiply. ________ × $5 = ________ Add. ________ + $75 = $125

Practice the Strategy

Mr. Evans bought the items listed. He had $5 left over. How much did Mr. Evans have to start with?

Items Purchased	
Toothpaste	$4
Toothbrush	$2
Floss	$1

1 Understand

What facts do you know?

What do you need to find?

2 Plan

3 Solve

4 Check

Is my answer reasonable? Explain.

Name

Apply the Strategy

Solve each problem by working backward.

Movie Costs	
Popcorn	$4
Drink	$3
Ticket	$8

1. Seth bought a movie ticket, popcorn, and a drink. After the movie, he played 4 video games that each cost the same amount. He spent a total of $19. How much did it cost to play each video game?

2. Students sold raffle tickets to raise money for a field trip. The first 20 tickets sold cost $4 each. To sell more tickets, they lowered the price to $2 each. If they raise $216, how many tickets did they sell in all?

Brain Builders

3. Jeanette's sister charges $5.50 per hour before 9:00 P.M. for babysitting and $8 per hour after 9:00 P.M. She finished babysitting at 11:00 P.M. and earned $38. At what time did she begin babysitting? Explain how you found your answer.

4. Processes & Practices 2 **Use Algebra** Work backward to find the value of the variable in the equation below. Show your work.

$$d + 7 \times 2 = 20$$

5. Allie collected 15 more cans of food than Peyton. Ling collected 8 more than Allie. Ling collected 72 cans of food. How many cans of food did Allie, Peyton, and Ling collect in all?

Review the Strategies

Use any strategy to solve each problem.

- Work backward.
- Make a table.
- Solve a simpler problem.
- Determine extra or missing information.

6. Rebecca sold 11 more magazine subscriptions than Chad. Laura sold 4 more than Rebecca. Laura sold 45 magazine subscriptions. How many magazine subscriptions did Chad sell?

____

7. **Processes & Practices** 8 **Look for a Pattern** Frankie is planning to buy a new MP3 player for $90. Each month he doubles the amount he saved the previous month. If he saves $3 the first month, in how many months will Frankie have enough money to buy the MP3 player?

____

8. The table shows the number of miles Michael ran each day over the past four days. How many more miles did he run on day 3 than on day 2? Determine if there is extra or missing information.

Day	Miles
1	5
2	2
3	7
4	3

____

____

9. Mrs. Stevens is delivering flowers to a local flower shop. She delivers the same number of flowers with each delivery. The flower shop has ordered 2,050 flowers and it will take 5 trips to deliver all the flowers. How many flowers will Mrs. Stevens have delivered after 4 trips?

____

10. Admission to a car show costs $5 for each ticket. After selling only 20 tickets, they decided to lower the price to $3 each. If they raise $217, how many tickets did they sell in all?

____

Name

Lesson 4

Problem Solving: Work Backward

Homework Helper

Need help? connectED.mcgraw-hill.com

Peyton and his friends built an outdoor game board in the shape of a rectangle that has a length of 4 feet and a width of 2 feet. If they cut a circular hole that has an area of 1 square foot, what is the area of the game board that does not include the hole?

1 Understand

What facts do you know?

the length and width of the game board

the area the hole takes up

What do you need to find?

the area of the game board, not including the hole

2 Plan

I can work backward to solve the problem.

3 Solve

Find the area of the game board. (Hint: area = length × width)

$4 \times 2 = 8$ square feet

Subtract the area of the hole from the area of the game board.

$8 - 1 = 7$ square feet

So, the game board that Peyton and his friends built has an area of 7 square feet, not including the hole.

4 Check

Is my answer reasonable? Explain.

area of game board + area of hole = total area

7 square feet + 1 square foot = 8 square feet

Problem Solving

Solve each problem by working backward.

1. The science club raised money to clean the beach. They spent $29 on trash bags and $74 on waterproof boots. They still have $47 left. How much did they raise?

____

2. Mr. Charles cut fresh roses from his garden and gave 10 roses to his neighbor. Then he gave half of what was left to his niece. He kept the remaining 14 roses. How many roses did he cut?

____

3. Processes & Practices 2 **Use Number Sense** A number is divided by 6. Then 8 is added to the quotient. Next 3 is subtracted from the sum. The result is 7. What is the number?

____

Brain Builders

4. Purchasing a movie combo pack of one popcorn and two drinks for $14 saves $4 compared to purchasing the items individually. If a drink is $5, what is the individual cost of a popcorn?

____

5. Ms. Houston's fifth-grade class of 20 students is going to a museum. The class raises $128 for the trip. Transportation to the museum costs $5 for each student. The museum sells small fossils for $4 each. How many fossils can the students buy with the money they have left?

____

Check My Progress

Vocabulary Check

State whether each sentence is *true* or *false*. If *false*, replace the underlined word or number to make a true sentence.

1. A combination of numbers and operations is called a **formula**.

 ____

2. The **numerical expression** of $(2 \times 4) + (3 \times 3)$ has a value of 33.

 ____

3. The **order of operations** is a set of rules to follow when more than one operation is used in an expression.

 ____

Concept Check

4. Find the value of $2 \times \{15 - [(12 \div 3) \times 2]\}$.

Write the expression.	____ × {15 − [(____ ÷ 3) × ____]}	
Divide 12 by 3.	2 × {15 − [____ × 2]}	parentheses 1st
Multiply.	2 × {15 − ____}	brackets 2nd
Subtract.	2 × ____	braces 3rd
Multiply.	____	

So, $2 \times \{15 - [(12 \div 3) \times 2]\}$ = ____.

Write each phrase as a numerical expression.

5. multiply 4 and 7, then subtract 5 ______

6. add 3 to the product of 10 and 4 ______

7. subtract 8 from the quotient of 15 and 3 ______

8. subtract 9 from 13, then multiply the result by 2 ______

Brain Builders

9. Tia and her five friends are going to the ice skating rink. Each person pays $5 for admission and $6 for food. Tia has a coupon for $10 off the total cost. Write and evaluate a numerical expression to find the total cost for admission and food.

______

10. Cameron has 2 video game holder stands. Each stand has 2 rows of 20 games and 2 rows of 24 DVDs. Write and evaluate a numerical expression to find the total number of games and DVDs Cameron's stands can hold.

______

11. **Test Practice** Arturo buys 3 containers of ice cream for $5 each and a cake that costs $8 to take to his friend's party. Which expression will allow you to find how much money Arturo spent on ice cream and cake?

Ⓐ $8 × 3 × $5

Ⓑ (3 × $5) + $8

Ⓒ (3 × $8) + $5

Ⓓ 3 × ($5 + $8)

Name

Lesson 5
Hands On
Generate Patterns

ESSENTIAL QUESTION
How are patterns used to solve problems?

Build It

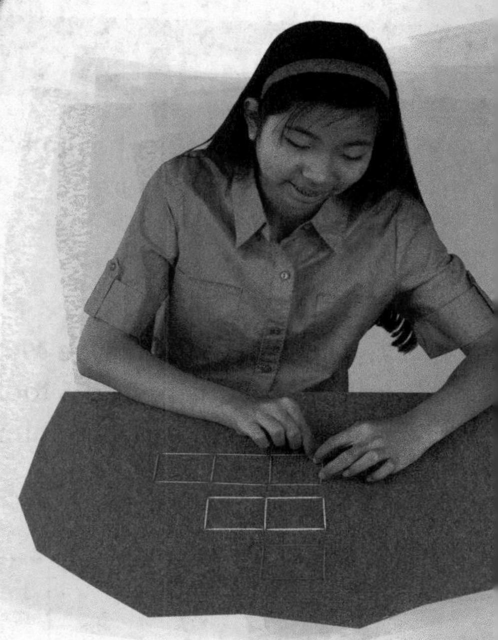

The pattern below is made from toothpicks. The first figure uses 4 toothpicks, the second figure uses 7 toothpicks, and the third figure uses 10 toothpicks. Assume the pattern continues.

Figure 1 Figure 2 Figure 3

1. Use toothpicks to model the fourth figure.

   How many toothpicks did you use? ________

2. Use toothpicks to model the fifth figure. Draw the result below.

   How many toothpicks did you use? ________

Complete the table to show the number of toothpicks needed if the pattern continues.

Figure Number	1	2	3	4	5	6	7
Number of Toothpicks	4	7	10				

What do you notice about the number of toothpicks needed for each new figure? ______

Talk About It

1. Using your rule, how many toothpicks would be needed for the eighth figure? ninth figure?

Figure 8 ______ Figure 9 ______

2. Processes & Practices 5 **Use Math Tools** Create a new pattern. Start with Figure 1 again. Add 6 toothpicks for each new figure as shown. Complete the table to show the number of toothpicks used for each figure.

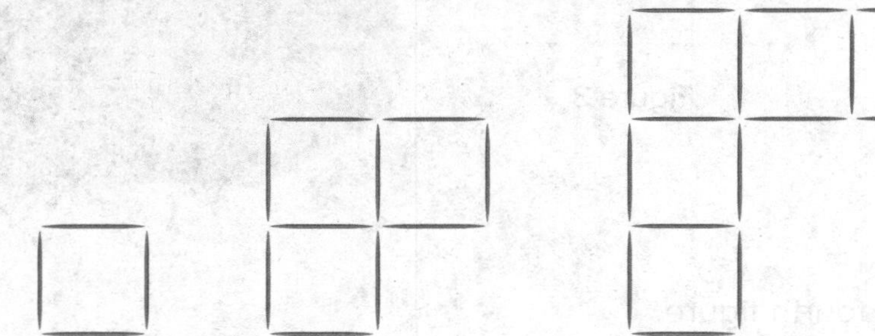

Figure 1 Figure 2 Figure 3

Figure Number	1	2	3	4	5	6	7
Number of Toothpicks	4	10	16				

3. Compare the pattern in Exercise 2 to the pattern in the activity on the previous page.

The number of toothpicks in Figure 2 for the pattern on the previous page is ______ less than the number of toothpicks in Figure 2 for the pattern in Exercise 2.

The number of toothpicks in Figure 3 for the pattern on the previous page is ______ less than the number of toothpicks in Figure 3 for the pattern in Exercise 2.

Name

Practice It

Processes &Practices 8 **Look for a Pattern** **For each pattern, find the number of toothpicks needed for the next figure.**

4.  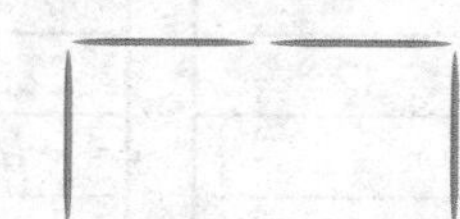 

Figure 1 Figure 2 Figure 3

Figure 4 needs ________ toothpicks.

How is this pattern different than the pattern in the activity on the first page of this lesson?

____

____

5.   

Figure 1 Figure 2 Figure 3

Figure 4 needs ________ toothpicks.

6.   

Figure 1 Figure 2 Figure 3

Figure 4 needs ________ toothpicks.

How does this pattern compare to the pattern for Exercise 5?

____

____

Apply It

7. The tables show the number of laps Tammi and Kelly swim each day. Complete each table if the patterns continue.

Compare the number of laps swam by each person on each day.

Tammi's Swimming Log	
Day	**Number of Laps**
1	0
2	6
3	12
4	
5	
6	

Kelly's Swimming Log	
Day	**Number of Laps**
1	0
2	3
3	6
4	
5	
6	

8. Two stores sell scented candles. Assume the pattern in the table below continues. Compare the price of candles sold by each store.

Store 1 Candles	
Number of Candles	**Cost ($)**
2	8
3	12
4	16

Store 2 Candles	
Number of Candles	**Cost ($)**
2	4
3	6
4	8

9. **Processes & Practices 4** **Model Math** Use Figure 1 of Exercise 5 to create a different pattern using toothpicks. Draw the pattern. How does your pattern compare to the pattern in Exercise 5?

Write About It

10. How can models be used to generate and analyze patterns?

Name

Lesson 5
Hands On: Generate Patterns

Homework Helper

Need help? connectED.mcgraw-hill.com

The pattern below is made from toothpicks. Figure 1 uses 4 toothpicks, Figure 2 uses 8 toothpicks, and Figure 3 uses 12 toothpicks. How many toothpicks will be needed for Figures 4, 5, 6, 7 and 8?

Figure 1

Figure 2

Figure 3

1. Use toothpicks to model Figure 4.

   Sixteen toothpicks were used.

Figure 4

2. Use toothpicks to model Figure 5.

   Twenty toothpicks were used.

Figure 5

3. Complete the table. The number of toothpicks increases by 4.

Figure Number	1	2	3	4	5	6	7	8
Number of Toothpicks	4	8	12	16	20	24	28	32

So, Figure 4 uses 16 toothpicks, Figure 5 uses 20 toothpicks, Figure 6 uses 24 toothpicks, Figure 7 uses 28 toothpicks, and Figure 8 uses 32 toothpicks.

Practice

Processes &Practices 8 **Look for a Pattern For each pattern, draw toothpicks to find the number of toothpicks needed for the next figure.**

1.

Figure 1 Figure 2 Figure 3

Figure 4 uses ________ toothpicks.

2.

Figure 1 Figure 2 Figure 3

Figure 4 uses ________ toothpicks.
How does this pattern compare to the pattern for Exercise 1?

____

____

Problem Solving

3. The tables show the height in centimeters each plant grew during a week. Assume the patterns continue. Compare the growth in height of each plant.

____

____

____

Plant A	
Day	Height (cm)
1	0
2	2
3	4
4	6
5	8
6	10
7	12

Plant B	
Day	Height (cm)
1	0
2	6
3	12
4	18
5	24
6	30
7	36

Name ....

Lesson 6
Patterns

ESSENTIAL QUESTION
How are patterns used to solve problems?

A **sequence** is a list of numbers that follow a specific pattern. Each number in the list is called a **term**.

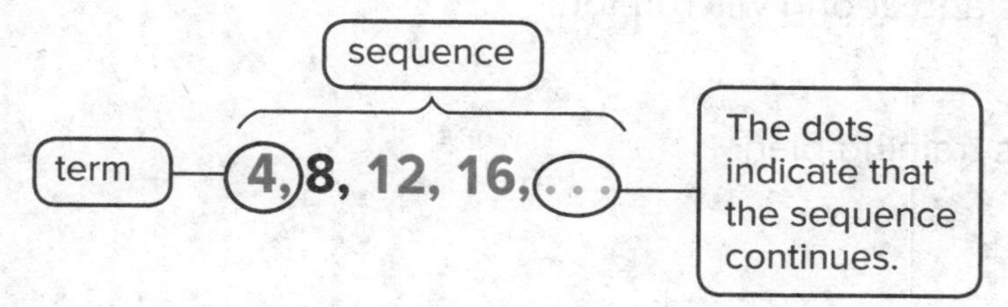

Math in My World

Example 1

Mary and her friends find a four-leaf clover during lunch. A four-leaf clover has four leaves. The table shows the total number of leaves for several four-leaf clovers. Extend the pattern to find the next three terms.

Number of Four-Leaf Clovers	1	2	3	4
Number of Leaves	4	8	12	16

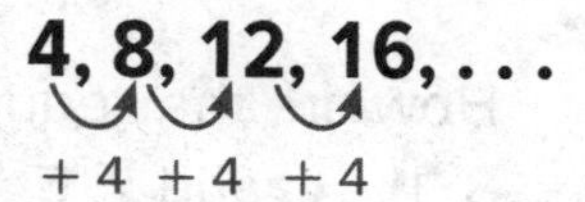

Each term in the sequence can be found by adding ________ to the previous term.

16 + 4 = ________   ________ + 4 = ________   ________ + 4 = ________

The next three terms are ________________.

Example 2 

Maria and Jeong are training to run a half-marathon. A half-marathon is about 13 miles. Their weekly training plans are shown in the table. Use the information to write a sequence to represent each person's weekly training plan. Then compare the plans.

Runner	Starting Miles	Training Plan
Maria	2	Add 2 miles per week for each of the next 4 weeks.
Jeong	4	Add 4 miles per week for each of the next 4 weeks.

Each week, Maria will run two more miles and Jeong will run four more miles than the previous week.

Write a sequence with 5 terms for Maria's training plan.

____

Write a sequence with 5 terms for Jeong's training plan.

____

Compare the training plans.

Each week, Jeong plans to run ____________ as many miles as Maria plans to run.

Guided Practice

1. Write the next three terms in the sequence 1, 4, 7, 10, ...

Each term in the sequence can be found

by adding ________ to the previous term.

10 + 3 = ________

________ + 3 = ________

________ + 3 = ________

The next three terms are ______________.

How are the sequences 2, 5, 8, 11, . . . and 2, 6, 18, 54, . . . alike? How are they different?

Name

Independent Practice

Algebra **Identify the pattern. Then write the next three terms in each sequence.**

2. 0, 7, 14, 21, . . .

3. 1,458, 486, 162, 54, . . .

4. 72, 66, 60, 54, . . .

5. 1, 3, 9, 27, . . .

6. 2, 4, 8, 16, . . .

7. 94, 88, 82, 76, . . .

8. 12, 24, 36, 48, . . .

9. 512, 256, 128, 64, . . .

10. 8, 13, 18, 23, . . .

11. 11, 24, 37, 50, . . .

12. 83, 75, 67, 59, . . .

13. 2, 8, 32, 128, . . .

Problem Solving

14. An amusement park offers discounted tickets after 4 P.M. Both ticket prices are shown to the right. Write the total cost of 1, 2, 3, and 4 tickets for each time period. Compare the cost of 4 tickets before 4 P.M. to 4 tickets after 4 P.M.

Admission Tickets	
Time	Cost ($)
Before 4 P.M.	45
After 4 P.M.	15

Brain Builders

15. Processes & Practices 3 **Which One Doesn't Belong?** Circle the sequence that does not belong with the other three. Explain your reasoning. Describe a real-world situation that could be modeled by the sequence that does not belong.

2, 5, 8, 11, ...

3, 6, 12, 24, ...

4, 14, 24, 34, ...

7, 12, 17, 22, ...

16. **Building on the Essential Question** How can we extend patterns?

Name ....

Lesson 6
Patterns

Homework Helper

Need help? connectED.mcgraw-hill.com

Tom is allowed to download 3 new songs each week. The table shows the total number of songs he can download for several weeks. Extend the pattern to find the next three terms.

Week	1	2	3	4
Number of Songs	3	6	9	12

3, 6, 9, 12, . . .

+3 +3 +3

Each term in the sequence can be found by adding 3 to the previous term.

$12 + 3 = 15$ $15 + 3 = 18$ $18 + 3 = 21$

The next three terms are 15, 18, and 21.

Practice

Algebra Identify the pattern. Then write the next three terms in each sequence.

1. 5, 10, 20, 40, ...

____

2. 63, 58, 53, 48, . . .

____

3. 192, 96, 48, 24, . . .

____

4. 4, 11, 18, 25, . . .

____

Problem Solving

5. Processes & Practices 1 **Make a Plan** Dino's Diner charges $4 for each sandwich. Carla's Café charges $8 for each sandwich. Write the total costs of 1, 2, 3, and 4 sandwiches for each restaurant. Then compare the total cost of 4 sandwiches at each restaurant.

____

____

____

____

Brain Builders

6. Processes & Practices 7 **Look for Patterns** Identify the pattern. Then find the missing terms.

81, ______ 9, ______, 1

____

Vocabulary Check

7. Fill in each blank with the correct word to complete each sentence.

A sequence is a list of numbers that follow a specific ______.

Each number in the list is called a ______.

8. **Test Practice** Which represents the next three terms in the sequence 8, 16, 24, 32, . . . ?

Ⓐ 36, 40, 44 Ⓒ 40, 48, 56

Ⓑ 64, 128, 256 Ⓓ 72, 216, 648

Name

Lesson 7 Hands On

Map Locations

ESSENTIAL QUESTION

How are patterns used to solve problems?

Draw It

You can use grid paper to represent locations on a map. From school, Marcia walks three blocks north to the library. Then she walks two blocks east to the park. Her home is located one block south of the park. Draw a map that shows these locations.

1. Use the blank grid to draw and label a dot in the lower left corner to represent the school.

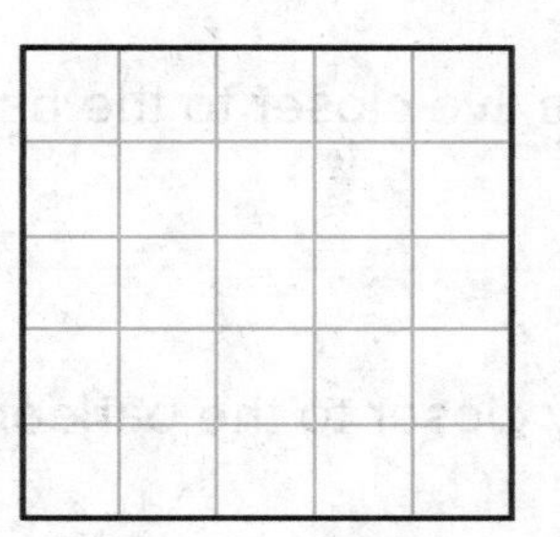

2. From the dot labeled "school", in what direction along the grid should you move to get to the library?

____

How many units should you move to get to the library?

____

3. Draw and label a dot to represent the library's location. From the dot labeled "library", in what direction along the grid should you move to get to the park?

____

How many units should you move to get to the park?

____

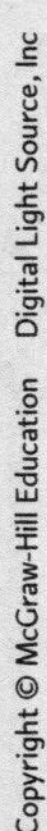

4 Draw and label a dot to represent the park's location. From the dot labeled "park", in what direction along the grid should you move to get to Marcia's home?

____

How many units should you move to get to Marcia's home?

____

Draw and label a dot to represent the location of Marcia's home.

Talk About It

1. On the grid provided, draw and label the locations from the Draw It Activity.

2. Does Marcia live closer to the park or to the library?

____

3. Is the library closer to the park or to Marcia's school?

____

4. Processes & Practices 1 **Make a Plan** Describe how Marcia could walk from her home to her school.

____

____

____

5. Write a problem that could represent locations of real-world objects. Use the grid to draw the map.

____

____

____

Name ....

Practice It

For Exercises 6–9, use the grid paper to draw a map of the given locations.

6. From the zoo entrance, Marco walks three units east to the gift shop. Then he walks four units north to the bear exhibit. The lion exhibit is located two units south and one unit west of the bear exhibit.

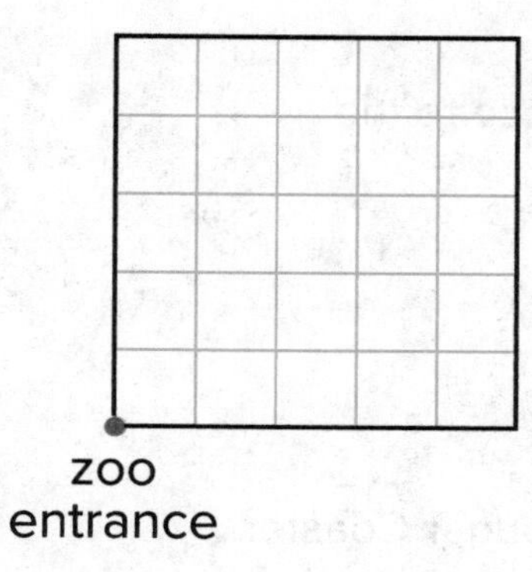

7. From the dining hall, a camper rides her bike four units north to the nature center. Then she rides her bike five units east and one unit south to her cabin. The campfire is one unit west and three units south of her cabin.

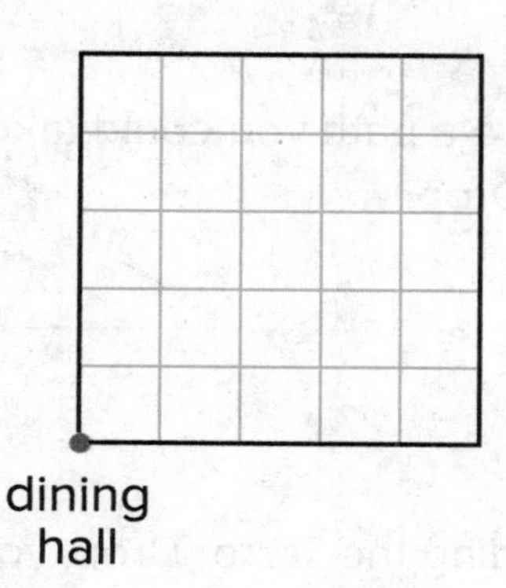

8. From the gymnasium entrance, Norah ran to the jump ropes that are located four units north and four units east. Then she ran two units west to the tumble mats. The skills challenge is one unit west and two units south of the tumble mats.

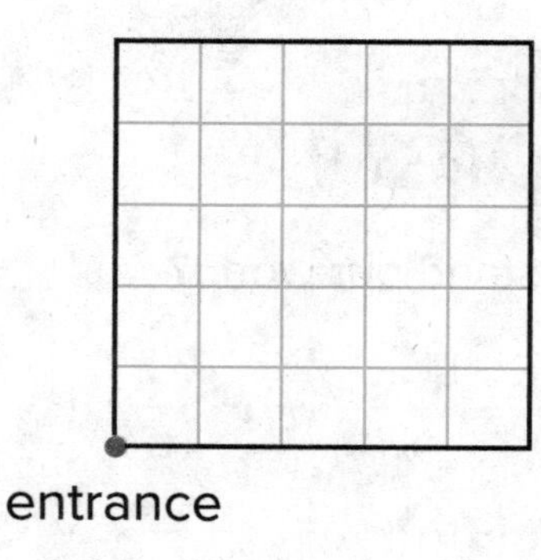

9. From the store entrance, Kyle walks to the toy section that is located four aisles north and two rows east. Then he walks two aisles south and two rows east to the boys' clothes. The cashier is three rows west and one aisle south of the boys' clothes.

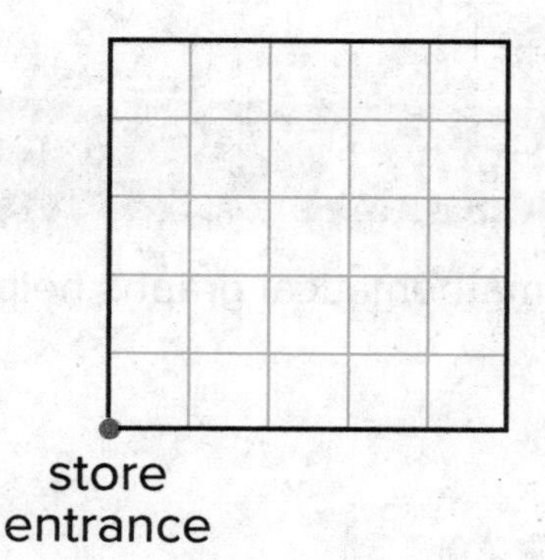

Apply It

Use the map of the amusement park below for Exercises 10–12. The walkways of the amusement park are represented by the vertical and horizontal lines on the map.

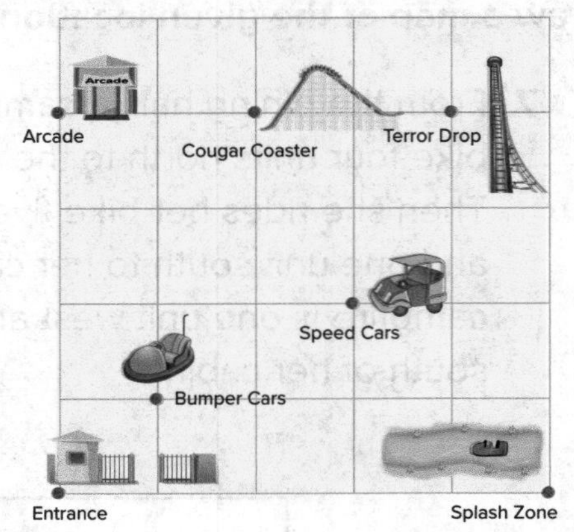

10. Describe a path you could take to get from the entrance to the Terror Drop.

____

____

11. After riding the Terror Drop, you decide to ride the Cougar Coaster. How many units do you need to move to get to the Cougar Coaster?

____

12. Processes & Practices 6 **Explain to a Friend** Lorenzo and Emily enter the park at the same time and head to two different attractions. Lorenzo walks to Speed Cars and Emily walks to the Cougar Coaster. Who walks farther? Explain to a friend.

____

____

Write About It

13. How do mathematical graphs help us better understand our world?

____

____

____

Name

MY Homework

Lesson 7

Hands On: Map Locations

Homework Helper

Need help? connectED.mcgraw-hill.com

You can use grid paper to represent locations on a map. The map shows locations of animals at an aquarium. The walkways are shown as the vertical and horizontal lines on the map. Describe how Dashiell can walk from the aquarium entrance to the giant octopus, penguins, and sharks, in that order.

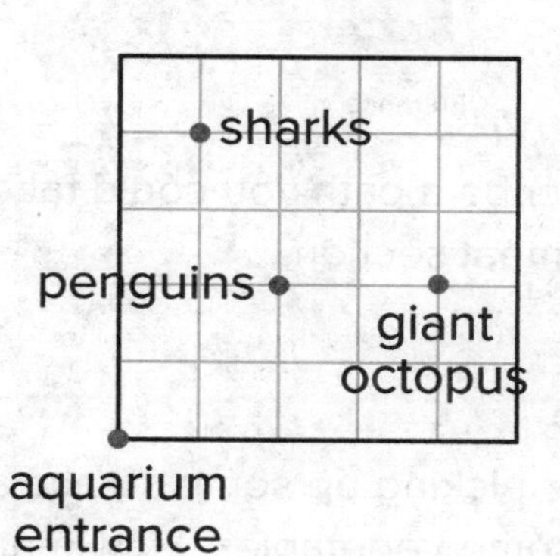

1. The aquarium entrance is located in the lower left corner of the map.

2. From the dot labeled "aquarium entrance", Dashiell walked 4 units to the right and then 2 units up to get to the giant octopus.

3. From the dot labeled "giant octopus", Dashiell walked 2 units to the left to get to the penguins.

4. From the dot labeled "penguins", Dashiell walked 2 units up and then 1 unit to the left to get to the sharks.

Practice

1. Refer to the Homework Helper. The squid exhibit is located three units east of the sharks. How many units north of the giant octopus is the squid exhibit?

____

Problem Solving

Use the map of Megamart below for Exercises 2–5. The aisles and rows of Megamart are represented by the vertical and horizontal lines on the map.

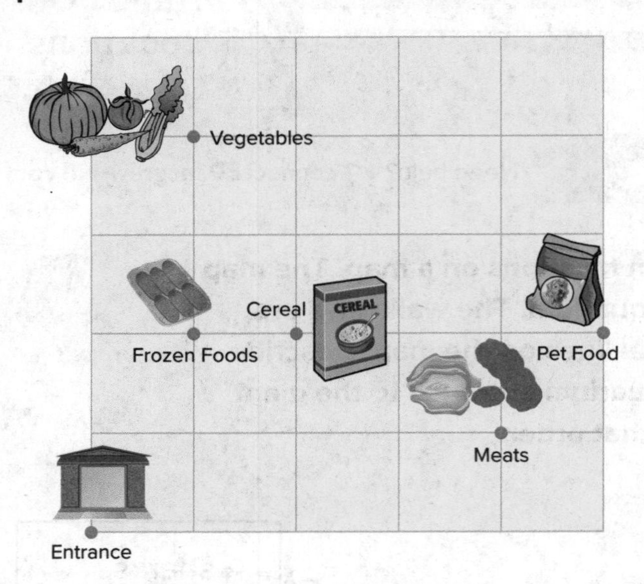

2. Describe a path you could take to get from the entrance to the meat section.

3. After picking up some steaks and hamburger, you decide to pick up some vegetables. How many total units do you need to move to get to the vegetables section? Explain.

4. **Processes & Practices** 5 **Use Math Tools** Rebecca and Lance enter Megamart and walk to two different sections. Rebecca walks to the cereal section and Lance walks to the frozen foods section. Who walks farther? Explain.

5. How many total units would you walk from the entrance to the pet food section? Explain.

Name ....

Lesson 8
Ordered Pairs

ESSENTIAL QUESTION
How are patterns used to solve problems?

A **coordinate plane** is formed when two perpendicular number lines intersect. One number line has numbers along the horizontal *x*-axis (across) and the other has numbers along the vertical *y*-axis (up and down). The point where the two axes intersect is the **origin.**

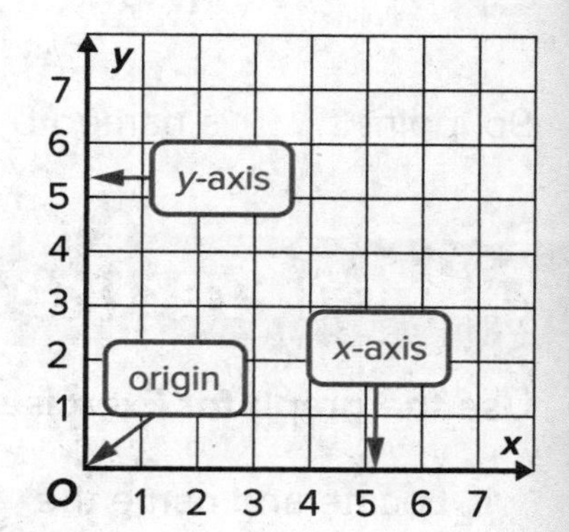

Math in My World

Example 1

Name the ordered pair for the location of Amy's house.

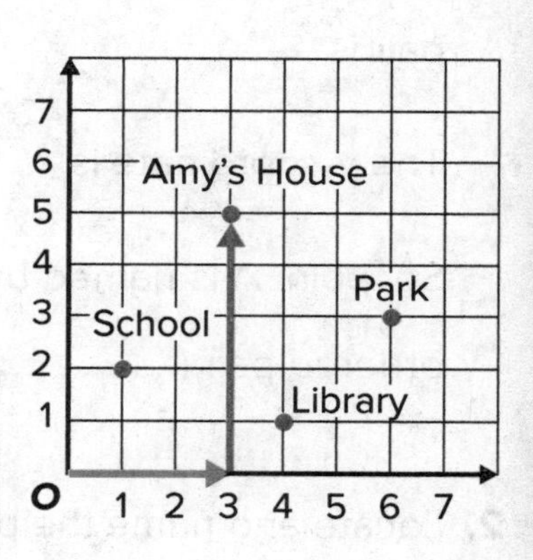

An **ordered pair** is a pair of numbers that is used to name a point.

The first number is the **x-coordinate** and corresponds to a number on the *x*-axis. → (3, 5) ← The second number is the **y-coordinate** and corresponds to a number on the *y*-axis.

1 Start at the origin (____, ____). Move right along the *x*-axis until you are under Amy's house. The *x*-coordinate of the ordered pair is ____.

2 Move up until you reach Amy's house. The *y*-coordinate is ____.

So, Amy's house is located at the ordered pair (____, ____).

Example 2

Name the point for the ordered pair (2, 3).

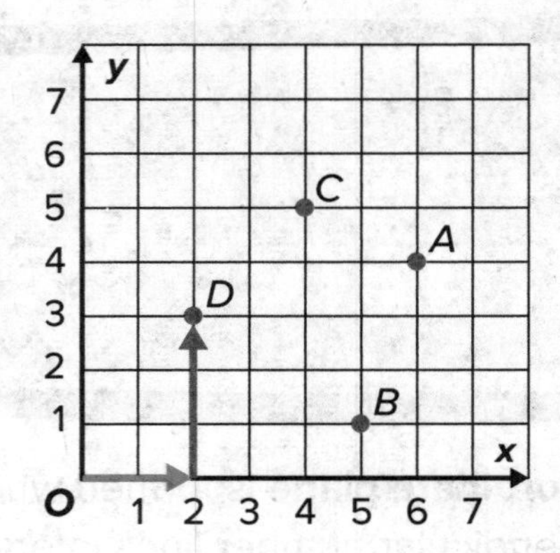

1 Start at the origin (_____, _____). Move right along the x-axis until you reach

_____, the x-coordinate.

2 Move up until you reach _____, the y-coordinate.

So, point _____ is named by the ordered pair (2, 3).

Guided Practice

Use the graph for Exercises 1 and 2.

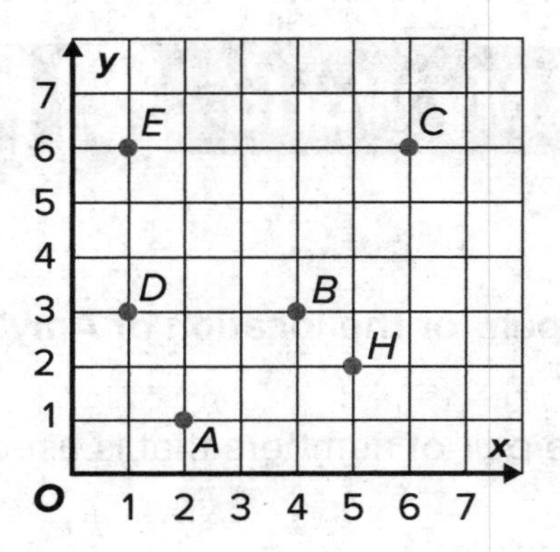

1. Locate and name the ordered pair for point A.

   The x-coordinate of the ordered pair is _____.

   The y-coordinate is _____.

   So, point A is named by the ordered pair (_____, _____).

2. Locate and name the point at (4, 3).

   Move _____ units to the right.

   Move up _____ units.

   So, point _____ is named by the ordered pair (4, 3).

Talk MATH

Are the points at (3, 8) and (8, 3) in the same location? Explain your reasoning.

Name ....

Independent Practice

Use the graph for Exercises 3–14.

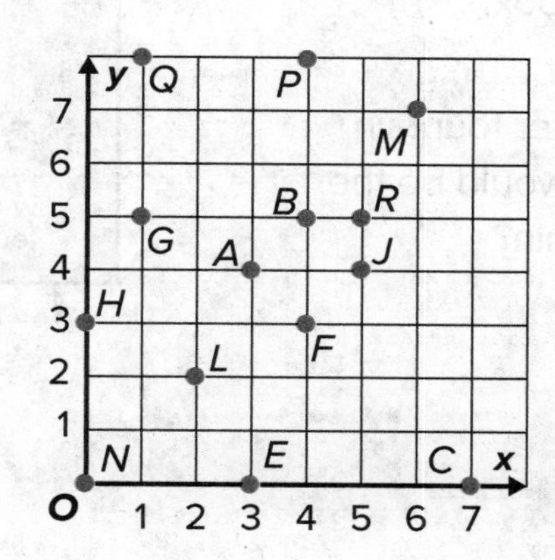

Locate and name each ordered pair.

3. *A* ________

4. *R* ________

5. *J* ________

6. *E* ________

7. *Q* ________

8. *N* ________

Locate and name each point.

9. (2, 2) ________

10. (0, 3) ________

11. (1, 5) ________

12. (6, 7) ________

13. (4, 8) ________

14. (7, 0) ________

Problem Solving

Use the map of the playground at the right for Exercises 15–20.

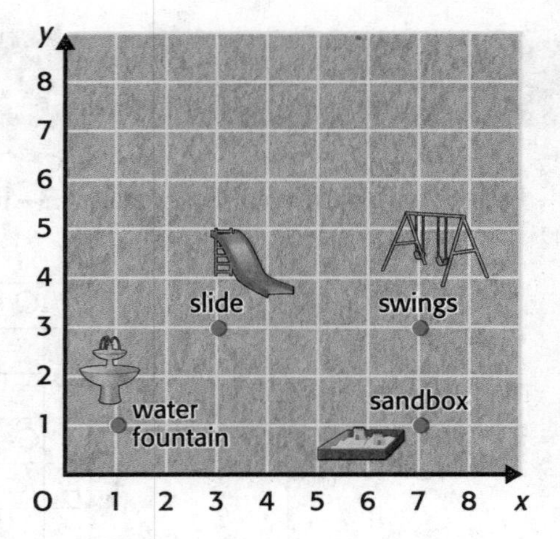

15. What is located at (7, 3)?

16. Write the ordered pair for the sandbox.

17. Suppose the *x*-coordinate of the water fountain was moved to the right 1 unit. What would be the new ordered pair of the water fountain?

Brain Builders

18. Write the ordered pair for the swings and the sandbox. Explain how the ordered pairs can be used to explain where the swings are when you are standing in the sandbox.

19. Cam walked 4 units down and 8 units left to reach the origin. What is the ordered pair of his starting point?

20. A new rock wall is going to be built 3 units up from the swings and 5 units to the left. Write the ordered pair for the rock wall.

21. Processes & Practices 1 **Make Sense of Problems** Name the ordered pair whose *x*-coordinate and *y*-coordinate are each located on an axis.

22. **Building on the Essential Question** How is the location of a point on a grid described?

Name ....

Lesson 8

Ordered Pairs

Homework Helper

Need help? connectED.mcgraw-hill.com

Name the ordered pair for point *A*.

Start at the origin (0, 0). Move right along the *x*-axis until you are under point *A*. The *x*-coordinate of the ordered pair is 3.

Move up until you reach point *A*. The *y*-coordinate is 5.

So, point *A* is named by the ordered pair (3, 5).

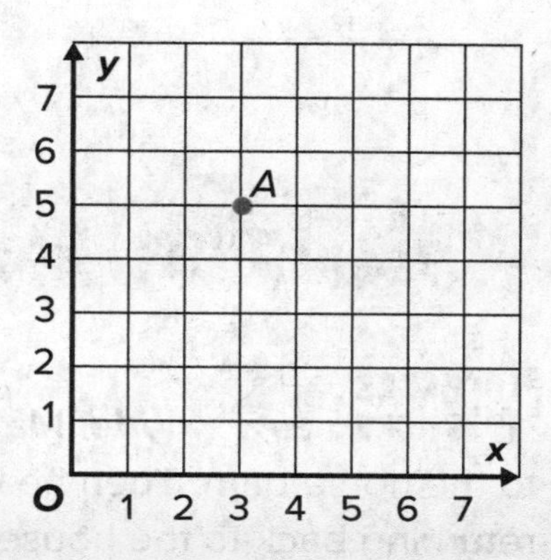

Practice

Use the graph for Exercises 1–6.
Locate and name each ordered pair.

1. *M* ________

2. *P* ________

3. *J* ________

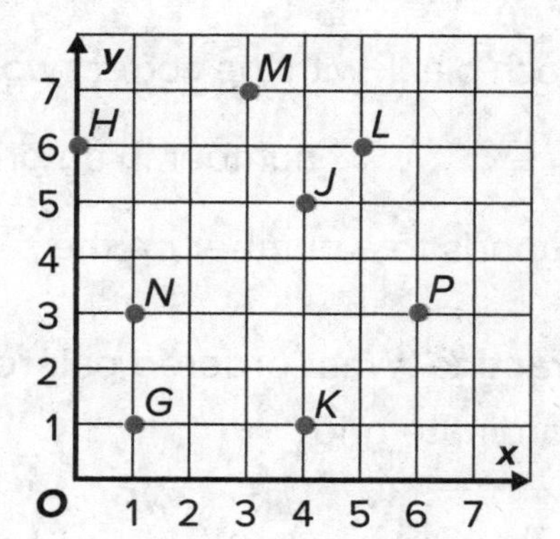

Locate and name each point.

4. (1, 3) ________

5. (5, 6) ________

6. (0, 6) ________

Problem Solving

Use the map for Exercises 7–10.

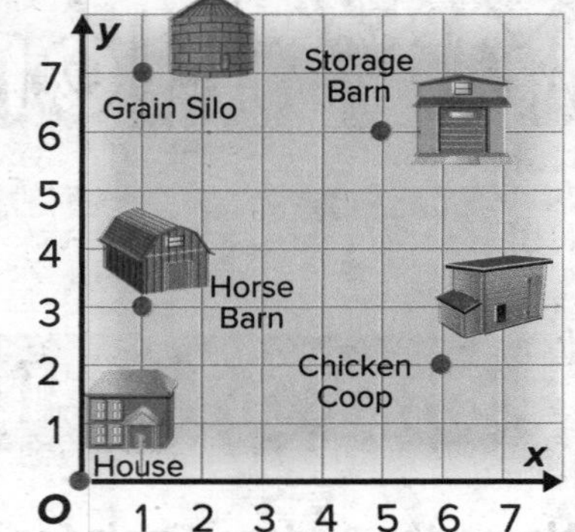

7. What ordered pair gives the location of the storage barn?

____

8. What is located at (1, 7)?

____

9. What is located at (6, 2)?

____

Brain Builders

10. Processes & Practices  **Model Math** Jimmy begins at the house. He walks to the horse barn. Then he walks 3 units up and 4 units right before returning back to the house. List the ordered pairs of the places Jimmy visits in the order he visits them.

____

____

Vocabulary Check

Vocab

11. Fill in each blank with the correct word to complete the sentence.

The ____________ number in an ordered pair is the *y*-coordinate and corresponds to a number on the ____________.

12. Test Practice What ordered pair represents point *D* on the coordinate grid?

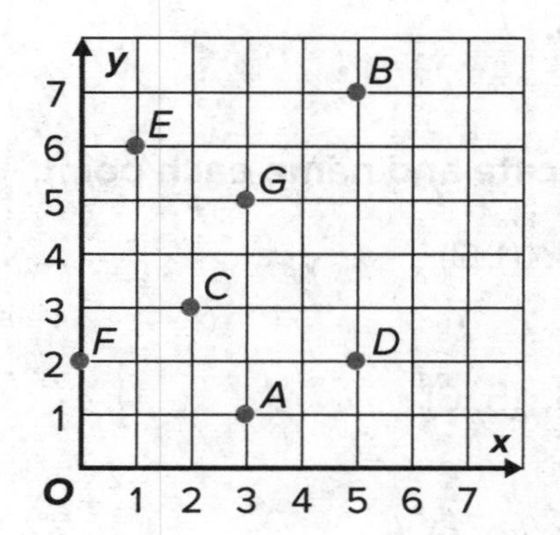

Ⓐ (5, 7)

Ⓑ (5, 2)

Ⓒ (2, 5)

Ⓓ (3, 1)

Name

Lesson 9
Graph Patterns

ESSENTIAL QUESTION
How are patterns used to solve problems?

Math in My World

Example 1

Tricia and her friends decided to rent bicycles to ride on their weekend trip. Bikes 'N More charges $5 for each hour and Adventure Bikes charges $10 for each hour. Find the cost of renting a bicycle from each store for 1, 2, 3, and 4 hours.

1. Complete the tables below.

Bikes 'N More				
Hours	1	2	3	4
Cost ($)				

Adventure Bikes				
Hours	1	2	3	4
Cost ($)				

2. Generate ordered pairs. Let each x-coordinate represent the number of ________. Let each y-coordinate represent the ______.

Bikes 'N More

(1, ____), (2, ____), (3, ____), (4, ____)

Adventure Bikes

(1, ____), (2, ____), (3, ____), (4, ____)

How much more would it cost to rent bicycles for 3 hours from Adventure Bikes than from Bikes 'N More? ________

Example 2

Refer to Example 1. Graph each set of ordered pairs on a coordinate plane. Label each set of ordered pairs. Does the difference in costs between the two stores increase or decrease as the number of hours increases?

The graph shows that the difference in costs between the two stores ___________ as the number of hours increases.

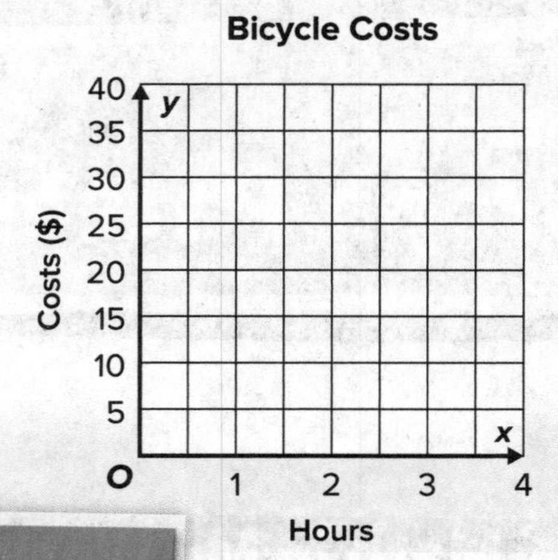

Guided Practice

1. Birdseed is sold in 8-pound bags and 24-pound bags at the local store. Find the weights of 1, 2, 3, and 4 bags of both sizes of birdseed.

Complete the tables below.

8-Pound Bag				
Bags	1	2	3	4
Weight (lb)				

24-Pound Bag				
Bags	1	2	3	4
Weight (lb)				

Generate ordered pairs. Let each x-coordinate represent the number of ________. Let each y-coordinate represent the ________.

8-Pound Bag

(1, ____), (2, ____), (3, ____), (4, ____)

24-Pound Bag

(1, ____), (2, ____), (3, ____), (4, ____)

How many more pounds of birdseed would there be if you purchased 2 bags of 24-pound birdseed than 2 bags of 8-pound birdseed?

____

Explain how you would graph two real-world patterns using ordered pairs.

Name

Independent Practice

2. Speedy Cab charges $2 per mile traveled, while Purple Cab charges $4 per mile traveled. Find the costs of traveling 1, 2, 3, and 4 miles for both cab companies. Then graph the results as ordered pairs.

Speedy Cab				
Miles	1	2	3	4
Cost ($)				

Purple Cab				
Miles	1	2	3	4
Cost ($)				

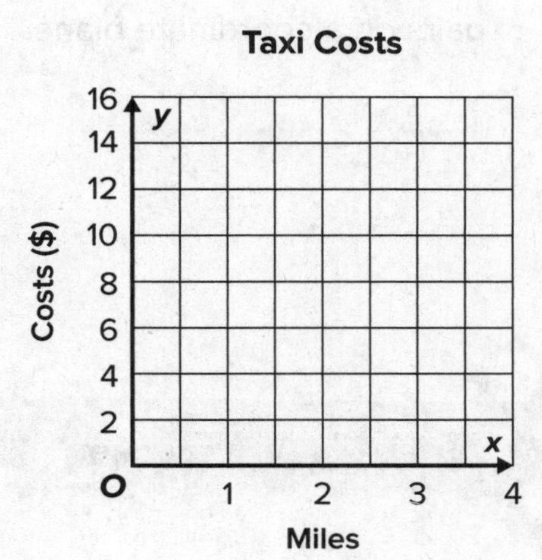

Does the difference in cost between the two taxi services increase or decrease as the number of miles increases?

____

3. Jennifer makes $9 per hour working for her neighbors after school each day. Carmen works for a local farmer and makes $3 per hour working after school each day. Find the total amount earned for each girl if they work 1, 2, 3, and 4 hours. Then graph the results as ordered pairs.

Jennifer's Hourly Wages				
Hours	1	2	3	4
Money Earned ($)				

Carmen's Hourly Wages				
Hours	1	2	3	4
Money Earned ($)				

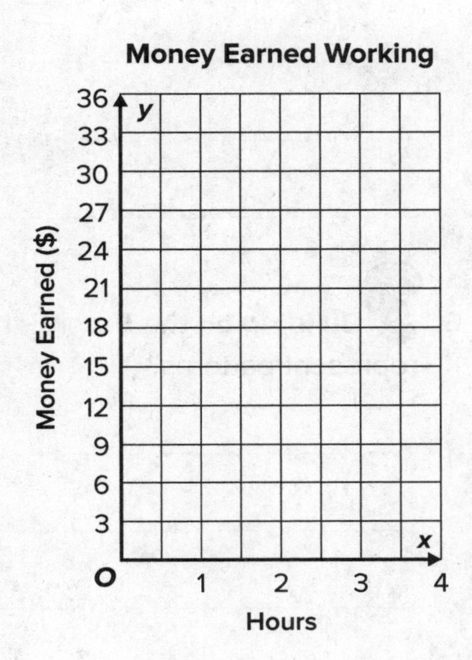

How much more money would Jennifer make if both girls worked 3 hours?

____

Problem Solving

4. Jason places books in his book bag to take home. Each book weighs 2 pounds. Mason has books that weigh 3 pounds each. Find the weights of 1, 2, 3, and 4 books for both Jason and Mason. How many more pounds would Mason carry if they both carried 4 books? Generate ordered pairs. Then graph the ordered pairs on a coordinate plane.

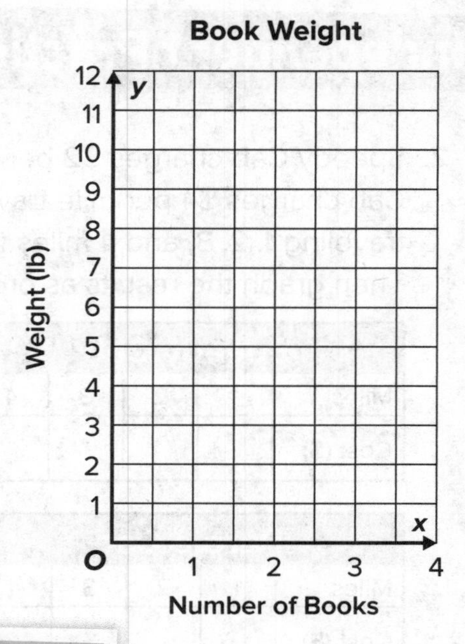

____

Brain Builders

5. Processes & Practices 1 **Plan Your Solution** Write a real-world problem for which you could compare patterns by graphing ordered pairs.

____

____

____

____

____

6. **Building on the Essential Question** How are graphs used to represent patterns?

____

____

____

Name

MY Homework

Lesson 9
Graph Patterns

Homework Helper

Need help? connectED.mcgraw-hill.com

Lance is helping his dad remodel their house. He cuts boards with lengths of 2 feet and 8 feet. Find the amount of material needed for 1, 2, 3, and 4 boards of each length. Then graph the results as ordered pairs on a coordinate plane. How many more feet of 8-foot boards would there be if Lance cuts 3 boards of both lengths?

Complete the tables below.

2-Foot Board				
Number of Boards	1	2	3	4
Length (ft)	2	4	6	8

8-Foot Board				
Number of Boards	1	2	3	4
Length (ft)	8	16	24	32

Generate ordered pairs. Let each x-coordinate represent the number of boards. Let each y-coordinate represent the total length in feet of the boards of that size.

2-Foot Board
(1, 2), (2, 4), (3, 6), (4, 8)

8-Foot Board
(1, 8), (2, 16), (3, 24), (4, 32)

Graph each set of ordered pairs on a coordinate plane.

If Lance cuts 3 boards of both lengths, he will have 6 feet of 2-foot boards and 24 feet of 8-foot boards.

$24 - 6 = 18$

So, there will be 18 more feet of 8-foot boards if he cuts 3 boards of both lengths.

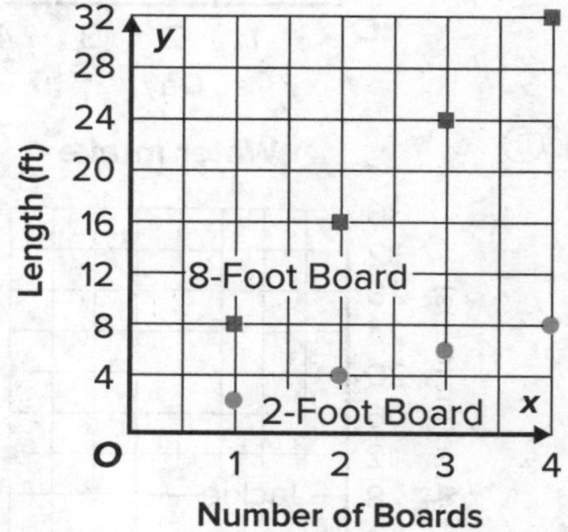

Brain Builders

1. Processes & Practices **4** **Model Math** Jarrett walks his puppy outside every day for 30 minutes. Angela walked her puppy every day for 90 minutes. Find the number of minutes that each puppy was walked for 1, 2, 3, and 4 days. Then graph the results as ordered pairs. How many more minutes does Angela spend walking her puppy over 2 days compared to Jarrett walking his puppy over 2 days?

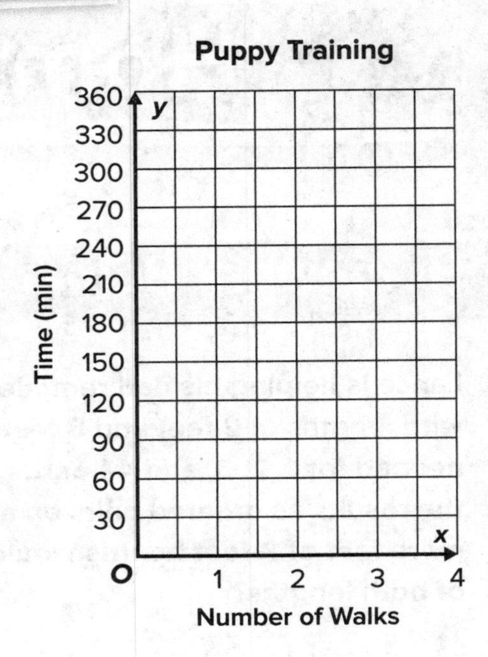

____

How many more minutes will Angela spend walking her puppy over 5 days compared to Jarrett?

____

2. Test Practice Emily drinks 6 cups of water every day, while Jackie drinks 8 cups of water every day. Which graph represents the total amount of water consumed by Emily and Jackie over a 4-day period?

Ⓐ

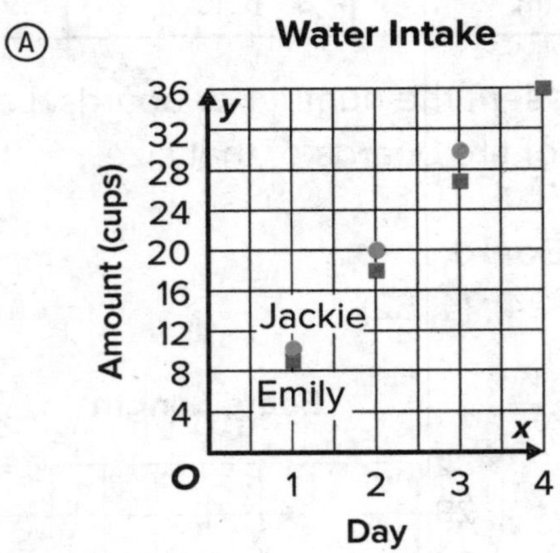

Ⓒ

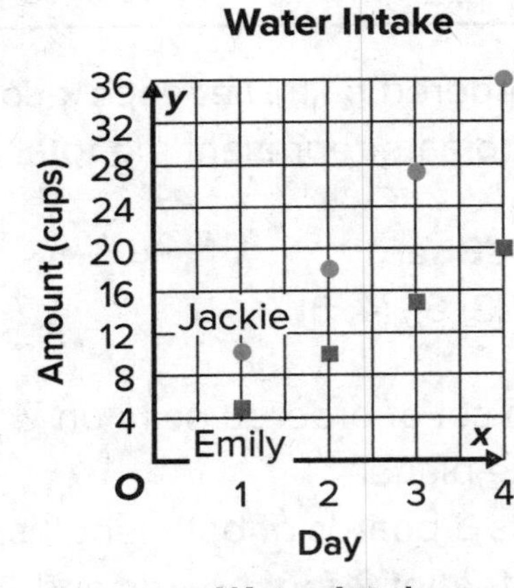

Ⓑ

Water Intake

Amount (cups): 36, 32, 28, 24, 20, 16, 12, 8, 4; y; Jackie; Emily; x; O; 1, 2, 3, 4; Day

Ⓓ

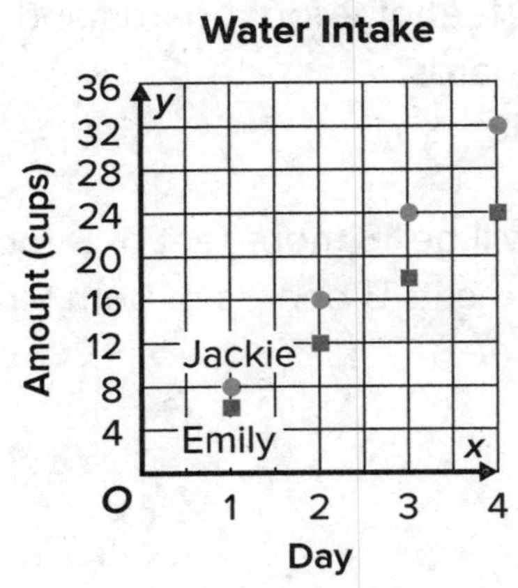

Review

Chapter 7

Expressions and Patterns

Vocabulary Check

Use context clues to write a description for each boldfaced vocabulary word.

1. The **numerical expression** $5 + 12$ represents the sum of 5 and 12.

2. Keira was asked why the **ordered pair** is important when plotting points on a graph.

3. Luis was given the expression $10 - (2 \times 6)$ to evaluate and asked to identify the **order of operations**.

4. The **coordinate plane** can be used to name ordered pairs.

5. Majorie was asked to name the ***x*-coordinate** on the coordinate plane for point *A*.

Concept Check

Evaluate each expression.

6. $(2 \times 2^2) \times (4 + 7) =$ ______

7. $10 \times [(7^2 + 3) - 9] =$ ______

8. $(21 \div 3) + (17 - 7) =$ ______

9. $\{[(66 \div 11) + 3] \times 2\} =$ ______

10. $6 \times [5 \times (3^3 - 17)] =$ ______

11. $\{[(18 - 3) + 3^2] - 14\} \times 3 =$ ______

Write each phrase as a numerical expression.

12. divide 18 by 3, then add 9 ______

13. subtract 5 from 13 then add the product of 3 and 7 ______

14. Compare the pair of numerical expressions without evaluating them.

Expression 1	Expression 2
12×2	$(12 \times 2) \times 2$

Both expressions contain the same multiplication expression.

Write the expression. ______

In Expression 2, the product is multiplied by ______.

So, Expression 2 is ______ times as large as Expression 1.

Locate and name each ordered pair.

15. *A* ______ **16.** *B* ______ **17.** *C* ______

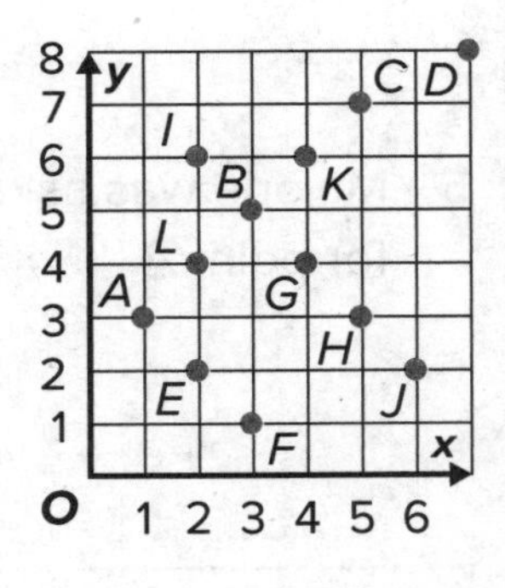

Locate and name each point.

18. (5, 3) ______ **19.** (4, 6) ______ **20.** (4, 4) ______

Problem Solving

21. A store display will have 6 rows, with 12 containers in each row. If 39 containers has been set up so far, explain how to find the number of containers that still need to be set up. Then solve the problem.

22. Fashion for All has jeans on sale for \$13. Designer Pants has jeans on sale for \$26. Write the total costs of 1, 2, 3, and 4 pairs of jeans for each store. Then compare the total cost of 4 pairs of jeans.

23. Glenn swam 2 laps every morning for 7 days. In addition to the laps he swam each morning, he swam 3 laps with his friends on Tuesday and Thursday. Write the expression that shows the number of laps he swam during the week. Evaluate the expression to find the total number of laps he swam that week.

Brain Builders

Use the graph for Exercises 24 and 25.

24. Before leaving the grocery store, Kenny wrote the *y*-coordinate first and the *x*-coordinate second when telling his sister where to meet him. To his surprise, she was at the right location. If his sister did not make an error, where did the siblings meet? Explain how you know.

25. Write the coordinates for the playground, the fountain, and the bank. Write a rule to describe the pattern for each coordinate.

26. **Test Practice** Refer to the graph for Exercises 24 and 25. If the *y*-coordinate of the grocery store was moved up 4 units, what would be the ordered pair of the grocery store?

Ⓐ (1, 5)   Ⓑ (5, 1)   Ⓒ (1, 7)   Ⓓ (5, 5)

Reflect

Chapter 7

Answering the ESSENTIAL QUESTION

Use what you learned about expressions and patterns to complete the graphic organizer.

ESSENTIAL QUESTION

How are patterns used to solve problems?

Real-World Example

Graph the Patterns

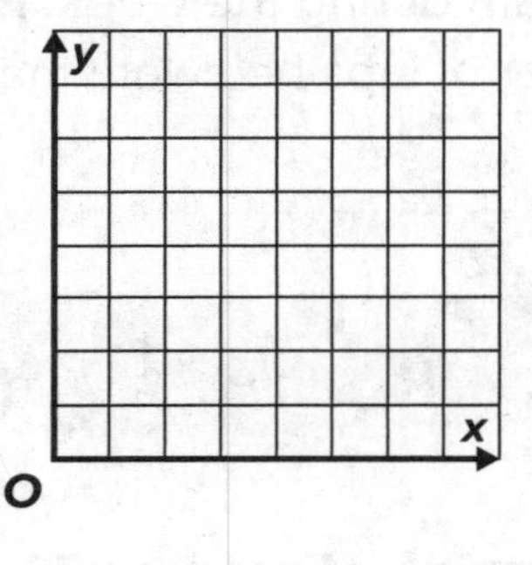

Compare the Patterns

Now reflect on the ESSENTIAL QUESTION  **Write your answer below.**

Name Date

Score

Performance Task

Brain Builders

Planting Community Gardens

Sierra and Mara have each developed a plan for planting new community gardens in their neighborhoods. Sierra will plant 1 garden the first year, 2 gardens the second year, 4 gardens the third year, and so on. Mara will plant 2 gardens the first year, 6 gardens the second year, 10 gardens the third year, and so on.

Show all your work to receive full credit.

Part A

Describe the pattern for each girl's community garden planting plan.

Part B

For each of the first six years, plot the number of community gardens that each girl plans to plant. Use a different mark for each of the girls.

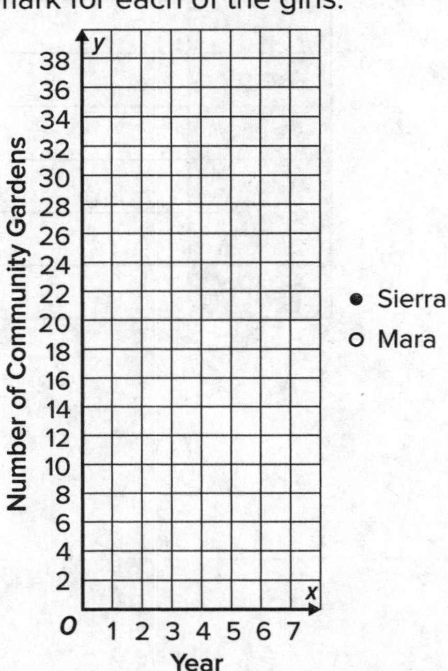

Part C

For the first several years Mara plants more gardens than Sierra. In what year will Sierra plant more gardens than Mara?

Part D

Fill in the table with the total number of gardens planted in that year by both girls.

Year 1	
Year 2	
Year 3	
Year 4	
Year 5	
Year 6	

Chapter

8 Fractions and Decimals

ESSENTIAL QUESTION

How are factors and multiples helpful in solving problems?

Let's Play Games and Sports!

Watch

Watch a video!

Name

MY Chapter Project

Fraction Party

1. In your group, discuss the different types of foods that could be cut into fractions. List the 5 foods you will include in your party plan in the table below.

2. In the table, write the number of students each food will feed. (This will be the number of students in your class.)

3. Decide what fraction each food will be cut into. Will you cut it into halves, thirds, fourths, or another fraction? You can use different fractions for different foods. In the table, list the fraction you will use to divide each food.

4. For each food, determine how many fractional pieces you will have and how many of the whole item you will need. For example:

Food	Number of Students	Fractional Part	Number of Fractional Pieces	Number of Whole Pieces
apple	29	thirds	30	10

5. Complete the table for your five foods.

Food	Number of Students	Fractional Part	Number of Fractional Pieces	Number of Whole Pieces

6. Below, list how many leftover parts you'll have for each food.

Name ....

Am I Ready?

Find the factors of each number.

1. 8 ________

2. 12 ________

3. 6 ________

4. 21 ________

5. 32 ________

6. 45 ________

Multiply or divide.

7. $5 \times 12 =$ ________

8. $64 \div 4 =$ ________

9. $56 \div 7 =$ ________

10. $9 \times 13 =$ ________

11. Claire brought three packages of sports drinks to the volleyball game. If each package contains 8 drinks, how many total drinks did Claire bring?

____

Graph each number on the number line provided.

12. 6.3

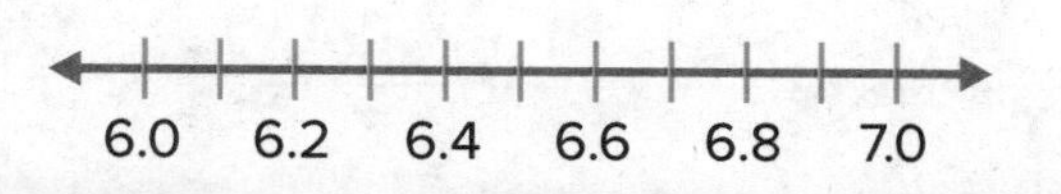

13. 1.8

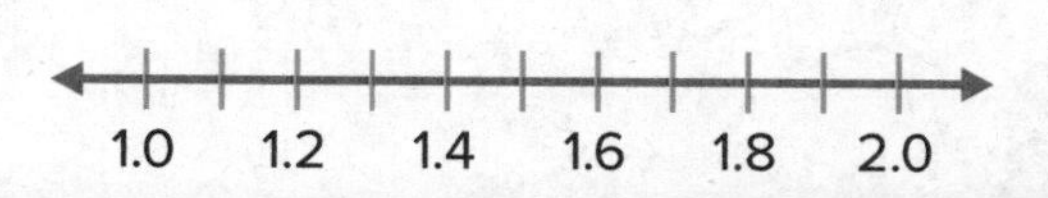

How Did I Do?

Shade the boxes to show the problems you answered correctly.

1	2	3	4	5	6	7	8	9	10	11	12	13

Name ....

MY Math Words

Vocab

Review Vocabulary

decimal   equivalent decimals   multiples   prime factorization

Making Connections

Use the chart below to identify the review vocabulary based on the examples provided. Then provide your own non-example. The first one is done for you.

Word	Example	Non-Example
equivalent decimals	5.25 = 5.250	3.05 > 3.005
	$2 \times 2 \times 3 \times 3 = 36$	
	twenty-one hundredths	
	3, 6, 9, 12, 15, 18, and 21 are multiples of 3.	

Explain why 0.5 and 0.50 express the same amount.

____

____

____

MY Vocabulary Cards

Processes & Practices

Lesson 8–2

common factor

12: 1, 2, 3, 4, 6, 12

30: 1, 2, 3, 5, 6, 10, 15, 30

common factors of 12 and 30: 1, 2, 3, 6

Lesson 8–5

common multiple

4: 4, 8, 12, 16, 20, 24, 28, 32, 36 . . .

6: 6, 12, 18, 24, 30, 36, 42 . . .

common multiples of 4 and 6: 12, 24, 36

Lesson 8–1

denominator

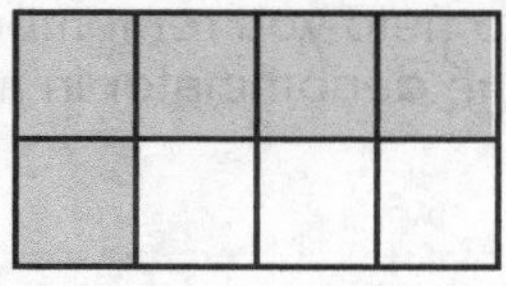

$\frac{5}{8} = 5 \div 8$

Lesson 8–3

equivalent fractions

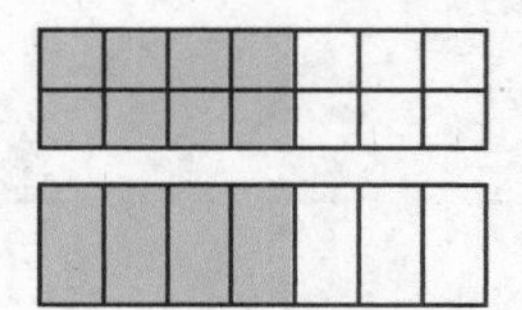

$\frac{8}{14} = \frac{4}{7}$

Lesson 8–1

fraction

$\frac{2}{3}$

Lesson 8–2

greatest common factor (GCF)

12: 1, 2, 3, 4, 6, 12

30: 1, 2, 3, 5, 6, 10, 15, 30

greatest common factor: 6

Lesson 8–6

least common denominator (LCD)

$\frac{2}{3} = \frac{2 \times 2}{3 \times 2} = \frac{4}{6}$

$\frac{1}{6} = \frac{1 \times 1}{6 \times 1} = \frac{1}{6}$

6 is the LCD

Lesson 8–5

least common multiple (LCM)

4: 4, 8, 12, 16, 20, 24, 28, 32, 36 . . .

6: 6, 12, 18, 24, 30, 36, 42 . . .

least common multiple of 4 and 6: 12

Ideas for Use

- Develop categories for the words. Sort them by category. Ask another student to guess each category.
- Design a crossword puzzle. Use the definition for each word as the clues.

A whole number that is a multiple of two or more numbers.

Use a thesaurus to write a synonym and an antonym for *common*.

A number that is a factor of two or more numbers.

When are factors useful?

Fractions that have the same value.

Give an example of two fractions that are equivalent.

The bottom number in a fraction. It represents the total number of equal parts.

Write a tip to help you remember which number is the denominator in a fraction.

The greatest of the common factors of two or more numbers.

Find the greatest common factor of 24 and 36. Show your work.

A number that represents equal parts of a whole or parts of a set.

Explain how fractions represent division.

The least multiple, other than 0, common to sets of multiples.

How can finding the LCM of two numbers be helpful in a real-world situation?

The least common multiple of the denominators of two fractions.

Explain why finding the LCD is important in working with fractions.

MY Vocabulary Cards

Processes & Practices

Lesson 8–5

multiple

7,	14,	21,
1×7	2×7	3×7
28,	35,	42,...
4×7	5×7	6×7

Lesson 8–1

numerator

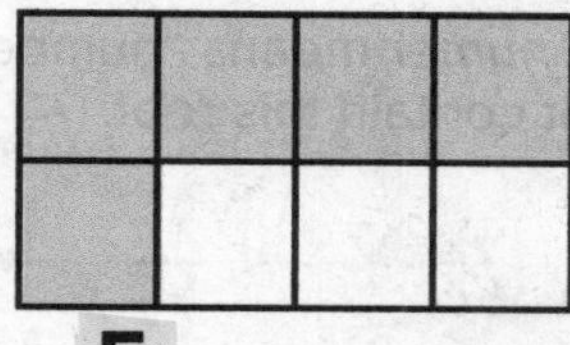

$\frac{5}{8} = 5 \div 8$

Lesson 8–3

simplest form

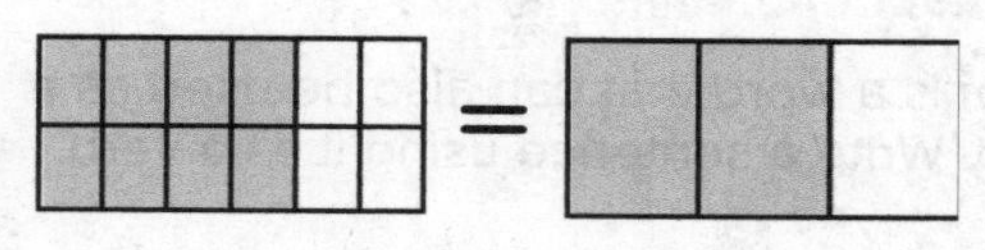

$\frac{8 \div 4}{12 \div 4} = \frac{2}{3}$

Ideas for Use

- Write a tally mark on each card every time you read the word in this chapter or use it in your writing. Challenge yourself to make at least 3 or 4 tally marks for each card.
- Write the names of lessons you would like to review on the front of each blank card. Write a few study tips on the back of each card.

The top number in a fraction. It tells how many of the equal parts are being used.

The Latin root *numer* means "number." Write two words that contain this root.

____

____

A multiple of a number is the product of that number and any whole number.

Explain why 32 is a multiple of 8. What is another multiple of 8?

____

____

A fraction in which the GCF of the numerator and the denominator is 1.

Form is a word that can also be used as a verb. Write a sentence using it as a verb.

____

____

MY Foldable

FOLDABLES® Follow the steps on the back to make your Foldable.

Fractions	Decimals	Models
$\frac{1}{2}$	0.5	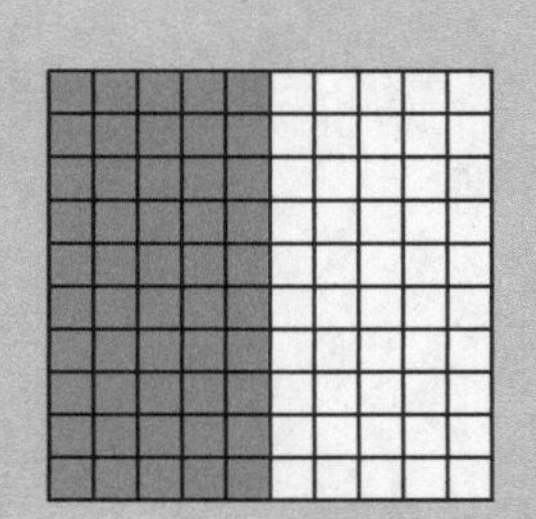

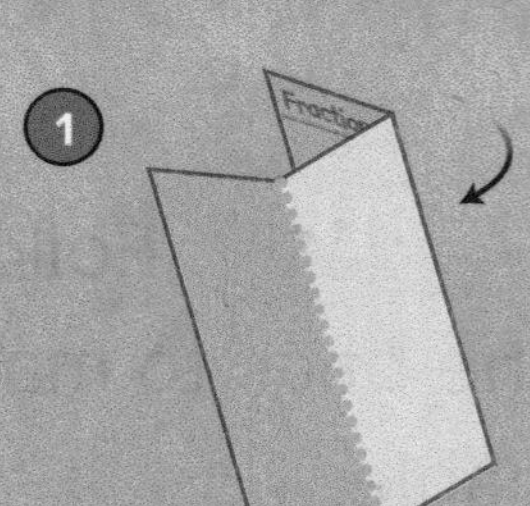
Fractions

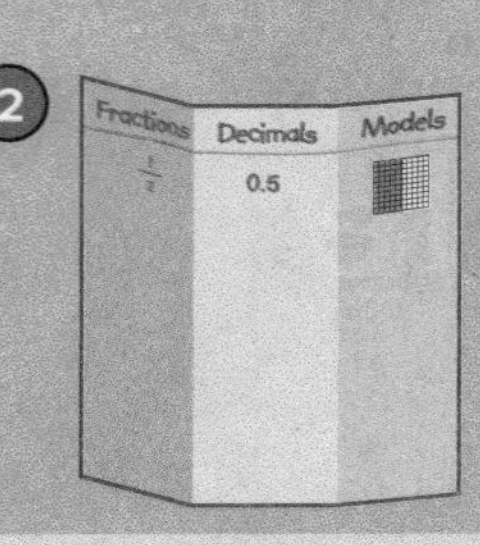
Fractions
Decimals
Models
0.5

Name

Lesson 1
Fractions and Division

ESSENTIAL QUESTION
How are factors and multiples helpful in solving problems?

A **fraction** is a number that names equal parts of a whole or parts of a set. A fraction represents division of the numerator by the denominator.

numerator ⟶ $\frac{1}{3} = 1 \div 3$
denominator ⟶

The **numerator** is the number of parts represented. The **denominator** represents the number of parts in the whole.

Math in My World

Example 1

Dylan, Drake, and Jade are sharing 2 small pizzas equally after their lacrosse game. How much does each person get?

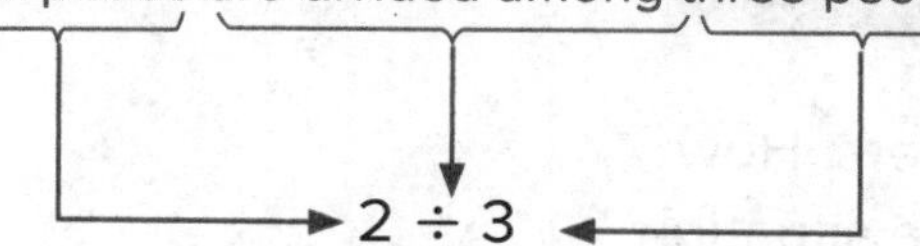

Each person gets $\frac{2}{3}$ of a pizza.

So, $2 \div 3 = \frac{\square}{\square}$

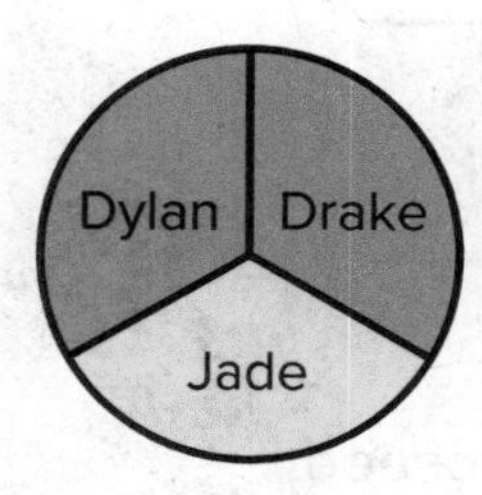

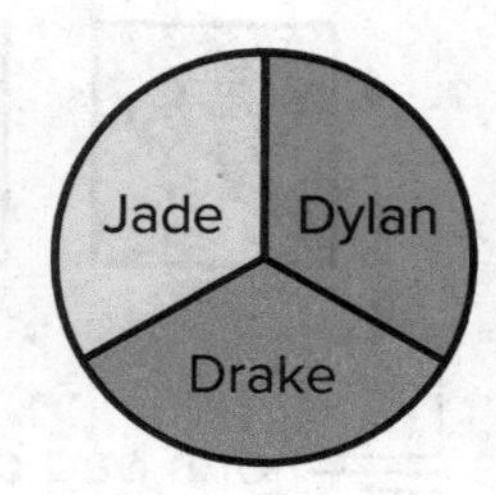

The amount of pizza each person gets is between what two whole numbers?

____

Example 2

Ray and Bailey are sharing 3 brownies equally. How much does each person get?

Three brownies are divided between two people.

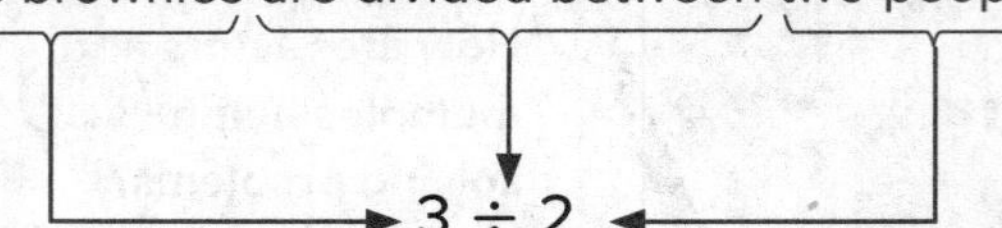

Each person gets $\frac{3}{2}$ of a brownie.

The models at the right show that each person gets 1 whole brownie. There is one brownie remaining, which can be divided equally between the two people.

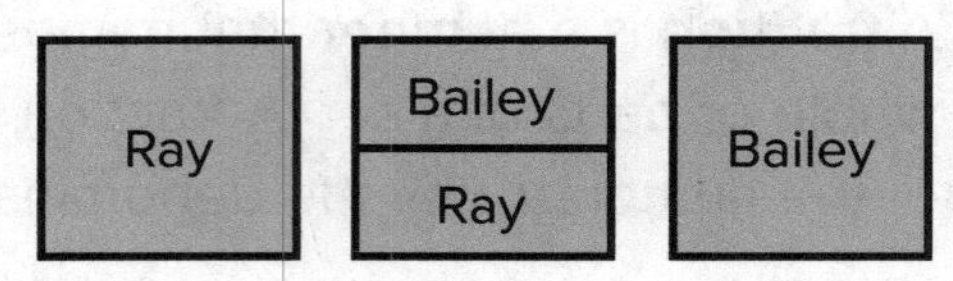

Each person gets $1\frac{1}{2}$ brownies.

So, $3 \div 2 = \square\frac{\square}{\square}$

The amount of brownies each person gets is between what two whole numbers?

____

Guided Practice

1. Two bags of birdseed are used to fill three bird feeders. How much birdseed does each feeder use? Represent the situation using the model. Then solve.

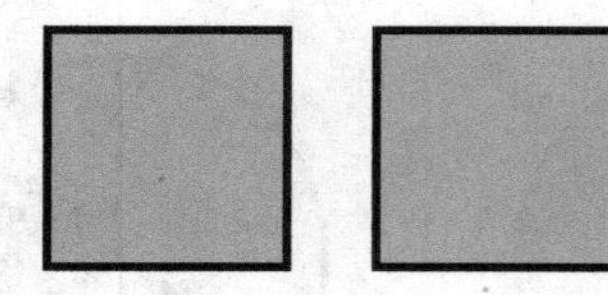

Each feeder uses $\frac{\square}{\square}$ of a bag of birdseed.

So, $2 \div 3 = \frac{\square}{\square}$.

Give an example of how a fraction represents a division situation in real life.

Name ____________________

Independent Practice

Represent each situation using a model. Then solve.

2. Four families equally share 5 pies. How much pie will each family receive?

Each family receives $\frac{\square}{\square}$, or $\square\frac{\square}{\square}$, of a pie.

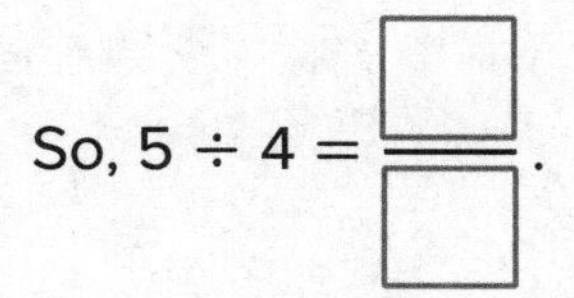

So, $5 \div 4 = \frac{\square}{\square}$.

The answer is between the whole numbers ______ and ______.

3. Six bags of soil are used to fill 5 flower pots. How much soil does each flower pot use? Between what two whole numbers does the answer lie?

Each flower pot uses $\frac{\square}{\square}$, or $\square\frac{\square}{\square}$, of a bag of soil.

So, $6 \div 5 = \frac{\square}{\square}$.

The answer is between the whole numbers ______ and ______.

4. Forty yards of fabric are used to make 9 school banners. How many yards of fabric does each banner use? Between what two whole numbers does the answer lie?

Each banner uses $\frac{\square}{\square}$, or $\square\frac{\square}{\square}$, yards of fabric.

So, $40 \div 9 = \frac{\square}{\square}$.

The answer is between the whole numbers ______ and ______.

Problem Solving

5. Processes & Practices 5 **Use Math Tools** Demont used 4 gallons of gasoline in three days driving to work. Each day he used the same amount of gasoline. How many gallons of gasoline did he use each day?

6. Suzanne made 2 gallons of punch to be divided equally among 10 people. How much of the punch did each person receive?

Brain Builders

7. The baseball team is selling 30 loaves of banana bread. Each loaf is sliced and equally divided into 12 large storage containers. If each slice is the same size, how many loaves of banana bread are in each container? Between what two whole numbers does the answer lie?

8. Processes & Practices 2 **Reason** You know that if $15 \div 3 = 5$, then $5 \times 3 = 15$. If you know that $7 \div 8 = \frac{7}{8}$, what can you conclude about the product of $\frac{7}{8}$ and 8? Explain.

9. **Building on the Essential Question** How can division be represented by using a fraction? Give a real-world example.

Name ______________________________

Lesson 1

Fractions and Division

Homework Helper

Need help? connectED.mcgraw-hill.com

Two truckloads of mulch are used to cover seven playground areas. Each playground receives the same amount of mulch. How much mulch does each playground receive?

Two truckloads are divided to cover seven playgrounds.

$2 \div 7$

1	2	3	4	5	6	7

1	2	3	4	5	6	7

Each playground receives $\frac{2}{7}$ of a truckload.

So, $2 \div 7 = \frac{2}{7}$.

This number is between the whole numbers 0 and 1.

Practice

1. Three pounds of potatoes make eight equal-size servings of mashed potatoes. How many pounds of potatoes are in each serving? Represent the situation with a model. Then solve.

   Each serving uses $\frac{\square}{\square}$ pound of potatoes.

   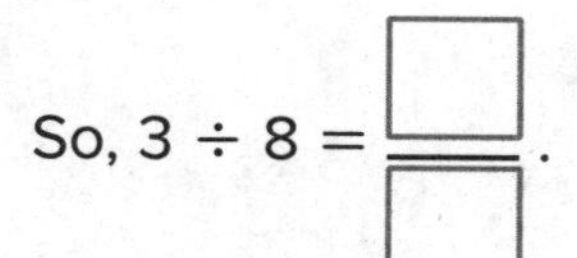

   So, $3 \div 8 = \frac{\square}{\square}$.

   The answer is between the whole numbers ______ and ______.

Problem Solving

2. One large submarine sandwich is divided equally among four people. How much of the sandwich did each person receive?

____

Brain Builders

3. Four gallons of paint are used to paint 20 chairs and 5 small tables. If each chair and table used the same amount of paint, how many gallons are used to paint each piece of furniture? Between what two whole numbers does your answer lie?

____

4. Processes & Practices 1 **Make Sense of Problems** Mrs. Larsen made 12 pillows from 16 yards of the same fabric. How much fabric was used to make each pillow? Between what two whole numbers does your answer lie?

____

How many feet of fabric are needed for each pillow?

____

Vocabulary Check

5. Fill in each blank with the correct word to complete the sentence.

The numerator is the __________ number in a fraction, while the denominator is the __________ number in a fraction.

6. **Test Practice** Elena drank 5 bottles of water over 7 volleyball practices. How much water did Elena drink each practice if she drank the same amount each time?

Ⓐ $\frac{2}{7}$ bottle

Ⓑ $\frac{2}{5}$ bottle

Ⓒ $\frac{5}{7}$ bottle

Ⓓ $\frac{7}{5}$ or $1\frac{2}{5}$ bottles

Name ....

Lesson 2
Greatest Common Factor

ESSENTIAL QUESTION
How are factors and multiples helpful in solving problems?

Factors shared by two or more numbers are called **common factors**. The greatest of the common factors of two or more numbers is the **greatest common factor (GCF)** of the numbers.

Math in My World

Example 1

Sevierville Middle School arranges their sports trophies in rows in a display case. There is an equal number of trophies in each row. Each row has only one kind of trophy. What is the greatest possible number of trophies in each row?

Trophies	
Type	**Number**
Volleyball	40
Football	24
Baseball	32

Write the prime factorization to find common factors.

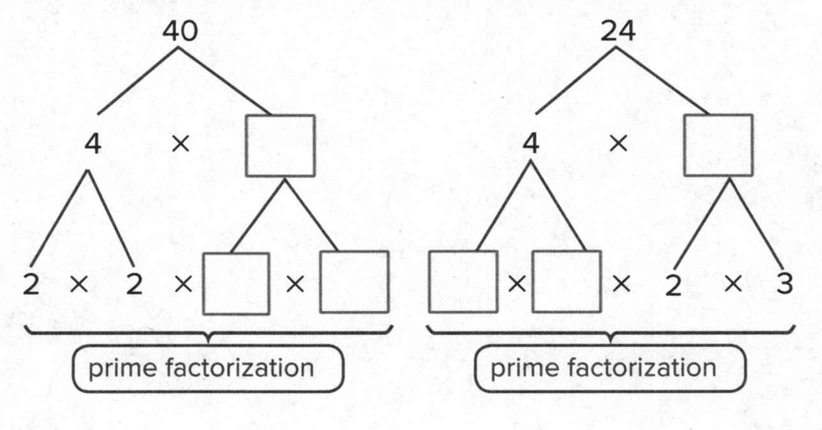

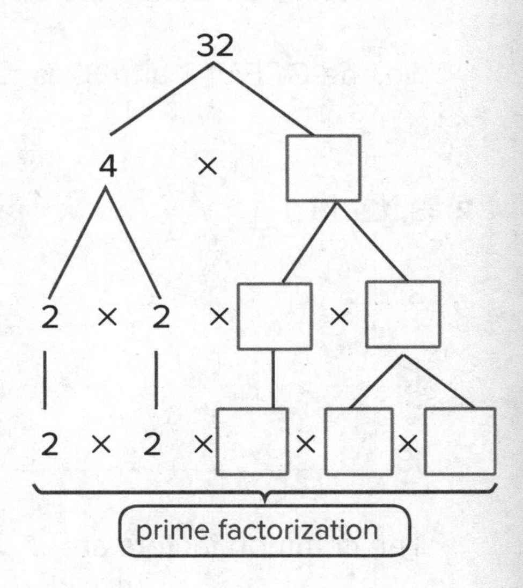

The common prime factors are 2, 2, and 2.

Multiply to find the GCF.

_____ × _____ × _____ or _____

So, the greatest number of trophies that could be placed in each row is _____.

Example 2 Tutor

Find the GCF of 60 and 54.

Make an organized list of the factors for each number. Then circle the common factors.

60: 1, 2, 3, 4, 5, 6, 10, 12, 15, 20, 30, 60

54: 1, 2, 3, 6, 9, 18, 27, 54

The common factors are ____, ____, ____, and ____.

So, the greatest common factor, or GCF, of 60 and 54 is ____.

Guided Practice

Find the GCF of each set of numbers.

Explain which method you prefer to find the GCF of two numbers.

1. 8, 32

8: ________________________

32: ________________________

The common factors are ____, ____, ____, and ____.

So, the GCF of 8 and 32 is ____.

2. 3, 12, 18

3: ________________________

12: ________________________

18: ________________________

The common factors are ____ and ____.

So, the GCF of 3, 12, and 18 is ____.

Name ....

Independent Practice

Find the GCF of each set of numbers.

3. 24, 60 ______

4. 12, 18 ______

5. 18, 42 ______

6. 30, 72 ______

7. 4, 10, 14 ______

8. 14, 35, 84 ______

9. 9, 18, 42 ______

10. 16, 52, 76 ______

Problem Solving

11. Processes & Practices 3 **Justify Conclusions** Annika is placing photos in a scrapbook. Each page will have only one size of photo. She also wants to place the same amount of photos on each page. What is the greatest number of photos that could be on each page? Justify your response.

Scrapbooking	
photo size	photos
Large	8
Medium	12
Small	16

12. Twelve pens and 16 pencils will be placed in bags with an equal number of each item. What is the most number of bags that can be made?

Brain Builders

13. Oliver has 14 chocolate chip cookies and 21 iced cookies. Oliver gives each of his friends an equal number of each type of cookie. What is the greatest number of friends with whom he can share the cookies? How many of each type of cookie will each friend receive?

14. Processes & Practices 3 **Which One Doesn't Belong?** What is the GCF of the four numbers shown? Circle the number that you would take away so that 8 will be the GCF of the remaining three numbers. Explain.

16 8 24 20

15. **Building on the Essential Question** How can you find the greatest common factor of two numbers?

Name ....

MY Homework

Lesson 2
Greatest Common Factor

Homework Helper

Need help? connectED.mcgraw-hill.com

The table shows the amount of money Ms. Ayala made over three days selling 4-inch × 6-inch prints at an arts festival. Each print costs the same amount. What is the most each print could have cost?

Ms. Ayala's Artwork	
Day	**Cost ($)**
Friday	60
Saturday	144
Sunday	96

Write the prime factorization to find common factors.

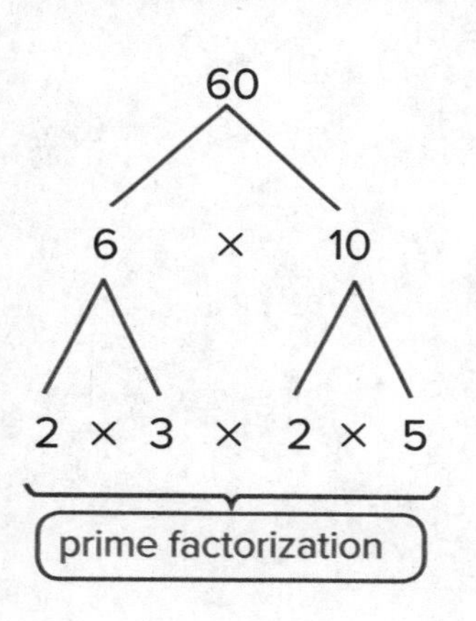

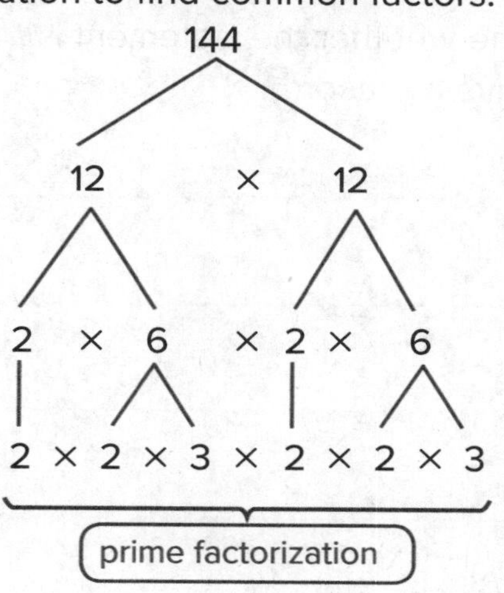

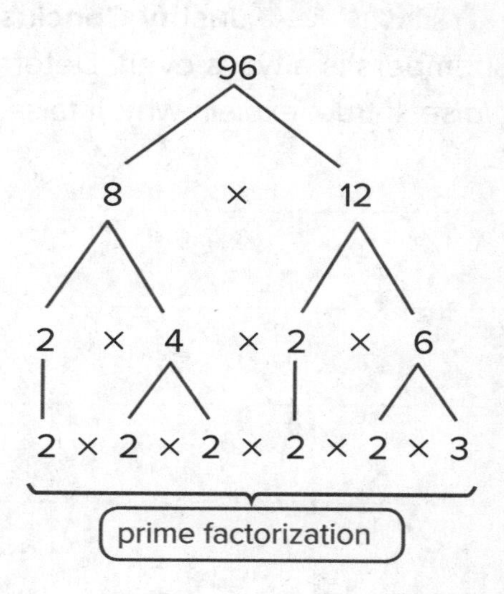

The common prime factors are 2, 2, and 3.

Multiply to find the GCF. $2 \times 2 \times 3 = 12$

So, the greatest cost of each print would be $12.

Practice

Find the GCF of each set of numbers.

1. 21, 30 ______

2. 12, 30, 72 ______

Problem Solving

3. A store sells bottles of juice in equal-size boxes. Garth bought 18 bottles, Rico bought 36 bottles, and Mai bought 45 bottles. What is the greatest number of bottles in each box? How many boxes did each person buy if each box contained the greatest number of bottles possible?

Brain Builders

4. Processes &Practices  **Justify Conclusions** The GCF of any two even numbers is always even. Determine whether the statement is *true* or *false*. If true, explain why. If false, give a reason.

Vocabulary Check

5. Circle the correct term that makes the sentence true.
The (greatest, least) of the common factors of two or more numbers is the (greatest, least) common factor of the numbers.

6. Test Practice Jeremiah will give away all of his sports cards to a number of his friends. What is the greatest number of friends he can give his cards to so that each friend will receive an equal number of baseball cards and football cards?

Sports Cards	
Type	**Number**
Baseball	32
Football	24

Ⓐ 4 friends

Ⓑ 8 friends

Ⓒ 12 friends

Ⓓ 16 friends

Name

Lesson 3
Simplest Form

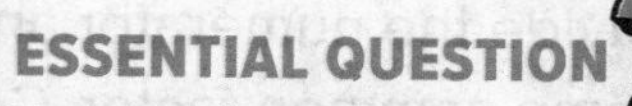

ESSENTIAL QUESTION
How are factors and multiples helpful in solving problems?

A fraction is written in **simplest form** when the GCF of the numerator and the denominator is 1. The simplest form of a fraction is one of its many equivalent fractions. **Equivalent fractions** are fractions that name the same number.

Math in My World

Example 1

Angie has a vertical jump height of 12 inches, and Holly has a vertical jump height of 22 inches. So, Angie's vertical jump is $\frac{12}{22}$ of the vertical jump of Holly. Write the fraction in simplest form.

1. List the factors. Circle the GCF.

12: 1, 2, 3, 4, 6, 12

22: 1, 2, 11, 22

The GCF of 12 and 22 is ______.

2. Divide both the numerator and the denominator by the GCF. This results in an equivalent fraction.

$\frac{12}{22} = \frac{12 \div 2}{22 \div 2} = \frac{\square}{\square}$

The GCF of 6 and 11 is ______.

So, Angie's vertical jump is $\frac{\square}{\square}$ of Holly's vertical jump.

Check Use models.

Shade 12 out of 22.

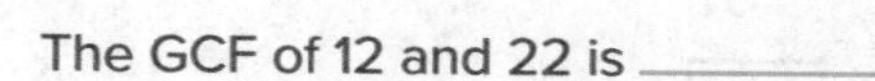

Shade 6 out of 11.

Example 2

Write $\frac{18}{30}$ in simplest form.

Divide the numerator and denominator by the same common factor. Continue dividing until the fraction is in simplest form.

Helpful Hint

Dividing the numerator and denominator by the same number is the same as dividing the fraction by 1. The result is an equivalent fraction.

$\frac{18}{30} = \frac{18 \div 2}{30 \div 2} = \frac{\square}{\square}$ Divide 18 and 30 by the common factor 2.

$= \frac{9 \div 3}{15 \div 3}$ Divide the numerator and denominator by the common factor 3.

$= \frac{\square}{\square}$

Since 3 and 5 have no common factors other than 1, stop dividing.

So, $\frac{18}{30}$ written in simplest form is $\frac{\square}{\square}$.

Guided Practice

Write each fraction in simplest form. If the fraction is already in simplest form, write *simplified*.

Explain how to find the simplest form of any fraction.

1. $\frac{4}{6}$

$\frac{4}{6} = \frac{4 \div 2}{6 \div 2} = \frac{\square}{\square}$

In simplest form, $\frac{4}{6} = \frac{\square}{\square}$.

2. $\frac{2}{12}$

$\frac{2}{12} = \frac{2 \div 2}{12 \div 2} = \frac{\square}{\square}$

In simplest form, $\frac{2}{12} = \frac{\square}{\square}$.

Name

Independent Practice

Write each fraction in simplest form. If the fraction is already in simplest form, write *simplified*.

3. $\frac{6}{8}$ ______

4. $\frac{6}{10}$ ______

5. $\frac{3}{18}$ ______

6. $\frac{2}{5}$ ______

7. $\frac{4}{16}$ ______

8. $\frac{12}{24}$ ______

9. $\frac{6}{25}$ ______

10. $\frac{21}{30}$ ______

11. $\frac{4}{11}$ ______

Algebra Find each unknown.

12. $\frac{8}{28} = \frac{m}{7}$

$m =$ ______

13. $\frac{12}{40} = \frac{b}{10}$

$b =$ ______

14. $\frac{9}{24} = \frac{3}{y}$

$y =$ ______

Problem Solving

15. Processes & Practices 1 **Plan Your Solution** The table shows the results of a survey about favorite movie theater snacks. Write a fraction in simplest form that compares the number of people who chose popcorn to the total number of people surveyed.

Favorite Movie Snack	
Snack	**Frequency**
Popcorn	24
Hot dog	12
Nachos	11
Chocolate	8
Licorice	5

____

16. Kara buys 24 bagels. Ten are whole wheat. What fraction of the bagels are whole wheat? Write in simplest form.

____

Brain Builders

17. Processes & Practices 3 **Which One Doesn't Belong?** Circle the fraction that is not in simplest form. Explain and simplify the fractions.

$\frac{9}{21}$ $\frac{7}{18}$ $\frac{3}{25}$ $\frac{12}{31}$

____

____

18. **Building on the Essential Question** Does dividing the numerator and denominator by a common factor guarantee that the result will be in simplest form? Explain.

____

____

Name

Lesson 3

Simplest Form

Homework Helper

Need help? connectED.mcgraw-hill.com

Amelia rode $\frac{8}{24}$ mile on the bike trail. Express in simplest form the fraction of the distance that Amelia rode on the bike trail.

Divide the numerator and denominator by the same common factor. Continue dividing until the fraction is in simplest form.

$\frac{8}{24} = \frac{8 \div 2}{24 \div 2}$ Divide 8 and 24 by the common factor 2.

$= \frac{4}{12}$ Simplify.

$= \frac{4 \div 4}{12 \div 4}$ Divide the numerator and denominator by the common factor 4.

$= \frac{1}{3}$ Simplify.

Since 1 and 3 have no common factors other than 1, stop dividing.

So, $\frac{8}{24}$ written in simplest form is $\frac{1}{3}$.

Practice

Write each fraction in simplest form. If the fraction is already in simplest form, write *simplified*.

1. $\frac{8}{9}$ ______________

2. $\frac{9}{18}$ ______________

3. $\frac{26}{32}$ ______________

Brain Builders

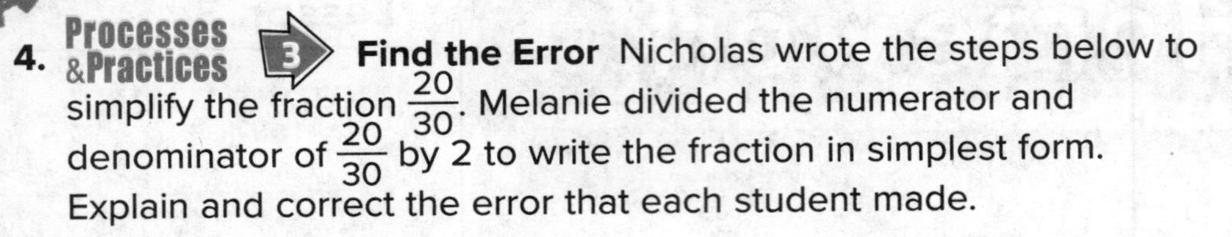

4. **Processes & Practices** 3 **Find the Error** Nicholas wrote the steps below to simplify the fraction $\frac{20}{30}$. Melanie divided the numerator and denominator of $\frac{20}{30}$ by 2 to write the fraction in simplest form. Explain and correct the error that each student made.

$$\frac{20}{30} = \frac{20 \div 5}{30 \div 6} = \frac{4}{5}$$

____

____

____

5. Mr. Rolloson's class had a board game tournament. Out of the 24 games played, Timeka won 10 games. Write the fraction of games Timeka lost in simplest form.

____

Vocabulary Check

6. Fill in the blank with the correct term or number to complete the sentence.

   A fraction is written in simplest form when the GCF of the numerator and the denominator is ________.

7. **Test Practice** Gil's aunt cut his birthday cake into 32 equal pieces. Eighteen pieces were eaten at his birthday party. What fraction of the cake was left?

   Ⓐ $\frac{7}{16}$ Ⓒ $\frac{7}{12}$

   Ⓑ $\frac{9}{16}$ Ⓓ $\frac{9}{14}$

Name ....

Lesson 4
Problem-Solving Investigation
STRATEGY: Guess, Check, and Revise

ESSENTIAL QUESTION
How are factors and multiples helpful in solving problems?

Learn the Strategy

The Bactrian camel has two humps, while the Dromedary camel has just one. Toby counted 20 camels with a total of 28 humps. How many camels of each type are there?

1 Understand

What facts do you know?

- Bactrian camels have two humps and Dromedary camels have one hump.
- There are ________ camels with ________ humps.

What do you need to find?

- How many ________ of each type are there?

2 Plan

I will guess, check, and revise to solve the problem.

3 Solve

Bactrian Camels	Dromedary Camels	Number of Humps	Check
7	13	$(7 \times 2) + (13 \times 1) = 27$	too low
8	12	$(8 \times 2) + (12 \times 1) = 28$	correct

So, there are ________ Bactrian camels and ________ Dromedary camels.

4 Check

Is my answer reasonable?

________ + ________ = 20 camels and ________ + ________ = 28 humps

Practice the Strategy

Conner spent $66 on rookie cards and Hall of Famer cards. How many of each type of card did he buy?

Baseball Card	Cost
Rookie	4 for $6
Hall of Famer	2 for $9

1 Understand

What facts do you know?

What do you need to find?

2 Plan

3 Solve

4 Check

Is my answer reasonable?

Name

Apply the Strategy

Guess, check, and revise to solve each problem.

1. Bike path A is 4 miles long. Bike path B is 7 miles long. If April biked a total of 37 miles, how many times did she bike each path?

2. Ruben sees 14 wheels on a total of 6 bicycles and tricycles. How many bicycles and tricycles are there?

3. A teacher is having three students take care of 28 goldfish during the summer. He gave some of them to Alaina. Then he gave twice as many to Miguel. He gave twice as many to Kira as he gave to Miguel. How many fish did each student get?

Brain Builders

4. Ticket prices for a science museum are $18 for adults and $12 for students. If $162 is collected from a group of 12 people, how many adults and students are in the group? How can you check the reasonableness of your answer?

5. Processes &Practices 1 **Keep Trying** Jerome bought 2 postcards and received $1.50 in change in quarters, dimes, and nickels. The nickels were worth $0.15. If he got 9 coins back, how many of each coin did he get?

6. A tour director collected $180 for 13 tour packages. Tour package A costs $18, tour package B costs $10, and tour package C costs $12. If 5 of the tour packages were A, how many of each tour package were sold?

Review the Strategies

Use any strategy to solve each problem.

- Guess, check, and revise.
- Work backward.
- Determine an estimate or exact answer.
- Make a table.

7. The sum of two numbers is 30. Their product is 176. What are the two numbers?

____

8. Processes &Practices 1 **Make a Plan** Howard downloaded 6 more songs than Erin. Lee downloaded 3 more than Howard. Lee downloaded 16 songs. How many songs did Erin download?

____

9. **Algebra** Work backward to find the value of the variable in the equation below.

$$d - 4 = 23$$

____

10. A bakery can make 175 pastries each day. The bakery has been sold out for 3 days in a row. Determine if an estimate or exact answer is needed. Then determine how many pastries were sold during the 3 days.

____

11. Charlotte bought packages of hot dogs for $4 each. Each package contains 8 hot dogs. If she spent $16 on hot dogs, how many hot dogs did she buy?

____

12. Mr. Thompson took his 5 children to the bowling alley. The cost for children 12 and older is $3.50. The cost for children under 12 is $2.25. He spent a total of $16.25. How many of his children are 12 and older?

____

Name

Lesson 4

Problem Solving: Guess, Check, and Revise

Homework Helper

Need help? connectED.mcgraw-hill.com

An automobile museum has cars and motorcycles displayed. There are a total of 19 cars and motorcycles in the collection and a total of 64 tires. How many cars and motorcycles are in the collection?

1 Understand

What facts do you know?

- There are a total of 19 cars and motorcycles.
- There are a total of 64 tires.

What do you need to find?

- how many cars and motorcycles are in the collection

2 Plan

Guess, check, and revise to solve the problem.

3 Solve

Number of Cars	Number of Motorcycles	Number of Tires	Check
14	5	$(14 \times 4) + (5 \times 2) = 66$	too high
13	6	$(13 \times 4) + (6 \times 2) = 64$	correct

So, there are 13 cars and 6 motorcycles.

4 Check

Is my answer reasonable?

yes; $52 \div 4 = 13$, $12 \div 2 = 6$, and $13 + 6 = 19$.

Real World

Problem Solving

Guess, check, and revise to solve each problem.

1. A cabin has room for 7 campers and 2 counselors. How many cabins are needed for a total of 49 campers and 14 counselors?

2. Marcus is selling lemonade and peanuts. Each cup of lemonade costs \$0.75, and each bag of peanuts costs \$0.35. On Saturday, Marcus sold 5 more cups of lemonade than bags of peanuts. He earned \$7.05. How many cups of lemonade did Marcus sell on Saturday?

Brain Builders

3. Processes & Practices 1 **Keep Trying** A letter to Europe from the United States costs \$0.94 to mail. A letter mailed within the United States costs \$0.44. Nancy mailed 5 letters for \$3.70, some to Europe and some to the United States. How many letters did she send to Europe? Explain how you can check your answer.

4. Alma counts 26 legs in a barnyard with horses and chickens. If there are 8 animals, how many are horses? Explain.

5. A volleyball league organizer collected \$2,040 for both divisions of volleyball teams. The Blue division costs \$160 per team and the Red division costs \$180 per team. How many teams will play in each division? Explain.

Check My Progress

Vocabulary Check

Draw lines to match each question with its correct answer.

1. Which is the greatest of the common factors of two or more numbers?
2. Which is a number that names equal parts of a whole or parts of a set?
3. A fraction that is written when the greatest common factor of the numerator and denominator is 1 is written in which form?

- fraction
- greatest common factor
- simplest form

Concept Check

Represent each situation using a model. Then solve.

4. Three gallons of paint are used to paint 16 wooden signs. How much paint did each sign use? Between what two whole numbers does the answer lie?

____

____

5. Refer to Exercise 4. How many wooden signs can be painted with one gallon of paint? Between what two whole numbers does the answer lie?

____

____

Find the GCF of each set of numbers.

6. 36, 90 ______

7. 16, 24, 40 ______

Write each fraction in simplest form. If the fraction is already in simplest form, write *simplified.*

8. $\frac{4}{14}$ ______

9. $\frac{15}{20}$ ______

10. $\frac{21}{35}$ ______

Problem Solving

11. Anja buys a magazine and a pizza. She spends $9.75. The pizza costs twice as much as the magazine. How much does the pizza cost? Guess, check, and revise to solve.

______

Brain Builders

12. In a pile of coins, there are 7 more quarters than nickels. If there is a total of $2.65 in coins, how many total coins are there? How many of each type of coin is in the pile? Guess, check, and revise to solve.

______

13. Test Practice A warehouse has shelves that can hold 8, 12, or 16 skateboards. Each shelf has sections holding the same number of skateboards. What is the greatest number of skateboards that can be put in a section?

Ⓐ 1 skateboard

Ⓑ 2 skateboards

Ⓒ 3 skateboards

Ⓓ 4 skateboards

Name ....

Lesson 5
Least Common Multiple

ESSENTIAL QUESTION
How are factors and multiples helpful in solving problems?

A **multiple** of a number is the product of the number and any other whole number (0, 1, 2, 3, 4, . . .). Multiples that are shared by two or more numbers are **common multiples**.

The **least common multiple (LCM)** is the least multiple, other than 0, common to sets of multiples.

Math in My World

Example 1

Tom hits golf balls at the driving range every 3 days, practices putting every 4 days, and goes golfing every 6 days. If he did all three activities today, in how many days will he do all three activities again?

Find and circle the LCM of 3, 4 and 6 by listing nonzero multiples of each number.

3: ______________________________

4: ______________________________

6: ______________________________

So, Tom will complete all three activities again in ________ days.

Check The number line shows on which days all three activities will be completed.

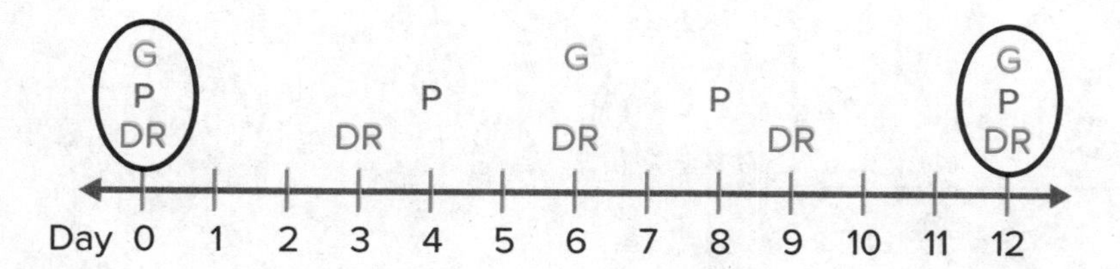

G = golf
P = putting
DR = driving range

Example 2

Find the LCM of 15 and 40.

Write the prime factorization of each number.

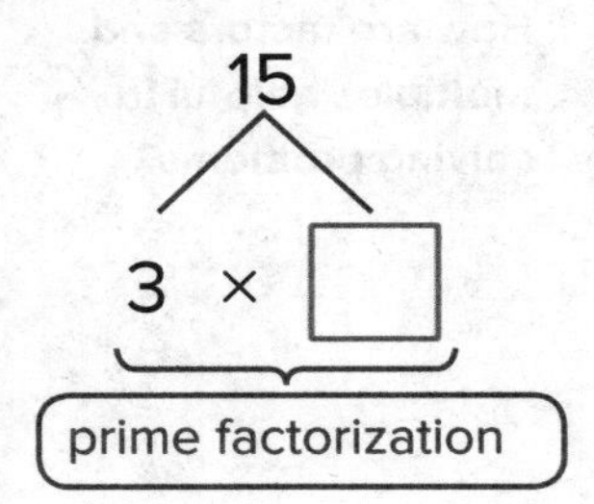

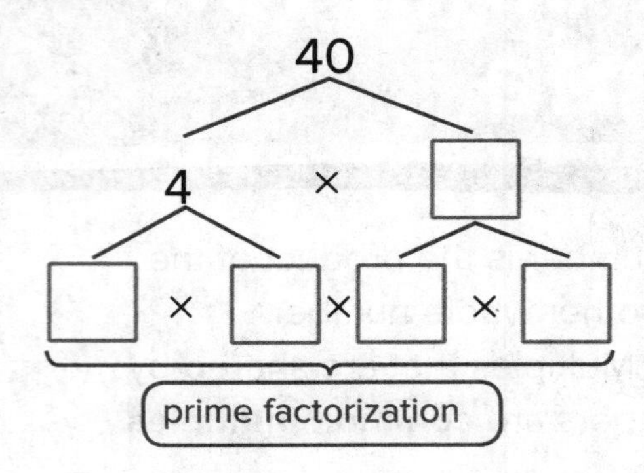

2 Find the common prime factors.

15 = ____ × ____

The only common prime factor is ____.

40 = ____ × ____ × ____ × ____

3 Find the product of the prime factors using each common prime factor only once and any remaining factors.

The LCM is ______ × ______ × ______ × ______ × ______, or ______.

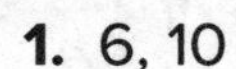

Guided Practice

Find the LCM of each set of numbers by first listing nonzero multiples of each number.

Talk MATH

Could the LCM of two numbers be one of the numbers? Explain.

1. 6, 10

6: ________________________________

10: ________________________________

So, the LCM of 6 and 10 is ________.

2. 3, 4

3: ________________________________

4: ________________________________

So, the LCM of 3 and 4 is ________.

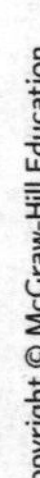

Name

Independent Practice

Find the LCM of each set of numbers.

3. 2, 13 ______

4. 7, 9 ______

5. 2, 10 ______

6. 12, 15 ______

7. 16, 20 ______

8. 3, 8 ______

9. 4, 8, 10 ______

10. 3, 9, 18 ______

11. 15, 25, 75 ______

12. 9, 12, 15 ______

13. 4, 7, 10 ______

14. 6, 7, 9 ______

Problem Solving

15. Juan goes to the bowling alley every 3 weeks. Percy goes to the bowling alley every 5 weeks. If Juan and Percy meet the first time they go bowling, how many weeks will it be before they see each other at the bowling alley again?

____

16. A full moon occurs about every 30 days. If the last full moon occurred on a Friday, how many days will pass before a full moon occurs again on a Friday?

____

Brain Builders

17. Processes & Practices 3 **Find the Error** Maria is finding the LCM of 6 and 8. Help find and correct her mistake. Then find the LCM of 6, 8 and 14.

$6 = 2 \times 3$
$8 = 2 \times 2 \times 2$
The LCM of 6 and 8 is 2.

____

18. Processes & Practices 4 **Model Math** Write a real-world problem in which it would be helpful to find the least common multiple of three numbers. Include the solution.

____

____

____

19. **Building on the Essential Question** How can I find the least common multiple of two or more numbers?

____

____

Name

MY Homework

Lesson 5

Least Common Multiple

Homework Helper

Need help? connectED.mcgraw-hill.com

Ben's Burgers gives away a free order of fries every 2 days, a free milkshake every 3 days, and a free hamburger every 4 days. If they gave away all three items today, in how many days will they give away all three items again?

Find and circle the LCM of 2, 3 and 4 by listing nonzero multiples of each number.

2: 2, 4, 6, 8, 10, (12) . . .

3: 3, 6, 9, (12), 15 . . .

4: 4, 8, (12), 16, 20 . . .

The least common multiple of 2, 3, and 4 is 12.

So, Ben's Burgers will give away all three items again in 12 days.

Check The number line shows on which days all three activities will be completed.

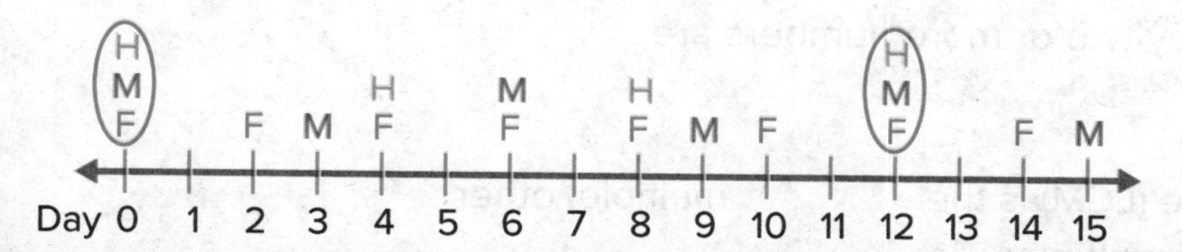

H = hamburger
M = milkshake
F = fries

Practice

Find the LCM of each set of numbers.

1. 7, 14 ________

2. 6, 15 ________

3. 5, 9, 15 ________

Brain Builders

4. The cycles for two different events are shown in the table. Each of these events happened in the year 2000. What is the next year in which both will happen? Explain.

Event	Cycle (yr)
Summer Olympics	4
United States Census	10

____

5. Processes & Practices 3 **Draw a Conclusion** Is the statement below *always*, *sometimes*, or *never* true? Give at least two examples to support your reasoning.

The LCM of two numbers is the product of the two numbers. Discuss your conclusion with a partner.

____

____

____

Vocabulary Check

Fill in each blank with the correct word(s) to complete each sentence.

6. Multiples that are shared by two or more numbers are

____________________.

7. The least common multiple (LCM) is the ________ multiple, other than 0, common to sets of multiples.

8. Test Practice Micah is buying items for a birthday party. If he wants to have the same amount of each item, what is the least number of packages of cups he needs to buy?

Party Supplies	
Item	**Number in Each Package**
Cups	6
Plates	8

Ⓐ 2 packages

Ⓑ 3 packages

Ⓒ 4 packages

Ⓓ 5 packages

Name ______________________________

Lesson 6
Compare Fractions

ESSENTIAL QUESTION
How are factors and multiples helpful in solving problems?

The **least common denominator (LCD)** is the LCM of the denominators of the fractions. You can use the LCD to compare fractions.

Math in My World

Trevor made 2 out of 3 goals and Tyler made 5 out of 6. Who made a greater fraction of goals?

Find the LCM of the denominators of $\frac{2}{3}$ and $\frac{5}{6}$.

3: ______________________________

6: ______________________________

The LCM of 3 and 6 is ________

Find equivalent fractions with a denominator of ________.

$\frac{2}{3} = \frac{2 \times 2}{3 \times 2} = \frac{\square}{\square}$ $\quad$ $\frac{5}{6} = \frac{5 \times 1}{6 \times 1} = \frac{\square}{\square}$

Compare the numerators.

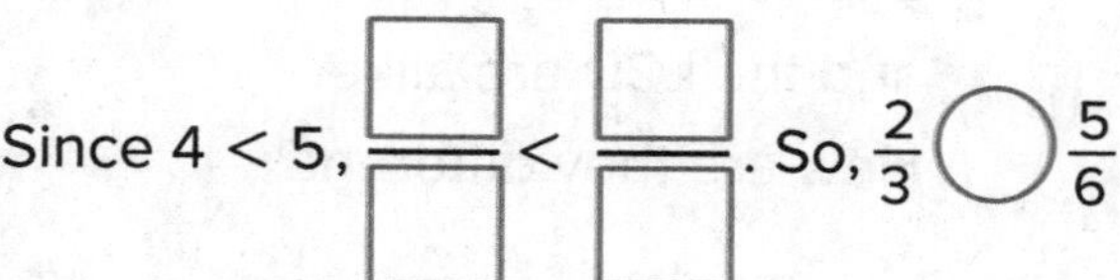

Since 4 < 5, $\frac{\square}{\square} < \frac{\square}{\square}$. So, $\frac{2}{3} \bigcirc \frac{5}{6}$.

So, ________ made a greater fraction of goals.

Helpful Hint
Multiplying the numerator and denominator by the same number is the same as multiplying the fraction by 1. The result is an equivalent fraction.

Check

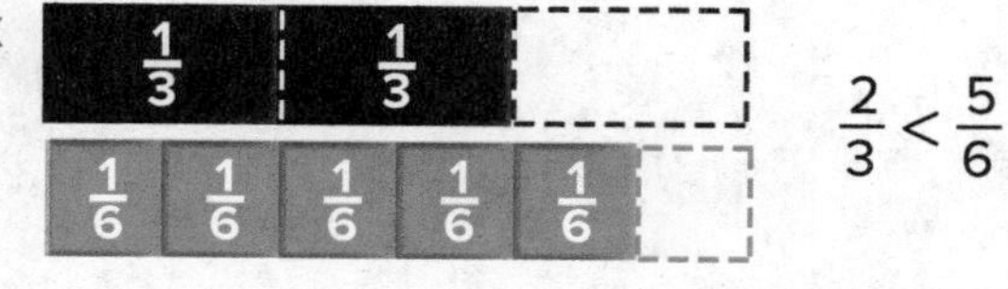

$\frac{2}{3} < \frac{5}{6}$

Example 2

Compare $\frac{3}{5}$ and $\frac{1}{2}$ using the least common denominator.

1 Find the LCM of the denominators.

5: ____________________

2: ____________________

The LCM of 2 and 5 is ________

2 Find equivalent fractions with a denominator of ________ .

$\frac{3}{5} = \frac{3 \times 2}{5 \times 2} = \frac{\square}{\square}$ $\qquad$ $\frac{1}{2} = \frac{1 \times 5}{2 \times 5} = \frac{\square}{\square}$

3 Compare the numerators.

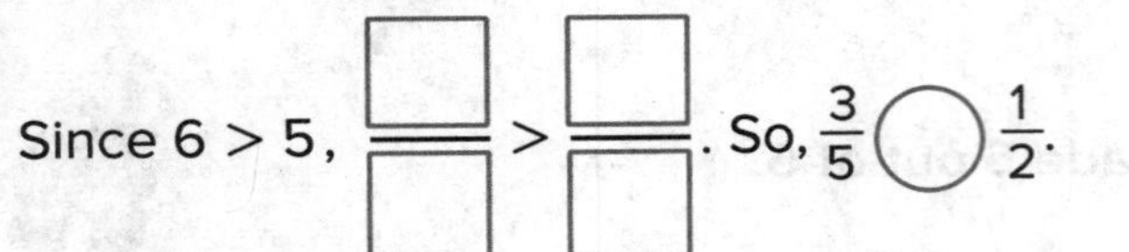

Since 6 > 5, $\frac{\square}{\square} > \frac{\square}{\square}$. So, $\frac{3}{5} \bigcirc \frac{1}{2}$.

Check The models show that $\frac{3}{5} > \frac{1}{2}$.

$\frac{1}{5}$	$\frac{1}{5}$	$\frac{1}{5}$		
$\frac{1}{2}$				

Guided Practice

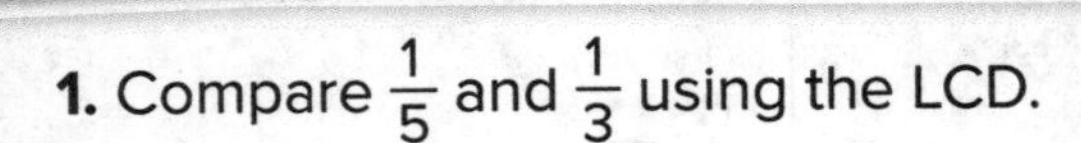

1. Compare $\frac{1}{5}$ and $\frac{1}{3}$ using the LCD.

5: ____________________

3: ____________________

$\frac{1}{5} = \frac{1 \times 3}{5 \times 3} = \frac{\square}{\square}$

$\frac{1}{3} = \frac{1 \times 5}{3 \times 5} = \frac{\square}{\square}$

So, $\frac{1}{5} \bigcirc \frac{1}{3}$.

Explain how the LCM and the LCD are alike. How are they different?

Name ....

Independent Practice

Processes & Practices 6 **Be Precise Compare each pair of fractions by drawing models or using the LCD. Use the symbols <, >, or =.**

2. $\frac{3}{4} \bigcirc \frac{7}{8}$

3. $\frac{2}{3} \bigcirc \frac{7}{10}$

4. $\frac{2}{3} \bigcirc \frac{7}{12}$

5. $\frac{1}{3} \bigcirc \frac{5}{9}$

6. $\frac{1}{4} \bigcirc \frac{1}{6}$

7. $\frac{2}{5} \bigcirc \frac{6}{15}$

8. $\frac{2}{3} \bigcirc \frac{3}{4}$

9. $\frac{1}{5} \bigcirc \frac{3}{15}$

10. $\frac{1}{6} \bigcirc \frac{1}{3}$

Algebra Find each unknown in the equations that shows equivalent fractions.

11. $\frac{3 \times m}{4 \times 5} = \frac{p}{20}$

$m =$ ________

$p =$ ________

12. $\frac{7 \times g}{8 \times k} = \frac{21}{24}$

$g =$ ________

$k =$ ________

13. $\frac{5}{6} = \frac{b}{48}$

$b =$ ________

Problem Solving

14. The amounts of water four runners drank are shown at the right. Who drank the most?

Runner	Amount (bottle)
Evita	$\frac{3}{5}$
Jack	$\frac{5}{8}$
Keisha	$\frac{3}{4}$
Sirjo	$\frac{5}{10}$

15. Processes &Practices 6 **Be Precise** A recipe calls for $\frac{5}{8}$ cup of brown sugar and $\frac{2}{3}$ cup of flour. Which ingredient has the greater amount?

16. A trail mix has $\frac{1}{2}$ cup of raisins and $\frac{2}{3}$ cup of peanuts. Which ingredient has the greater amount?

Brain Builders

17. Processes &Practices 2 **Use Number Sense** Jenny multiplies the numerator of a fraction by m and the denominator of the fraction by p to create an equivalent fraction. Is m greater than, less than, or equal to p? Explain.

18. **Building on the Essential Question** What is one way to compare fractions with unlike denominators? Include an example to support your reasoning.

Name ....

MY Homework

Lesson 6

Compare Fractions

Homework Helper

Need help? connectED.mcgraw-hill.com

Compare $\frac{4}{5}$ and $\frac{5}{6}$ using the least common denominator.

Find the LCM of the denominators.

5: 5, 10, 15, 20, 25, 30 . . .
6: 6, 12, 18, 24, 30 . . . ← The LCM of 5 and 6 is 30.

2 Find equivalent fractions with a denominator of 30.

$\frac{4}{5} = \frac{4 \times 6}{5 \times 6} = \frac{24}{30}$ $\quad \frac{5}{6} = \frac{5 \times 5}{6 \times 5} = \frac{25}{30}$

3 Compare the numerators.

Since $24 < 25$, then $\frac{24}{30} < \frac{25}{30}$. So, $\frac{4}{5} < \frac{5}{6}$.

Check The models show that $\frac{4}{5} < \frac{5}{6}$.

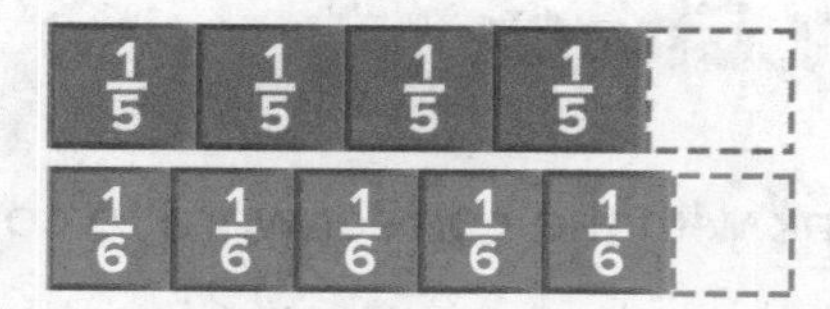

Practice

Compare each pair of fractions by drawing models or using the LCD. Use the symbols <, >, or =.

1. $\frac{3}{4}$ ◯ $\frac{7}{8}$

2. $\frac{1}{3}$ ◯ $\frac{3}{9}$

3. $\frac{3}{4}$ ◯ $\frac{2}{3}$

Problem Solving

4. A survey showed that $\frac{7}{15}$ of a class liked soccer and $\frac{2}{5}$ liked baseball. Which sport was liked less?

____

Brain Builders

5. The fifth graders were given sandwiches for lunch during their field trip. Nathan ate $\frac{5}{6}$ of his sandwich, Leroy ate $\frac{7}{8}$ of his sandwich, and Sofia ate $\frac{3}{4}$ of her sandwich. Who ate the greatest amount of their sandwich? Explain.

____

6. Processes & Practices 2 **Use Symbols** Replace each ■ with a number to make $\frac{■}{24} > \frac{1}{4} > \frac{4}{■}$ a true statement.

____

Vocabulary Check

Vocab

7. Fill in the blank with the correct word to complete the sentence.
The least common denominator (LCD) is the least common multiple of the ______________ of the fractions.

8. Test Practice Eighteen out of 24 of Emil's CDs are country music. Five out of 8 of Imani's CDs are country music. Which is a true statement?

Ⓐ Half of each CD collection consists of country music.

Ⓑ Less than half of each CD collection consists of country music.

Ⓒ Emil has a greater fraction of country music than Imani.

Ⓓ Imani has a greater fraction of country music than Emil.

Name ______

Lesson 7
Hands On

Use Models to Write Fractions as Decimals

ESSENTIAL QUESTION
How are factors and multiples helpful in solving problems?

You can use models to write fractions as equivalent decimals.

Draw It

Tools

Use a model to write $\frac{1}{2}$ as a decimal.

1 Write $\frac{1}{2}$ as an equivalent fraction with a denominator of 10.

$$\frac{1}{2} = \frac{1 \times 5}{2 \times 5} = \frac{\square}{\square}$$

2 Shade a model of $\frac{\square}{\square}$ using the grid.

How many tenths are shaded? ______

The model shows ______ *tenths* or ______.

So, $\frac{1}{2}$ = ______.

Helpful Hint

Multiplying $\frac{1}{2}$ by $\frac{5}{5}$ is the same as multiplying $\frac{1}{2}$ by 1. The result is an equivalent fraction.

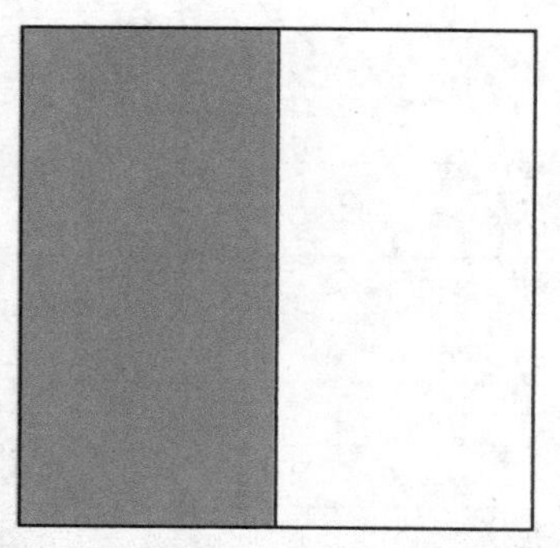

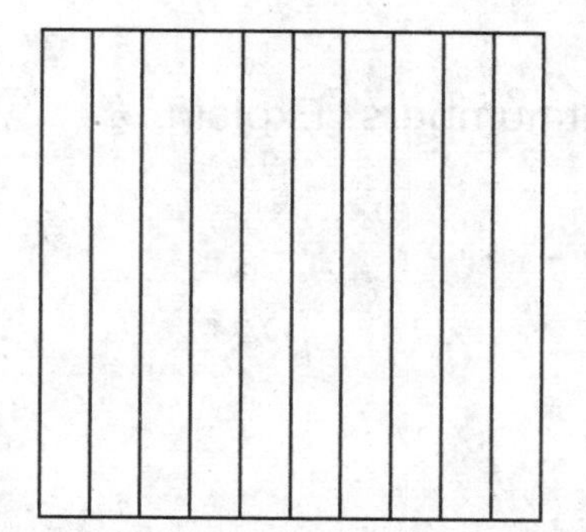

Try It

Use a model to write $\frac{3}{4}$ as a decimal.

Write $\frac{3}{4}$ as a fraction with a denominator of 100.

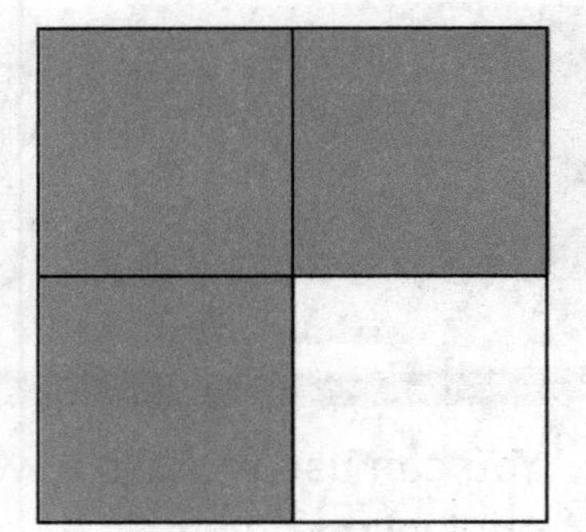

$\frac{3}{4} = \frac{3 \times 25}{4 \times 25} = \frac{\square}{\square}$

Shade a model of $\frac{\square}{\square}$ using the 10-by-10 grid.

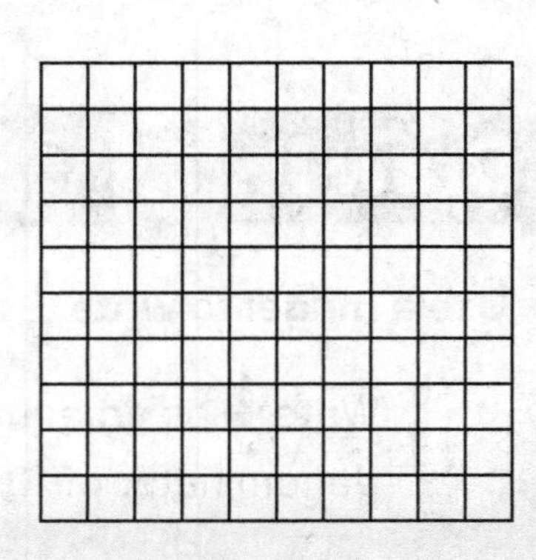

How many squares out of the 100 are shaded? ______

The model shows ______ *hundredths* or ______.

So, $\frac{3}{4} =$ ______.

Talk About It

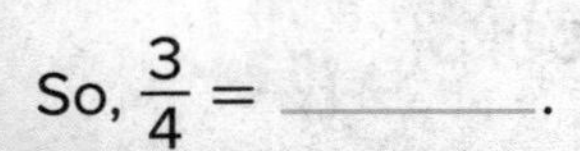

1. **Processes & Practices 4** **Model Math** In the first activity, how would it change if $\frac{1}{2}$ was written as a fraction with a denominator of 100? Would the result be the same? Explain.

______

______

______

2. Do $\frac{3}{5}$ and 0.6 represent equivalent numbers? Explain.

______

______

______

Name ________________

Practice It

Processes & Practices **Use Math Tools** **Shade each model. Then write each fraction as a decimal.**

3. $\frac{1}{4}$ = ________

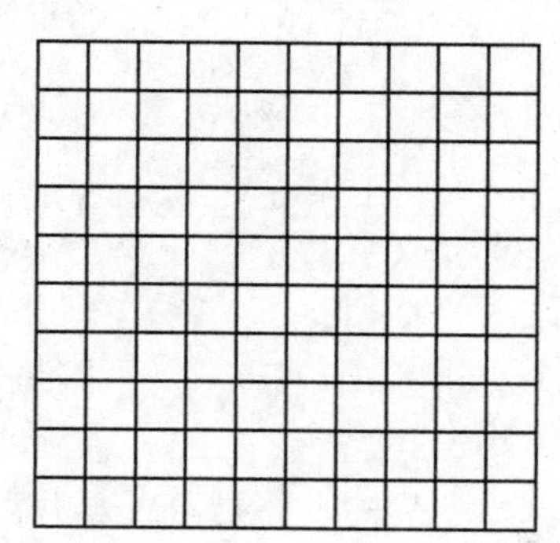

4. $\frac{3}{20}$ = ________

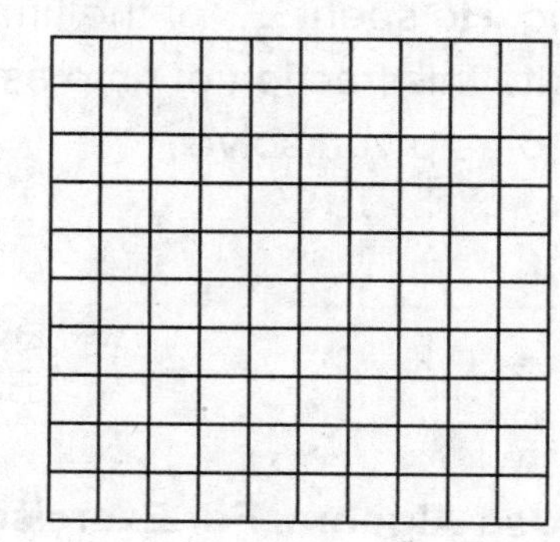

5. $\frac{2}{5}$ = ________

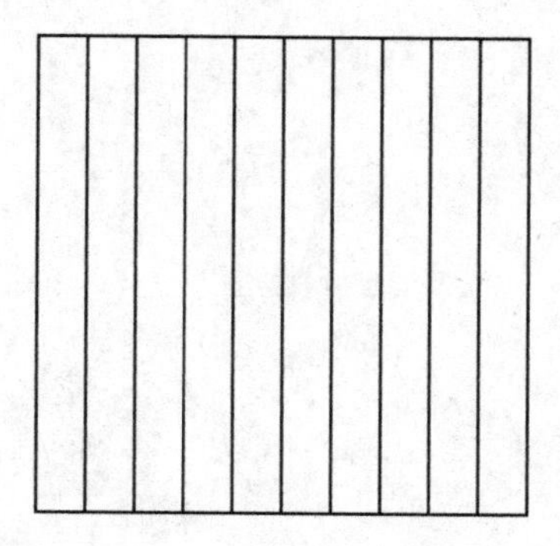

6. $\frac{3}{5}$ = ________

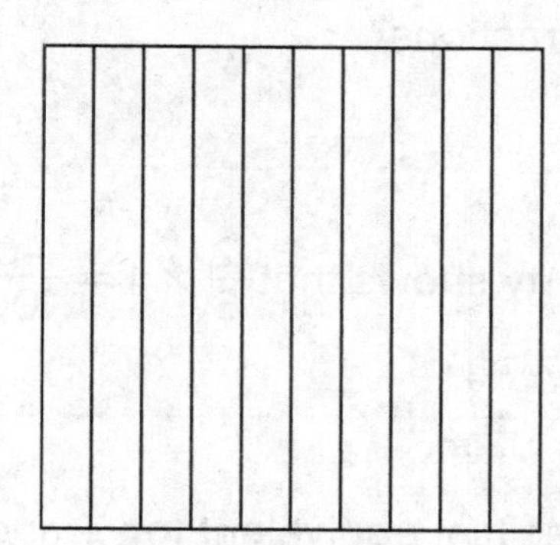

7. $\frac{7}{10}$ = ________

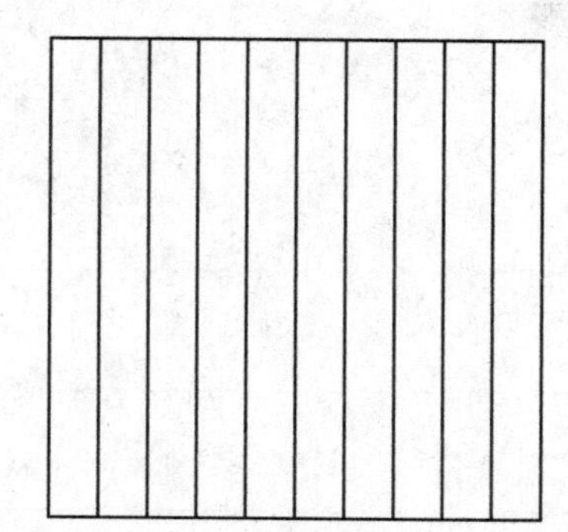

8. $\frac{8}{25}$ = ________

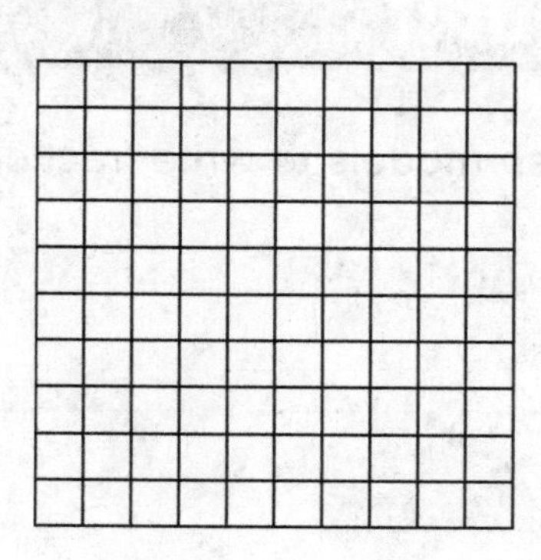

Apply It

9. Juanita practiced shooting 25 free throws at basketball practice. She made $\frac{17}{25}$ of the attempts. Write the fraction of attempts made as a decimal. Use models to help you solve.

____

10. Travis spent 20 minutes getting ready for school in the morning. He spent $\frac{9}{20}$ of the time eating breakfast. Write this fraction of time as a decimal. Use models to help you solve.

____

Use Algebra For Exercises 11–13, refer to the equation $\frac{2 \times p}{5 \times q} = \frac{40}{100}$.

11. What must be true about p and q if the equation shows equivalent fractions?

____

12. What property shows that $\frac{2}{5} \times 1 = \frac{40}{100}$?

____

13. Write the decimal equivalent for $\frac{2}{5}$ and $\frac{40}{100}$.

____

Write About It

14. How can I use models to write fractions as decimals?

____

____

____

Name

MY Homework

Lesson 7

Hands On: Use Models to Write Fractions as Decimals

Homework Helper

Need help? connectED.mcgraw-hill.com

Use a model to write $\frac{7}{20}$ as a decimal.

Helpful Hint

Multiplying the numerator and denominator by the same number is the same as multiplying the fraction by 1. The result is an equivalent fraction.

Write $\frac{7}{20}$ as a fraction with a denominator of 100.

$$\frac{7}{20} = \frac{7 \times 5}{20 \times 5} = \frac{35}{100}$$

The model represents $\frac{35}{100}$.

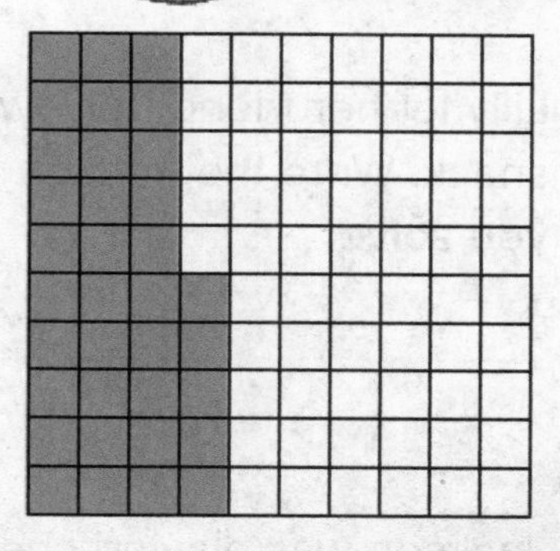

There are 35 squares out of the 100 that are shaded.

The model shows 35 *hundredths* or 0.35.

So, $\frac{7}{20} = 0.35$.

Practice

Shade each model. Then write each fraction as a decimal.

1. $\frac{9}{10} =$ ______

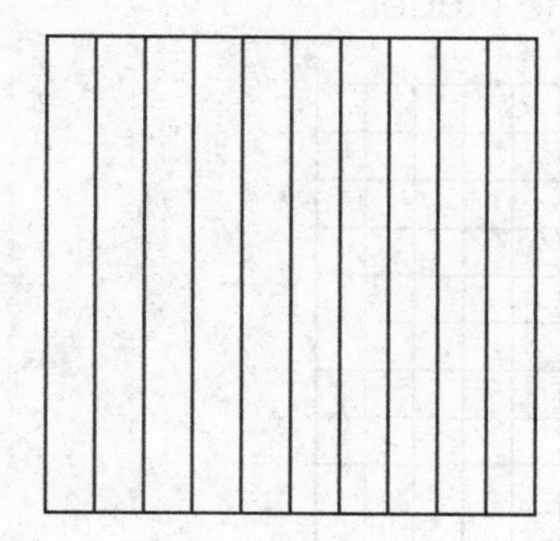

2. $\frac{11}{20} =$ ______

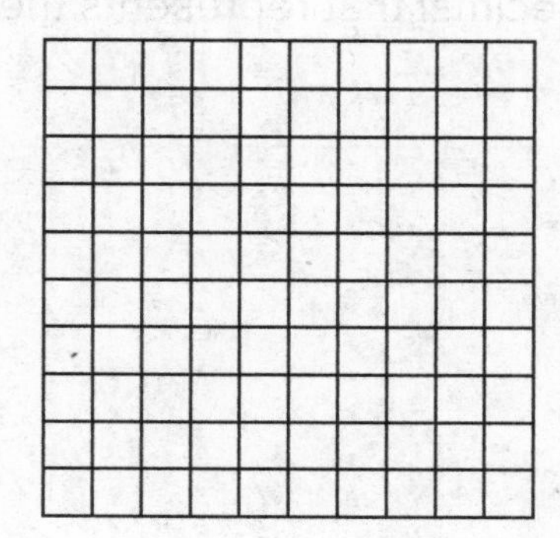

Problem Solving

3. Terrell hit a total of 10 home runs during the baseball season. He hit $\frac{4}{5}$ of the home runs during the first half of the season. Write the fraction of home runs hit during the first half of the season as a decimal. Draw models to help you solve.

____

4. Processes &Practices 5 **Use Math Tools** Bradley and his family drove to visit a museum. They drove $\frac{9}{25}$ of the way and stopped to get gasoline. Write the fraction of the distance traveled as a decimal. Draw models to help you solve.

____

5. Lilly let her friend borrow $\frac{1}{10}$ of the money in her purse to buy a snack. Write the fraction as a decimal. Draw models to help you solve.

____

6. Jackson was playing chess. Out of all the games he played, he won $\frac{7}{25}$ of the time. Write this fraction as a decimal. Draw models to help you solve.

____

7. Write the decimal that represents the shaded portion of the model.

____

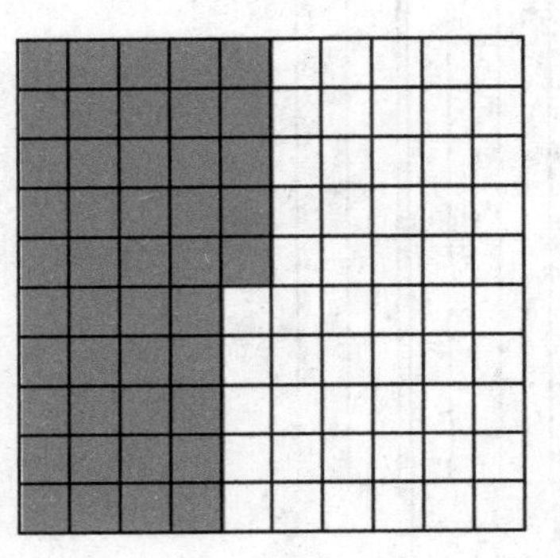

Name ....

Lesson 8
Write Fractions as Decimals

ESSENTIAL QUESTION
How are factors and multiples helpful in solving problems?

Math in My World

Example 1

The average weight of a tennis racquet is $\frac{2}{5}$ pound. Write this weight as a decimal.

Write $\frac{2}{5}$ as a decimal.

Write $\frac{2}{5}$ as an equivalent fraction with a denominator of 10.

$$\frac{2}{5} = \frac{2 \times 2}{5 \times 2} = \frac{\square}{\square}$$

Since 5 × 2 = 10, multiply 2 × 2 to obtain 4.

Write the fraction with a denominator of 10 as a decimal.

$\frac{2}{5}$ = ________ $\frac{4}{10}$ means *four tenths,* or 0.4.

You can use a place-value chart to read the decimal.

Ones	Tenths	Hundredths
0.	4	

Read the decimal. It reads four ____________.

The average weight of a tennis racquet is ________ pound.

Example 2

Write $\frac{3}{4}$ as a decimal.

Multiplying the numerator and denominator by the same number is the same as multiplying the fraction by 1. The result is an equivalent fraction.

Write $\frac{3}{4}$ as an equivalent fraction with a denominator of 100.

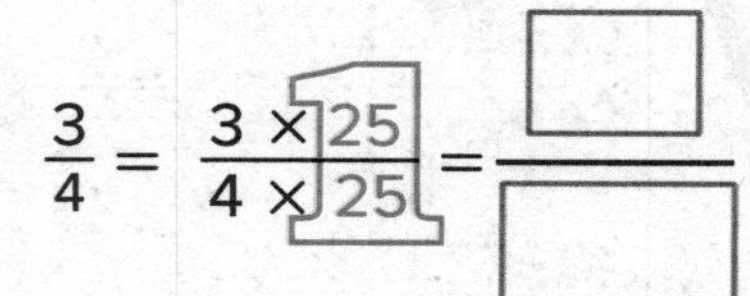

$\frac{3}{4} = \frac{3 \times 25}{4 \times 25} = \frac{\square}{\square}$ Since 4 × 25 = 100, multiply 3 × 25 to obtain 75.

Write the fraction with a denominator of 100 as a decimal.

$\frac{3}{4} =$ ________ $\frac{75}{100}$ means *seventy-five hundredths,* or 0.75.

Read the decimal as ______________________________.

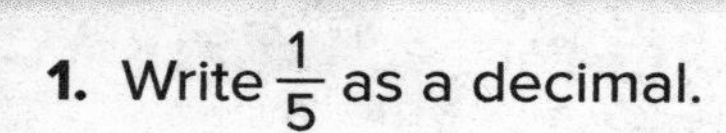

Guided Practice

1. Write $\frac{1}{5}$ as a decimal.

   $\frac{1}{5} = \frac{1 \times 2}{5 \times 2} = \frac{\square}{\square}$

   So, $\frac{1}{5} =$ ________

   Read the decimal as ____________.

Explain how to write a fraction as a decimal using equivalent fractions.

2. Write $\frac{11}{25}$ as a decimal.

   $\frac{11}{25} = \frac{11 \times 4}{25 \times 4} = \frac{\square}{\square}$

   So, $\frac{11}{25} =$ ________

   Read the decimal as ______________________________.

Name

Independent Practice

Write each fraction as a decimal.

3. $\frac{8}{10}$ = ________

4. $\frac{1}{20}$ = ________

5. $\frac{17}{20}$ = ________

6. $\frac{4}{25}$ = ________

7. $\frac{1}{10}$ = ________

8. $\frac{8}{25}$ = ________

9. $\frac{14}{25}$ = ________

10. $\frac{1}{4}$ = ________

11. $\frac{7}{20}$ = ________

12. $\frac{1}{25}$ = ________

13. $\frac{9}{10}$ = ________

14. $\frac{9}{25}$ = ________

Algebra **Find each unknown.**

15. $\frac{g}{20} = 0.65$

$g =$ ________

16. $0.7 = \frac{7}{w}$

$w =$ ________

17. $\frac{n}{50} = 0.18$

$n =$ ________

Problem Solving

18. The smallest known female spider is $\frac{23}{50}$ millimeter long. The smallest male spider is $\frac{37}{100}$ millimeter long. Write each fraction as a decimal.

19. Processes & Practices 4 **Model Math** Evan drank $\frac{2}{25}$ gallon of water throughout the day. Write $\frac{2}{25}$ as a decimal.

Brain Builders

20. At hockey practice, Savannah made 19 out of 20 shots. Write the fraction of shots Savannah missed as a decimal.

21. Processes & Practices 4 **Find the Error** Juliana claims that $\frac{18}{25}$ is equivalent to $\frac{72}{10}$ since they are both written as 0.72 in decimal form. Find her error and correct it.

22. **Building on the Essential Question** What is the relationship between fractions and decimals? Include a fraction and decimal model to help illustrate the relationship.

Name

MY Homework

Lesson 8

Write Fractions as Decimals

Homework Helper

Need help? connectED.mcgraw-hill.com

The average length of a honeybee is $\frac{4}{5}$ inch. Write $\frac{4}{5}$ as a decimal.

Write $\frac{4}{5}$ as an equivalent fraction with a denominator of 10.

$\frac{4}{5} = \frac{4 \times 2}{4 \times 2} = \frac{8}{10}$ Since $5 \times 2 = 10$, multiply 4×2 to obtain 8.

Write the fraction with a denominator of 10 as a decimal.

$\frac{4}{5} = 0.8$

Read the decimal as *eight tenths*.

Practice

Write each fraction as a decimal.

1. $\frac{1}{2} =$ ______

2. $\frac{11}{20} =$ ______

3. $\frac{13}{20} =$ ______

4. $\frac{6}{10} =$ ______

5. $\frac{13}{25} =$ ______

6. $\frac{14}{20} =$ ______

Problem Solving

7. Courtney hit the bullseye $\frac{3}{5}$ of the time when playing darts. Write $\frac{3}{5}$ as a decimal.

____

8. Processes & Practices 2 **Use Number Sense** Yesterday, it rained $\frac{9}{20}$ inch. Write $\frac{9}{20}$ as a decimal.

____

Brain Builders

9. Camille shaded $\frac{18}{30}$ of a model. Write the decimal that represents the unshaded portion of the model.

____

10. Paolo missed 3 out of 15 questions on a quiz, or $\frac{3}{15}$. Write the fraction of questions Paulo answered correctly as a decimal.

____

11. Bishon needs to sell at least 0.3 of his collection of sports cards to make room for new cards. Is $\frac{3}{20}$ of his collection enough? Explain.

____

12. **Test Practice** Emilia bought $\frac{9}{12}$ pound of sliced salami at the deli counter. Which of the following decimals did the scale show?

Ⓐ 0.25 pound

Ⓑ 0.34 pound

Ⓒ 0.70 pound

Ⓓ 0.75 pound

Review

Chapter 8
Fractions and Decimals

Vocabulary Check

Use the word bank below to complete each sentence.

common factor	common multiple
denominator	equivalent fractions
fraction	greatest common factor (GCF)
least common multiple (LCM)	least common denominator (LCD)
multiple	numerator
simplest form	

1. The ______________________ is the least multiple, other than 0, common to sets of multiples.

2. Fractions that name the same number are ______________________.

3. The top number of a fraction is called the ______________________.

4. When the numerator and the denominator have no common factor greater than 1, the fraction is written in ______________________.

5. The bottom number of a fraction is called the ______________________.

6. A whole number that is a factor of two or more numbers is called a(n) ______________________.

7. The greatest of the common factors of two or more numbers is the ______________________ of the numbers.

8. The ______________________ is the least common multiple of the denominators of the fractions.

Concept Check

Find the GCF of each set of numbers.

9. 11, 44 ______

10. 12, 21, 30 ______

Write each fraction in simplest form. If the fraction is already in simplest form, write *simplified.*

11. $\frac{3}{36}$ ______

12. $\frac{25}{30}$ ______

Find the LCM of each set of numbers.

13. 4, 9 ______

14. 5, 7, 10 ______

Compare each pair of fractions using models or the LCD. Use the symbols <, >, or =.

15. $\frac{2}{5}$ ◯ $\frac{3}{10}$

16. $\frac{1}{5}$ ◯ $\frac{1}{4}$

17. $\frac{3}{4}$ ◯ $\frac{18}{24}$

Write each fraction as a decimal.

18. $\frac{3}{10} =$ ______

19. $\frac{19}{50} =$ ______

20. $\frac{5}{10} =$ ______

21. $\frac{1}{5} =$ ______

22. $\frac{14}{25} =$ ______

23. $\frac{3}{25} =$ ______

Problem Solving

24. Three bags of packing peanuts are used to fill 2 boxes. How many bags of packing peanuts does each box use? Between which two whole numbers does the answer lie?

25. A van has room for 6 students and 2 teachers. How many vans are needed for a total of 48 students and 16 teachers?

Brain Builders

26. The table shows the number of each type of toy in a store. The toys will be placed on shelves so that each shelf has the same number of toys on it. There is only one type of toy per shelf. How many shelves are needed for each type of toy so that it has the greatest number of toys? How many toys will be on each shelf?

Toy	Amount
Dolls	45
Footballs	105
Small cars	75

27. Test Practice In a typical symphony orchestra, 16 out of every 100 musicians are violinists. What fraction of the orchestra are violinists?

Ⓐ $\frac{4}{25}$

Ⓑ $\frac{1}{4}$

Ⓒ $\frac{21}{50}$

Ⓓ $\frac{21}{25}$

Reflect

Chapter 8

Answering the ESSENTIAL QUESTION

Use what you learned about fractions and decimals to complete the graphic organizer.

Write an Example

Real-World Example

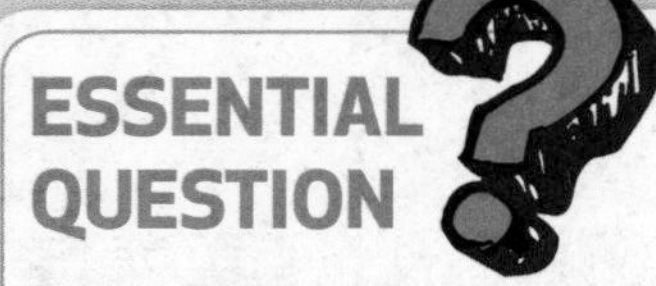

ESSENTIAL QUESTION

How are factors and multiples helpful in solving problems?

Vocabulary

Prime Factorization

Now reflect on the ESSENTIAL QUESTION  **Write your answer below.**

Name

Date

Score

Performance Task

Brain Builders

Free Throw Averages

Middle Valley's basketball coach is looking at data from his team's first five games. The table shows how many times his team had free throws and how many times those free throws were made for a point.

Game	Free Throw Attempts	Points
1	20	7
2	20	9
3	25	12
4	25	10
5	10	6

Show all your work to receive full credit.

Part A

For each game, what fraction of the free throw attempts was turned into points? Write your answer as a fraction in simplest form and as a decimal.

Game	Fraction in Simplest Form	Fraction as a Decimal
1		
2		
3		
4		
5		

Part B

List the games in order from the team's best performance in free throw attempts to their worst.

Part C

How many free throw attempts did the team have in all for the five games? How many points were scored directly from those free throw attempts? Use this to find the decimal that represents the fraction of the points scored from the free throw attempts in the first five games.

Part D

Another team had the exact same average of points scored from free throw attempts in five games. However, their number of attempts was only 50. How many points did they score from their free throws for the five games? Explain.

Chapter

Add and Subtract Fractions

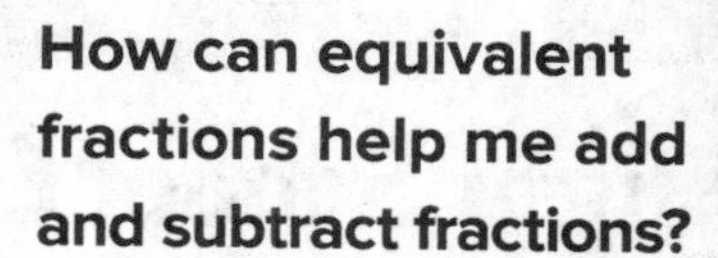

ESSENTIAL QUESTION

How can equivalent fractions help me add and subtract fractions?

Our Oceans

Watch a video!

Watch

Name ....

MY Chapter Project

Flight School Contest

Day 1

1. Work together to make a paper airplane. Search your library or the Internet to get ideas on different airplane designs. Build your airplane so it flies as far as possible.

2. Test your airplane. Make changes to your airplane's design until your group is happy with its performance.

3. Think of an addition or subtraction question about the contest that the class can answer. *Sample questions: What is the difference between the longest and shortest flight distance? What is the total distance flown by all the planes?*

Question: ____

____

____

Day 2

1. Each team will have three chances to fly their airplane, measure the largest distance flown, and record the distance using mixed numbers below.

Largest distance: ____

2. Answer the class questions designated by your teacher.

____

____

____

____

____

____

Name

Am I Ready?

Write each fraction in simplest form.

1. $\frac{4}{8}$ = ________

2. $\frac{4}{12}$ = ________

3. $\frac{15}{20}$ = ________

4. $\frac{4}{24}$ = ________

5. $\frac{16}{30}$ = ________

6. $\frac{24}{40}$ = ________

7. Marcela made 4 out of 16 free throws. Write the fraction of free throws she made in simplest form.

____

Write each improper fraction as a mixed number.

8. $\frac{10}{7}$ = ________

9. $\frac{3}{2}$ = ________

10. $\frac{14}{6}$ = ________

11. $\frac{22}{4}$ = ________

12. $\frac{30}{7}$ = ________

13. $\frac{41}{8}$ = ________

14. A recipe for potato casserole calls for $\frac{7}{4}$ cups of cheese. Write the fraction as a mixed number.

____

How Did I Do?

Shade the boxes to show the problems you answered correctly.

1	2	3	4	5	6	7	8	9	10	11	12	13	14

Name

MY Math Words

Vocab

Review Vocabulary

factors
greatest common factor (GCF)
least common multiple (LCM)
mixed numbers
multiples

Making Connections

Use the review words to complete the diagram below which relates factors and multiples of the numbers 9 and 12.

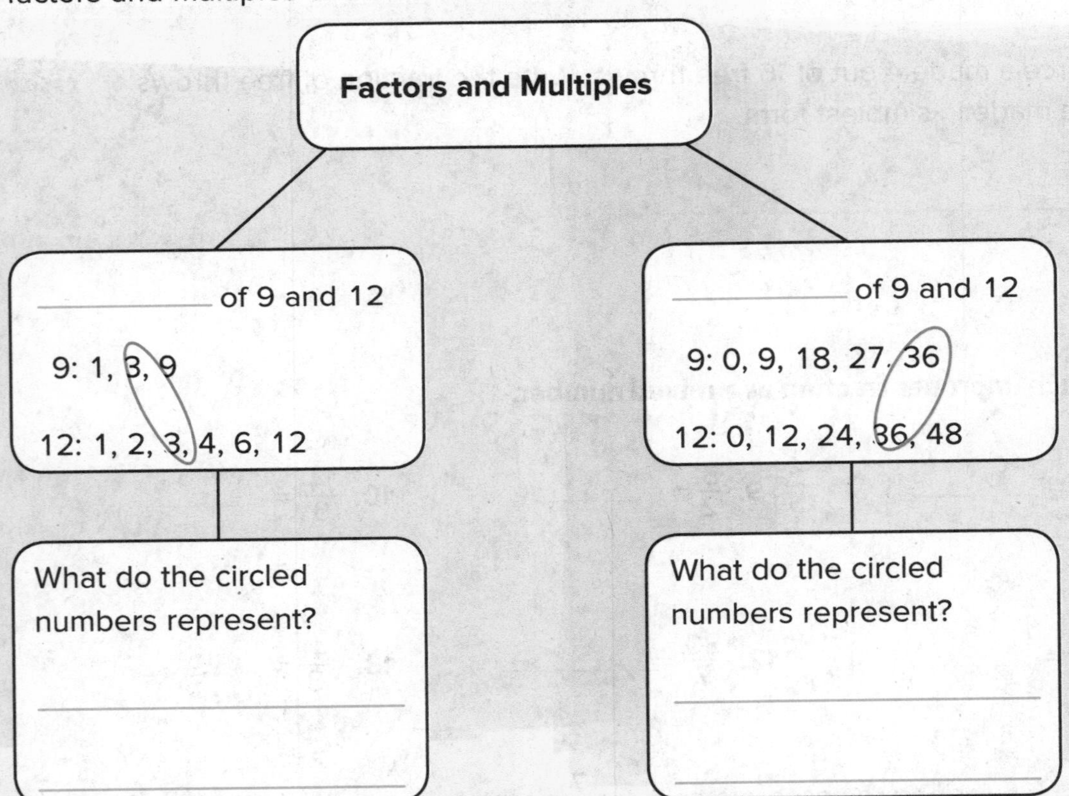

How can greatest common factors and least common multiples help you work with fractions?

MY Vocabulary Cards

Lesson 9-2

like fractions

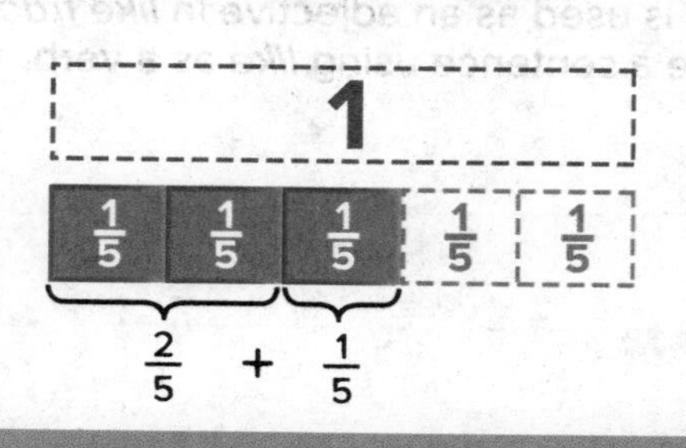

Lesson 9-4

unlike fractions

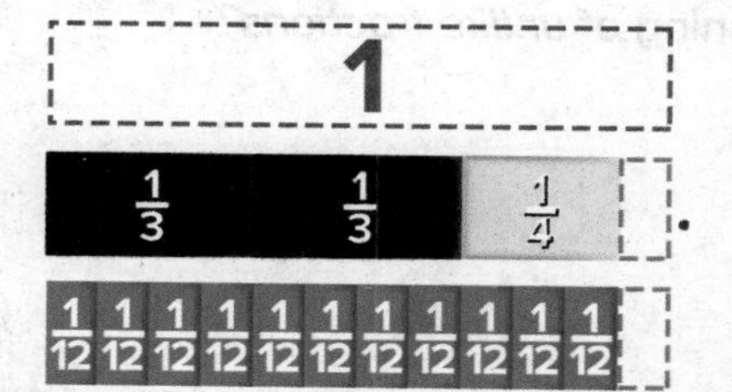

Ideas for Use

- Practice your penmanship! Write this chapter's essential question and each word in cursive.
- Use the back of the card to write or draw examples to help you answer the essential question.

Fractions that have different denominators.

How can the prefix *un-* help you remember the meaning of *unlike fractions*?

Fractions that have the same denominator.

***Like* is used as an adjective in *like fractions*. Write a sentence using *like* as a verb.**

MY Foldable

FOLDABLES® Follow the steps on the back to make your Foldable.

Like Denominators

$\frac{3}{8} + \frac{1}{8}$

1

2

3

Unlike Denominators

$\frac{3}{10} + \frac{1}{5}$

1

2

3

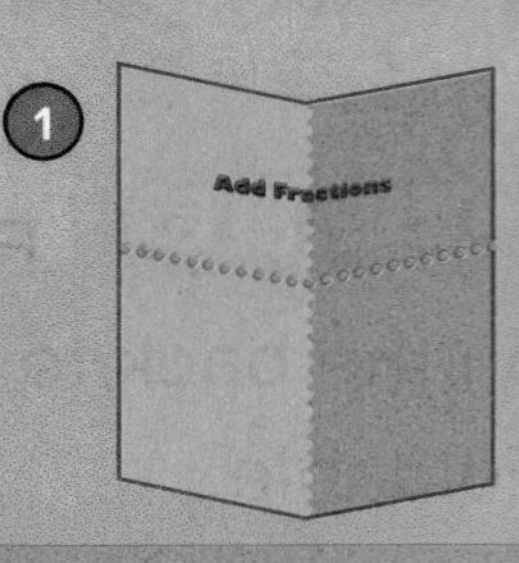

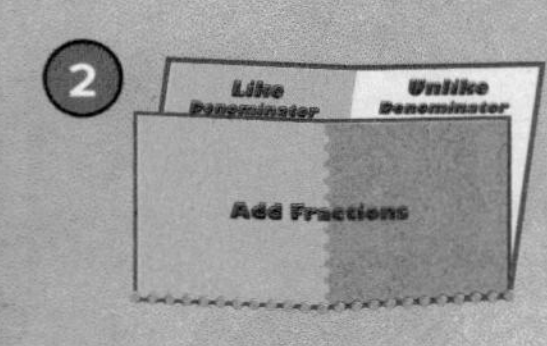

Add Fractions

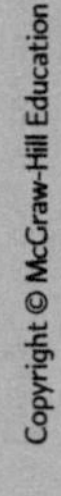

Name ....

Lesson 1
Rounding Fractions

ESSENTIAL QUESTION
How can equivalent fractions help me add and subtract fractions?

You can use benchmark fractions, such as $\frac{1}{2}$, to round a fraction to 0, $\frac{1}{2}$, or 1.

Math in My World

Example 1

A poison dart frog is 2 inches long. This is equal to $\frac{2}{12}$ foot. Is $\frac{2}{12}$ closest to 0, $\frac{1}{2}$, or 1?

Mark off 12 equal increments from 0 to 1.
Graph $\frac{2}{12}$ on a number line. You know $\frac{1}{2} = \frac{6}{12}$.

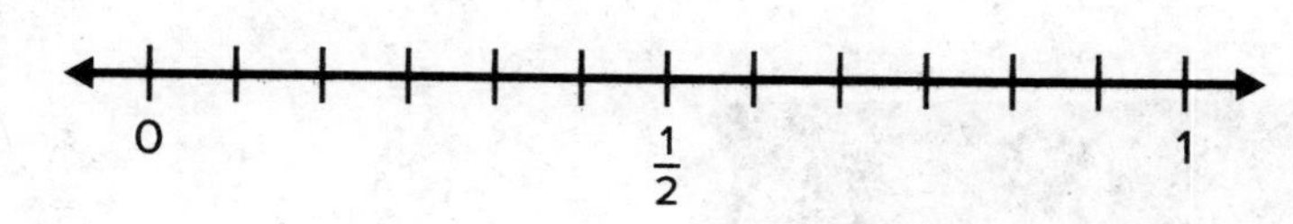

Which number, 0, $\frac{1}{2}$, or 1, is $\frac{2}{12}$ closest to on the number line? ______

So, the length of a poison dart frog is closest to ______ feet.

Key Concept Rounding Fractions

Round Down

If the numerator is much smaller than the denominator, round the fraction down to 0.

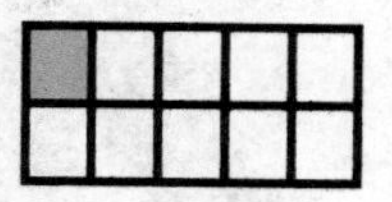

$\frac{1}{10}$ rounds to 0.

Round to $\frac{1}{2}$

If the numerator is about half of the denominator, round the fraction to $\frac{1}{2}$.

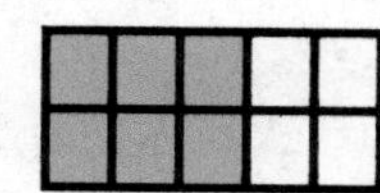

$\frac{6}{10}$ rounds to $\frac{1}{2}$.

Round Up

If the numerator is almost as large as the denominator, round the fraction up to 1.

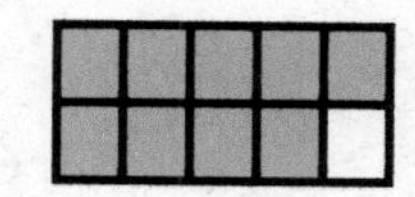

$\frac{9}{10}$ rounds to 1.

Example 2

Round $\frac{4}{9}$ to 0, $\frac{1}{2}$, or 1.

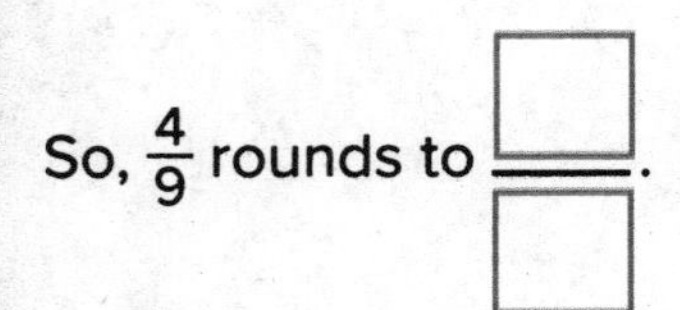

The number $4\frac{1}{2}$ is half of 9.

The numerator in $\frac{4}{9}$ is ________,

which is very close to $4\frac{1}{2}$.

So, $\frac{4}{9}$ rounds to $\frac{\square}{\square}$.

Helpful Hint

The denominator names the number of units. When rounding $\frac{4}{9}$, the number line should be divided into 9 equal sections.

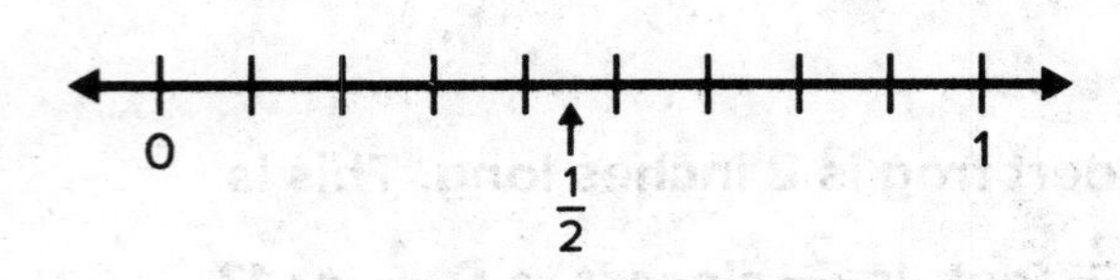

Check Graph $\frac{4}{9}$ on a number line.

$\frac{1}{2}$ is halfway between $\frac{4}{9}$ and $\frac{5}{9}$ since half of 9 is $4\frac{1}{2}$.

Guided Practice

Graph each fraction on the number line. Then state whether each fraction is closest to 0, $\frac{1}{2}$, or 1.

Talk MATH

Explain how to round fractions in your own words.

1. $\frac{5}{6} \approx$ ________

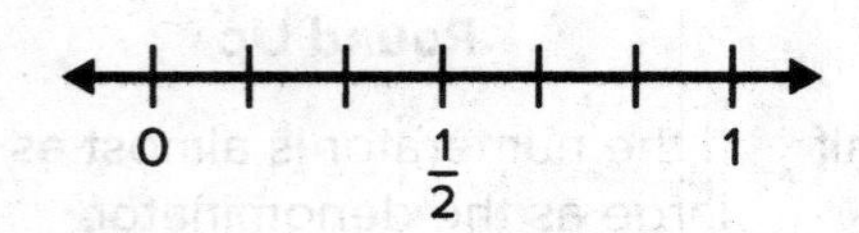

2. $\frac{5}{8} \approx$ ________

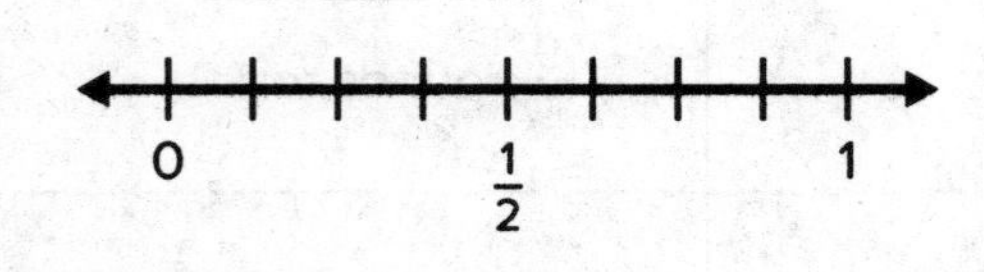

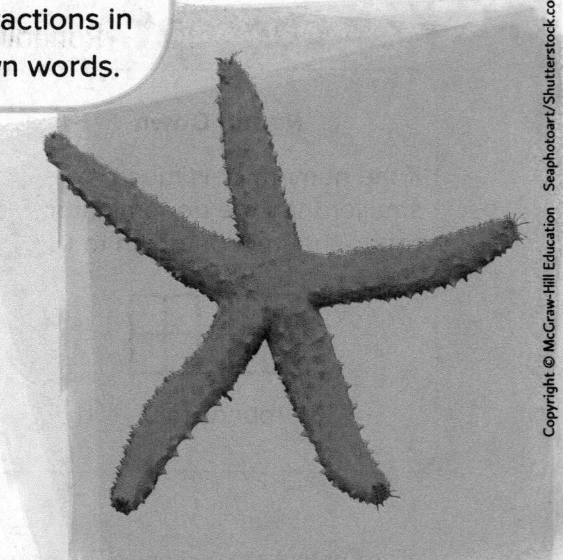

Name ________________

Independent Practice

Round each fraction to 0, $\frac{1}{2}$, or 1. Use a number line if needed.

3. $\frac{1}{8} \approx$ ______

4. $\frac{5}{9} \approx$ ______

5. $\frac{7}{8} \approx$ ______

6. $\frac{3}{7} \approx$ ______

7. $\frac{5}{11} \approx$ ______

8. $\frac{4}{5} \approx$ ______

9. $\frac{1}{9} \approx$ ______

10. $\frac{6}{7} \approx$ ______

11. $\frac{2}{5} \approx$ ______

12. $\frac{3}{8} \approx$ ______

13. $\frac{1}{5} \approx$ ______

14. $\frac{15}{16} \approx$ ______

Problem Solving

15. Processes & Practices 6 **Be Precise** Round the length of the ribbon to 0 inches, $\frac{1}{2}$ inch, or 1 inch.

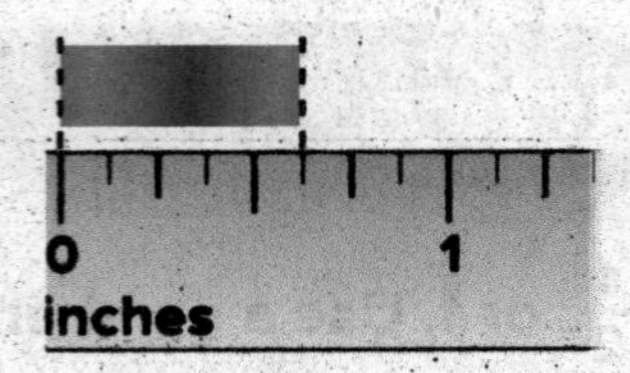

16. Corri has finished about $\frac{3}{5}$ of her daily chores. Has she finished about half of her chores or almost all of them?

Brain Builders

17. Peter has read about $\frac{12}{15}$ of his book. Ling has read about $\frac{7}{15}$ of her book. Has each student read about half of their book or almost all of their book?

18. Processes & Practices 3 **Which One Doesn't Belong?** Circle the fraction that does not belong with the other three. Explain your reasoning. Write a new fraction that would belong with the other three.

$\frac{2}{11}$ $\frac{8}{15}$ $\frac{7}{13}$ $\frac{5}{12}$

19. **Building on the Essential Question** Describe two different ways of rounding fractions.

Name

Lesson 1

Round Fractions

Homework Helper

Need help? connectED.mcgraw-hill.com

Round $\frac{10}{11}$ to 0, $\frac{1}{2}$, or 1.

Compare the numerator and denominator. The numerator, 10, is close to 11.

So, $\frac{10}{11}$ rounds to 1.

Check Graph $\frac{10}{11}$ on a number line.

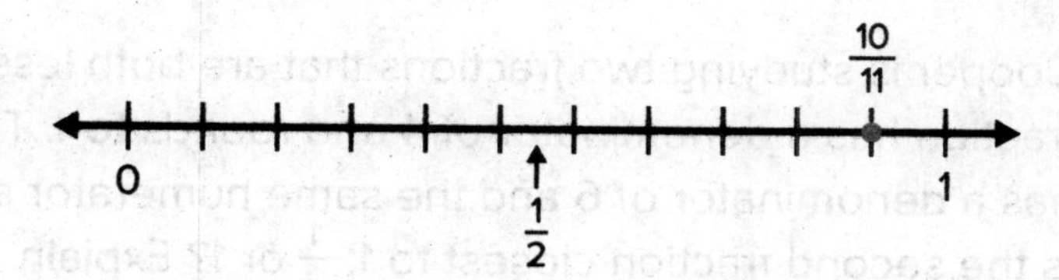

$\frac{1}{2}$ is halfway between $\frac{5}{11}$ and $\frac{6}{11}$ since half of 11 is $5\frac{1}{2}$.

Practice

Round each fraction to 0, $\frac{1}{2}$, or 1. Use a number line if needed.

1. $\frac{5}{9} \approx$ ______

2. $\frac{1}{14} \approx$ ______

3. $\frac{12}{13} \approx$ ______

4. $\frac{2}{13} \approx$ ______

5. $\frac{9}{11} \approx$ ______

6. $\frac{9}{17} \approx$ ______

Problem Solving

7. Kevin ate $\frac{5}{12}$ of a pizza. Which is a better estimate for the amount of pizza that he ate: about half of the pizza or almost all of the pizza?

8. Processes &Practices 1 **Make Sense of Problems** Darius has mowed $\frac{1}{5}$ of his backyard. Which is a better estimate for how much of the lawn he has left to mow: almost all of the lawn or about half of the lawn?

Brain Builders

9. Cooper is studying two fractions that are both less than 1. The first fraction has a denominator of 4 and rounds to 1. The second fraction has a denominator of 6 and the same numerator as the first fraction. Is the second fraction closest to 1, $\frac{1}{2}$ or 1? Explain.

10. Savannah is making a quilt with squares that have side lengths of $\frac{15}{16}$ foot each. Are the side lengths of the squares closer to $\frac{1}{2}$ foot or 1 foot long? Draw a diagram to support your answer.

11. **Test Practice** Samantha shaded $\frac{3}{7}$ of her design.

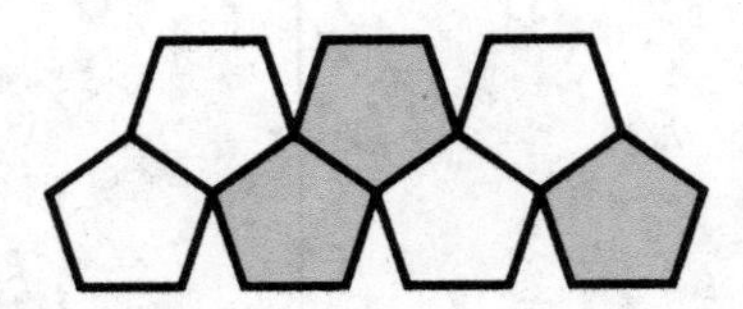

Which number is the best estimate for the shaded part of her design?

Ⓐ 0

Ⓑ $\frac{1}{7}$

Ⓒ $\frac{1}{2}$

Ⓓ 1

Name ....

Lesson 2
Add Like Fractions

ESSENTIAL QUESTION
How can equivalent fractions help me add and subtract fractions?

Like fractions have the same denominator.

Math in My World

Example 1

The length across the bell, or top, of a mushroom jellyfish is about $\frac{5}{6}$ foot. If two mushroom jellyfish were placed side by side, what would be the combined length?

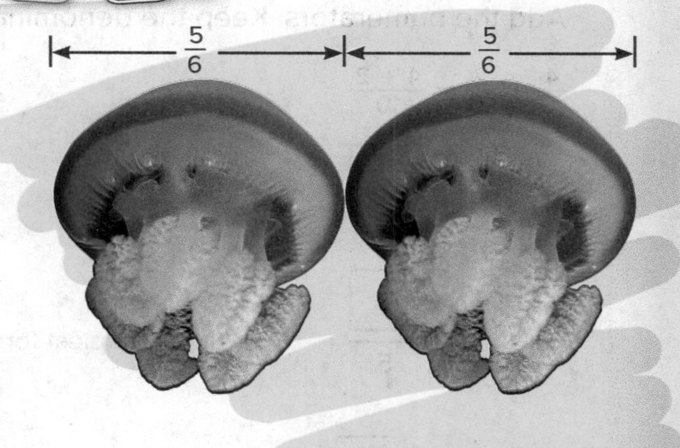

Find $\frac{5}{6} + \frac{5}{6}$.

One Way **Use models.**

Place two sets of five $\frac{1}{6}$-tiles side by side.

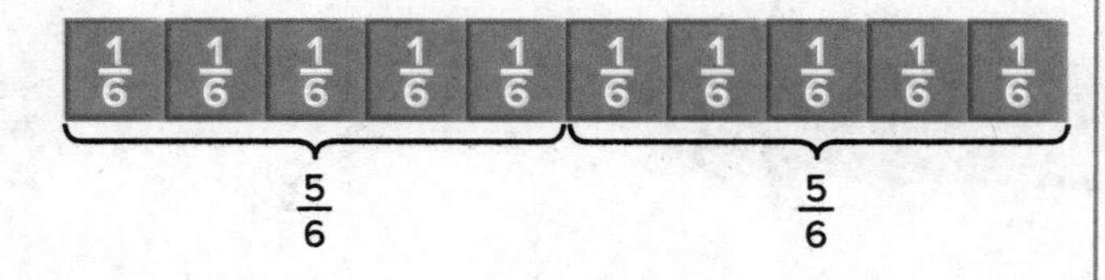

There are ______ $\frac{1}{6}$-tiles all together.

This shows the fraction $\frac{10}{6}$, or $1\frac{2}{3}$.

The combined length would be $\square\frac{\square}{\square}$ feet.

Another Way **Add the numerators. Keep the denominator.**

$$\frac{5}{6} + \frac{5}{6} = \frac{5+5}{6}$$

$$= \frac{10}{6} \quad 5 + 5 = 10$$

$$= 1\frac{2}{3} \quad \text{Write as a mixed number in simplest form.}$$

So, $\frac{5}{6} + \frac{5}{6} = \square\frac{\square}{\square}$

Example 2

The table shows how much of a book Teddy read each day. What fraction of the book did Teddy read all together on Tuesday and Thursday?

Day	Fraction Read
Monday	$\frac{1}{10}$
Tuesday	$\frac{4}{10}$
Wednesday	$\frac{3}{10}$
Thursday	$\frac{2}{10}$

On Tuesday, Teddy read $\frac{\square}{\square}$ of the book.

On Thursday, he read $\frac{\square}{\square}$ of the book.

Add the numerators. Keep the denominator.

$$\frac{4}{10} + \frac{2}{10} = \frac{4+2}{10}$$

$$= \frac{\square}{10}$$ Add. $4 + 2 = 6$

$$= \frac{\square}{5}$$ Write in simplest form.

So, Teddy read $\frac{\square}{\square}$ of the book on Tuesday and Thursday.

Guided Practice

Add. Write each sum in simplest form.

1. $\frac{1}{7} + \frac{3}{7} = \frac{\square}{\square}$

2. $\frac{1}{4} + \frac{1}{4} = \frac{\square}{\square} = \frac{\square}{\square}$

Describe a real-world problem that can be solved by adding like fractions.

Name

Independent Practice

Add. Write each sum in simplest form.

3. $\frac{1}{6} + \frac{1}{6} =$ ______

4. $\frac{5}{8} + \frac{3}{8} =$ ______

5. $\frac{2}{9} + \frac{3}{9} =$ ______

6. $\frac{4}{7} + \frac{2}{7} =$ ______

7. $\frac{2}{6} + \frac{2}{6} =$ ______

8. $\frac{2}{10} + \frac{5}{10} =$ ______

9. $\frac{3}{8} + \frac{1}{8} =$ ______

10. $\frac{3}{4} + \frac{1}{4} =$ ______

11. $\frac{4}{9} + \frac{5}{9} =$ ______

Algebra Find each unknown.

12. $\frac{1}{3} + \blacksquare = \frac{2}{3}$

The unknown is ______.

13. $\frac{5}{12} + \frac{4}{12} = \frac{\blacksquare}{4}$

The unknown is ______.

14. $\frac{3}{10} + \frac{4}{\blacksquare} = \frac{7}{10}$

The unknown is ______.

Problem Solving

15. Terri painted $\frac{5}{12}$ of the fence. Rey painted $\frac{4}{12}$ of the fence. How much of the fence did they paint all together? Write in simplest form.

16. Meagan walked $\frac{4}{10}$ mile to the park. She walked the same distance home. How much did she walk all together? Write in simplest form.

Brain Builders

17. Processes &Practices 2 **Use Number Sense** It rained $\frac{2}{8}$ inch in one hour. It rained $\frac{3}{8}$ inch each hour for the next two hours. Find the total amount of rain. Write in simplest form.

18. Processes &Practices 3 **Justify Conclusions** Select two fractions whose sum is $\frac{3}{4}$, and whose denominators are both the same, but are not 4. Justify your selection.

19. **Building on the Essential Question** When adding like fractions, why are the numerators added but not the denominators? When are equivalent fractions used when adding two like fractions?

Name ....

MY Homework

Lesson 2
Add Like Fractions

Homework Helper

Need help? connectED.mcgraw-hill.com

Find $\frac{3}{10} + \frac{3}{10}$. Write the sum in simplest form.

$\frac{3}{10} + \frac{3}{10} = \frac{3+3}{10}$ Add the numerators. Keep the denominator.

$= \frac{6}{10}$ Add. $3 + 3 = 6$

$= \frac{3}{5}$ Write in simplest form.

So, $\frac{3}{10} + \frac{3}{10} = \frac{3}{5}$.

Check The models show that $\frac{3}{10} + \frac{3}{10} = \frac{6}{10}$, or $\frac{3}{5}$.

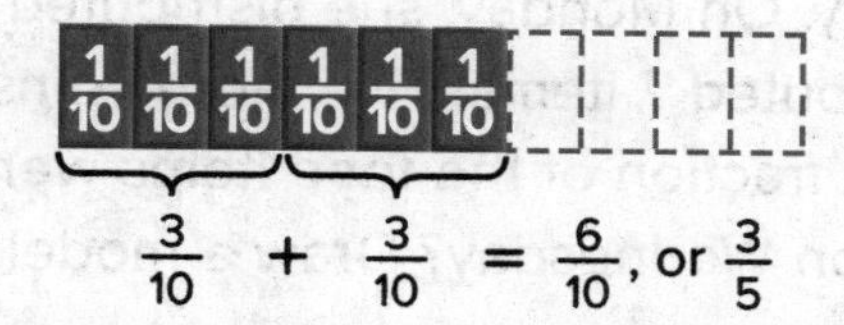

Practice

Add. Write each sum in simplest form.

1. $\frac{7}{10} + \frac{2}{10} =$ ______

2. $\frac{13}{16} + \frac{2}{16} =$ ______

3. $\frac{4}{5} + \frac{1}{5} =$ ______

4. $\frac{7}{15} + \frac{2}{15} =$ ______

5. $\frac{9}{20} + \frac{3}{20} =$ ______

6. $\frac{5}{8} + \frac{1}{8} =$ ______

Problem Solving

The table gives the fraction of each type of parade float used in a recent parade. Use the table to answer Exercises 7 and 8.

Type of Parade Float	Fraction
Sports Team	$\frac{6}{18}$
Radio Station	$\frac{5}{18}$
High School	$\frac{3}{18}$
Dance Group	$\frac{4}{18}$

7. What fraction of the floats were from either a dance group or a radio station? Write in simplest form.

____

8. What fraction of the floats were not from a sports team? Write in simplest form.

____

Brain Builders

9. Processes & Practices 3 **Draw a Conclusion** Sherry was in charge of distributing 25 food items that were donated to the local food pantry. On Monday, she distributed 8 items. On Tuesday, she distributed 7 items. Five more items were distributed on Wednesday. What fraction of the food items were distributed by the end of the day on Wednesday? Draw a model to help solve the problem.

____

What fraction of the food items were distributed each day?

____

Vocabulary Check

Complete the sentence with the correct vocabulary word(s).

10. The fractions in the expression $\frac{1}{3} + \frac{1}{3}$ are examples of ______________.

11. **Test Practice** Gina is working on a jigsaw puzzle. She completed $\frac{1}{10}$ of the puzzle in the morning, $\frac{1}{10}$ in the afternoon, and $\frac{2}{10}$ in the evening. In simplest form, what fraction of the puzzle is completed?

Ⓐ $\frac{2}{5}$ Ⓑ $\frac{3}{5}$ Ⓒ $\frac{2}{10}$ Ⓓ $\frac{3}{10}$

Name

Lesson 3
Subtract Like Fractions

ESSENTIAL QUESTION
How can equivalent fractions help me add and subtract fractions?

Math in My World

Example 1

About $\frac{7}{10}$ of Earth's surface is covered by oceans. The Pacific Ocean is the largest ocean and covers about $\frac{3}{10}$ of Earth's surface. How much of Earth's surface is covered by oceans other than the Pacific Ocean?

Find $\frac{7}{10} - \frac{3}{10}$.

One Way **Use Models.**

Place seven $\frac{1}{10}$-fraction tiles.

$\frac{1}{10}$	$\frac{1}{10}$	$\frac{1}{10}$	$\frac{1}{10}$	X	X	X			

Remove three of the tiles.

There are ______ tiles left, which represent $\frac{4}{10}$, or $\frac{2}{5}$.

Another Way **Subtract the numerators. Keep the denominator.**

$$\frac{7}{10} - \frac{3}{10} = \frac{7-3}{10}$$

$$= \frac{4}{10} \qquad 7 - 3 = 4$$

$$= \frac{\square}{\square} \qquad \text{Write in simplest form.}$$

So, $\frac{\square}{\square}$ of Earth's surface is covered by oceans other than the Pacific Ocean.

Example 2

The table shows the amount of rainfall several cities received in a recent month. How much more rain did Centerville receive than Brushton? Write in simplest from.

City	Rainfall (in.)
Spring Valley	$\frac{1}{10}$
Clarksburg	$\frac{6}{10}$
Centerville	$\frac{9}{10}$
Brushton	$\frac{3}{10}$

Subtract the numerators. Keep the denominator the same.

$\frac{9}{10} - \frac{3}{10} = \frac{9-3}{10}$

$= \frac{6}{10}$ $\quad 9 - 3 = 6$

$= \frac{\square}{\square}$ $\quad$ Write in simplest form.

So, $\frac{\square}{\square}$ inch more rain fell in ______________ than in ______________.

Guided Practice

Subtract. Write each difference in simplest form.

1. $\frac{5}{7} - \frac{3}{7} = \frac{\square}{\square}$

2. $\frac{3}{5} - \frac{2}{5} = \frac{\square}{\square}$

3. $\frac{6}{9} - \frac{3}{9} = \frac{\square}{\square}$

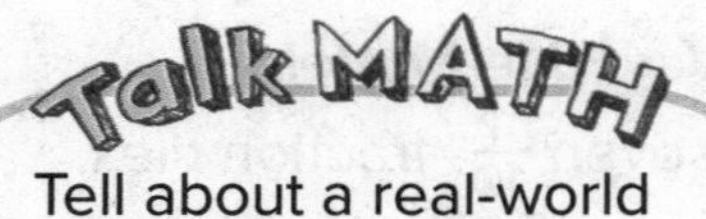

Tell about a real-world situation in which you would find $\frac{3}{4} - \frac{1}{4}$.

Name

Independent Practice

Subtract. Write each difference in simplest form.

4. $\frac{5}{6} - \frac{3}{6} =$ ______

5. $\frac{2}{3} - \frac{1}{3} =$ ______

6. $\frac{3}{5} - \frac{1}{5} =$ ______

7. $\frac{6}{7} - \frac{5}{7} =$ ______

8. $\frac{5}{9} - \frac{2}{9} =$ ______

9. $\frac{6}{8} - \frac{4}{8} =$ ______

10. $\frac{3}{4} - \frac{1}{4} =$ ______

11. $\frac{9}{12} - \frac{3}{12} =$ ______

12. $\frac{4}{5} - \frac{2}{5} =$ ______

Algebra **Find each unknown.**

13. $\frac{5}{9} - \frac{1}{9} = b$

$b =$ ______

14. $\frac{6}{8} - \frac{h}{8} = \frac{1}{8}$

$h =$ ______

15. $\frac{3}{4} - \frac{2}{4} = \frac{w}{4}$

$w =$ ______

Problem Solving

16. Processes & Practices 1 **Make Sense of Problems** A bucket was $\frac{7}{10}$ full with water. After Vick washed the car, the bucket was only $\frac{3}{10}$ full. What fraction of the water in the bucket did Vick use to wash the car? Write in simplest form.

17. Roshanda bought $\frac{5}{8}$ pound of ham and $\frac{7}{8}$ pound of roast beef. How much more roast beef than ham did she buy? Write in simplest form.

18. Chris spent $\frac{5}{6}$ hour drawing and $\frac{2}{6}$ hour reading. How much more time did he spend drawing than reading? Write in simplest form.

Brain Builders

19. Processes & Practices 2 **Use Number Sense** Explain the relationship between two like fractions whose difference is close to one-half.

20. **Building on the Essential Question** Describe two methods you can use to subtract like fractions.

Name ________________

Lesson 3

Subtract Like Fractions

Homework Helper

Need help? connectED.mcgraw-hill.com

Carla volunteers at a veterinarian's office. Of her time spent there, $\frac{6}{10}$ is spent grooming dogs and $\frac{1}{10}$ is spent answering phone calls. How much more of her time is spent grooming dogs than answering phone calls? Write the difference in simplest form.

Find $\frac{6}{10} - \frac{1}{10}$.

$\frac{6}{10} - \frac{1}{10} = \frac{6-1}{10}$ Subtract the numerators. Keep the denominator the same.

$= \frac{5}{10}$ $6 - 1 = 5$

$= \frac{1}{2}$ Write in simplest form.

So, $\frac{6}{10} - \frac{1}{10} = \frac{1}{2}$.

Carla spends $\frac{1}{2}$ more of her time grooming dogs than answering phone calls.

Check The models show that $\frac{6}{10} - \frac{1}{10} = \frac{5}{10}$ or $\frac{1}{2}$.

$\frac{1}{10}$	$\frac{1}{10}$	$\frac{1}{10}$	$\frac{1}{10}$	$\frac{1}{10}$	X				

Practice

Subtract. Write each difference in simplest form.

1. $\frac{3}{6} - \frac{1}{6} =$ ______

2. $\frac{7}{9} - \frac{3}{9} =$ ______

3. $\frac{7}{8} - \frac{2}{8} =$ ______

Problem Solving

The table shows the results of a survey of 28 students and their favorite tourist attractions. Use the table to answer Exercises 4 and 5.

Place	Fraction of Students
Mt. Rushmore	$\frac{14}{28}$
Grand Canyon	$\frac{8}{28}$
Statue of Liberty	$\frac{6}{28}$

4. What fraction of students prefer Mt. Rushmore over the Grand Canyon? Write in simplest form.

5. Suppose four students change their minds and choose the Statue of Liberty instead of the Grand Canyon. What part of the class now prefers Mt. Rushmore over the State of Liberty? Write in simplest form.

Brain Builders

6. Processes &Practices 1 **Make Sense of Problems** On a class trip to the museum, $\frac{5}{8}$ of the students saw the dinosaurs and $\frac{7}{8}$ of the students saw either the jewelry collection or the dinosaurs. What fraction of the students only saw the jewelry collection? Write in simplest form.

7. The Indian Ocean is $\frac{2}{10}$ of the area of the world's oceans. What fraction represents the area of the remaining oceans that make up the world's oceans? Explain. Write in simplest form.

8. Test Practice The pictures on the right show how much sausage and pepperoni pizza was left at the end of one day. Which pizza had more left? How much more?

Sausage

Pepperoni

Ⓐ pepperoni; $\frac{7}{8}$

Ⓑ pepperoni; $\frac{4}{8}$

Ⓒ sausage; $\frac{3}{8}$

Ⓓ sausage; $\frac{11}{8}$

Name ______________________________

Lesson 4
Hands On 
Use Models to Add Unlike Fractions

ESSENTIAL QUESTION
How can equivalent fractions help me add and subtract fractions?

Unlike fractions have different denominators. Before you can add unlike fractions, one or both of the fractions must be renamed so that they have a common denominator.

Build It 

To finish building a birdhouse, Jordan uses two boards. One is $\frac{1}{2}$ foot long and the other is $\frac{1}{4}$ foot long. What is the total length of the boards?

1. Model each fraction using fraction tiles and place them side by side.

$\frac{1}{2}$		$\frac{1}{4}$
$\frac{1}{4}$	$\frac{1}{4}$	$\frac{1}{4}$

2. Find fraction tiles that will match the length of the combined tiles. Line them up below the model.

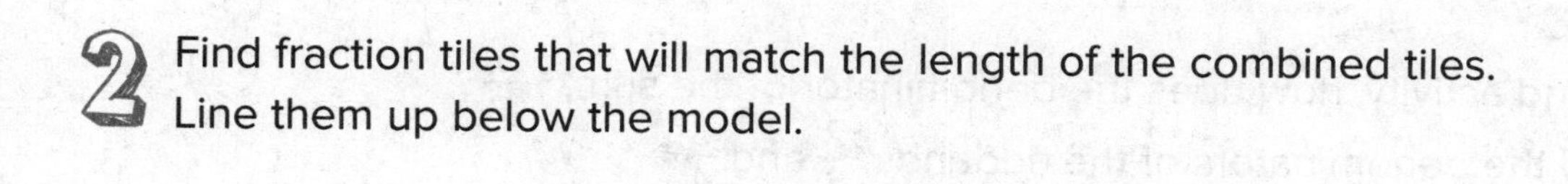

3. Count. There are ________ of the $\frac{1}{4}$-fraction tiles in all. This represents the fraction $\frac{3}{4}$.

So, $\frac{1}{2} + \frac{1}{4} = \frac{\square}{\square}$. The total length of the boards is $\frac{\square}{\square}$ foot.

Try It

Muna's family ate $\frac{2}{3}$ of a strawberry pie and Brendan's family ate $\frac{3}{4}$ of a different strawberry pie. How much did they eat all together?

1. Model each fraction using fraction tiles and place them side by side.

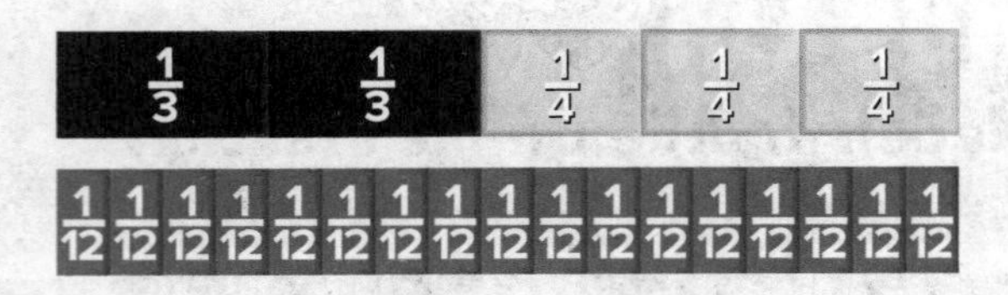

2. Find fraction tiles that will match the length of the combined tiles. Line them up below the model.

3. Count. There are ______ of the $\frac{1}{12}$-fraction tiles in all. This represents the fraction $\frac{17}{12}$, or $1\frac{5}{12}$.

So, $\frac{2}{3} + \frac{3}{4} = 1\frac{5}{12}$.

They ate $1\frac{5}{12}$ strawberry pies all together.

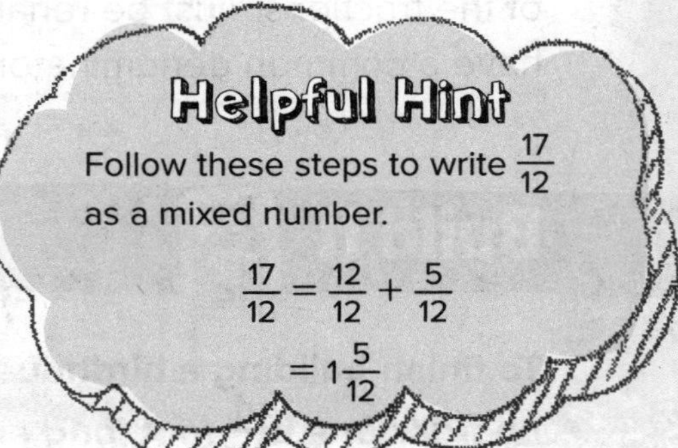

Talk About It

1. In the first activity, how does the denominator of the sum, $\frac{3}{4}$, compare to the denominators of the addends, $\frac{1}{2}$ and $\frac{1}{4}$?

____

____

2. In the second activity, how does the denominator of the sum, $1\frac{5}{12}$, compare to the denominators of the addends, $\frac{2}{3}$ and $\frac{3}{4}$?

____

____

3. **Processes & Practices** 1 **Explain to a Friend** Use your answers from Exercises 1 and 2 to predict the denominator of the sum of $\frac{1}{3}$ and $\frac{1}{4}$. Explain.

____

____

Practice It

Find the sum using fraction tiles. Write in simplest form. Draw the models.

4. $\frac{2}{3} + \frac{1}{6} =$ ______

5. $\frac{3}{8} + \frac{1}{4} =$ ______

6. $\frac{3}{10} + \frac{1}{5} =$ ______

7. $\frac{5}{8} + \frac{1}{4} =$ ______

8. $\frac{1}{3} + \frac{1}{4} =$ ______

9. $\frac{3}{4} + \frac{1}{6} =$ ______

Real World

Apply It

Processes & Practices 5 **Use Math Tools Draw models to solve Exercises 10–12.**

10. Alana walks east $\frac{4}{9}$ mile to school every morning. On Saturday, she walked to a friend's house, which is $\frac{1}{3}$ mile farther east from her school. How far did Alana walk to her friend's house on Saturday?

____

11. Mr. Hawkins gets two bonus payments every year. This year, his first bonus was $\frac{1}{4}$ of his salary. His second bonus was $\frac{1}{8}$ of his salary. What was the fraction of his salary for his combined bonus payments?

____

12. At a gas station, Kurt asked for directions to the nearest town. The attendant told him to go $\frac{5}{6}$ mile south and then $\frac{1}{4}$ mile east. How far does Kurt have to drive to get to the town?

____

13. Processes & Practices 4 **Model Math** Write a real-world problem that can be solved by adding unlike fractions.

____

____

Write About It

14. How can I use models to add fractions?

____

____

____

Name ______________________________

MY Homework

Lesson 4

Hands On: Use Models to Add Unlike Fractions

Homework Helper

Need help? connectED.mcgraw-hill.com

Find the sum of $\frac{2}{3}$ and $\frac{1}{2}$.

1. Model each fraction using fraction tiles and place them side-by-side.

2. Find fraction tiles that will match the length of the combined tiles. Line them up below the model.

$\frac{2}{3}$

$\frac{1}{3}$	$\frac{1}{3}$	$\frac{1}{2}$

$\frac{1}{6}$	$\frac{1}{6}$	$\frac{1}{6}$	$\frac{1}{6}$	$\frac{1}{6}$	$\frac{1}{6}$	$\frac{1}{6}$

$\frac{7}{6}$

3. Count. There are seven of the 1 _ 6 -fraction tiles in all.

So, $\frac{2}{3} + \frac{1}{2} = \frac{7}{6}$ or $1\frac{1}{6}$.

Helpful Hint

Follow these steps to write 7 _ 6 as a mixed number.

7 _ 6 $= \frac{6}{6} + \frac{1}{6} = 1\frac{1}{6}$

Practice

Find the sum using the fraction tiles shown. Write in simplest form.

1. $\frac{1}{2} + \frac{1}{4} =$ ______

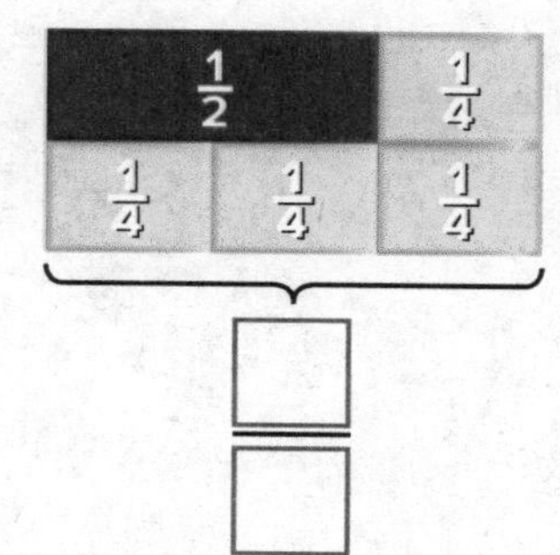

2. $\frac{1}{2} + \frac{1}{6} =$ ______

$\frac{1}{2}$			$\frac{1}{6}$
$\frac{1}{6}$	$\frac{1}{6}$	$\frac{1}{6}$	$\frac{1}{6}$

☐/☐ or ☐/☐

Problem Solving

Processes &Practices **Use Math Tools Draw models to solve Exercises 3–6. Write in simplest form.**

3. After school, Maurice walks $\frac{1}{3}$ mile to the park and then walks $\frac{1}{2}$ mile to his house. How far does Maurice walk from school to his house?

____

4. Ricki took a survey in the fifth grade and found that $\frac{2}{3}$ of the students ride the bus to school, and $\frac{1}{4}$ of the students walk. What fraction of the fifth grade students either ride the bus or walk to school?

____

5. Elizabeth made an English muffin pizza using $\frac{1}{4}$ cup of cheese and $\frac{3}{8}$ cup of sausage. How many cups of toppings did she use?

____

6. Craig and Alissa are building a sandcastle on the beach. They each have a bucket. Craig's bucket holds $\frac{1}{2}$ pound of sand and Alissa's bucket holds $\frac{9}{10}$ pound of sand. How much sand can Craig and Alissa collect together at one time?

____

Vocabulary Check

7. Complete the sentence with the correct vocabulary word(s).

Fractions that have different denominators are called ______________.

Name ....

Lesson 5 Add Unlike Fractions

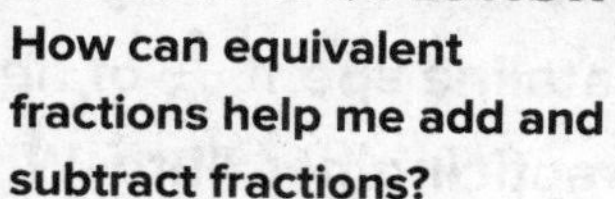

ESSENTIAL QUESTION
How can equivalent fractions help me add and subtract fractions?

Math in My World

Example 1

In the morning, an octopus swam for $\frac{1}{3}$ hour.

In the afternoon, the octopus swam for $\frac{1}{4}$ hour. How long did the octopus swim all together?

Find $\frac{1}{3} + \frac{1}{4}$.

Write equivalent, like fractions using the least common denominator, LCD. The LCD of $\frac{1}{3}$ and $\frac{1}{4}$ is 12.

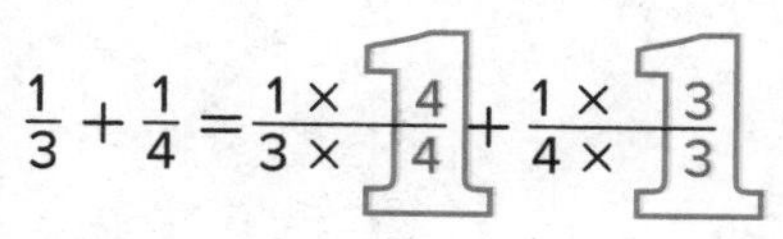

$\frac{1}{3} + \frac{1}{4} = \frac{1 \times 4}{3 \times 4} + \frac{1 \times 3}{4 \times 3}$ Write equivalent fractions using the LCD.

$= \frac{4}{12} + \frac{3}{12}$ Multiply.

$= \frac{4 + 3}{12}$ or $\frac{\_\_\_}{\_\_\_}$ Add like fractions.

Helpful Hint
The least common denominator, LCD, is the least common multiple of the denominators.

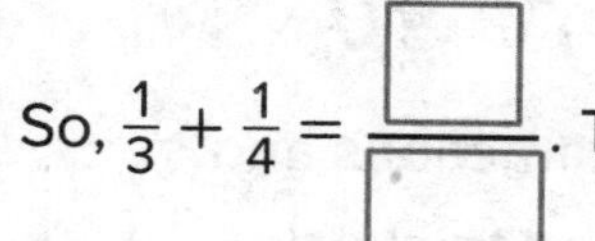
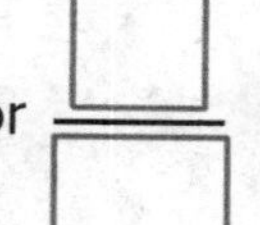

So, $\frac{1}{3} + \frac{1}{4} = \frac{\_\_\_}{\_\_\_}$. The octopus swam for $\frac{\_\_\_}{\_\_\_}$ hour all together.

Check The models show that $\frac{1}{3} + \frac{1}{4} = \frac{\_\_\_}{\_\_\_}$.

$\frac{1}{3}$				$\frac{1}{4}$		
$\frac{1}{12}$	$\frac{1}{12}$	$\frac{1}{12}$	$\frac{1}{12}$	$\frac{1}{12}$	$\frac{1}{12}$	$\frac{1}{12}$

You can use benchmark fractions such as $\frac{1}{2}$ to check an answer for reasonableness.

Example 2

Catalina spent $\frac{1}{10}$ of her free time reading and $\frac{4}{5}$ of her free time practicing her flute. What fraction of her free time did she spend reading and practicing her flute?

Add $\frac{1}{10}$ and $\frac{4}{5}$.

Estimate $\frac{1}{10} + \frac{4}{5} \approx 0 + 1$, or 1

Write equivalent, like fractions using the LCD. The LCD is 10.

$$\frac{1}{10} + \frac{4}{5} = \frac{1}{10} + \frac{4 \times 2}{5 \times 2}$$

$$= \frac{\square}{10} + \frac{\square}{10}$$ Multiply.

$$= \frac{\square + \square}{10} \text{ or } \frac{\square}{10}$$ Add like fractions.

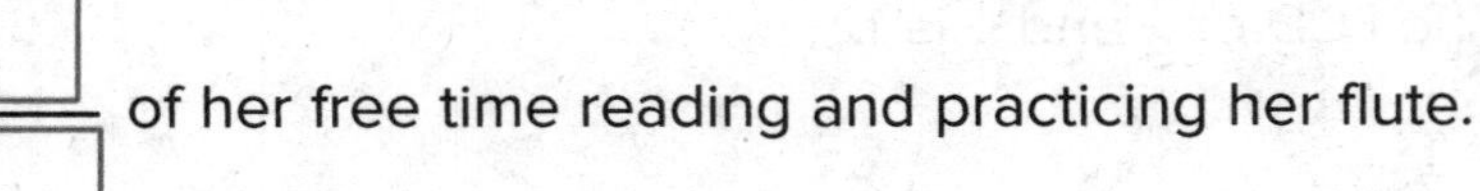

Catalina spent $\frac{\square}{\square}$ of her free time reading and practicing her flute.

Check $\frac{9}{10} \approx 1$

Guided Practice

Add. Write each sum in simplest form.

1. $\frac{2}{5} + \frac{1}{2} = \frac{\square}{\square}$

2. $\frac{3}{4} + \frac{1}{8} = \frac{\square}{\square}$

How can benchmark fractions and number sense be used to check answers for reasonableness?

Name

Independent Practice

Add. Write each sum in simplest form.

3. $\frac{1}{3} + \frac{1}{5} =$ ________

4. $\frac{1}{2} + \frac{1}{5} =$ ________

5. $\frac{5}{12} + \frac{1}{4} =$ ________

6. $\frac{2}{3} + \frac{1}{6} =$ ________

7. $\frac{1}{2} + \frac{1}{4} =$ ________

8. $\frac{5}{8} + \frac{1}{16} =$ ________

9. $\frac{3}{5} + \frac{3}{10} =$ ________

10. $\frac{5}{8} + \frac{3}{16} =$ ________

11. $\frac{3}{5} + \frac{3}{20} =$ ________

Algebra **Find each unknown.**

12. $\frac{7}{12} + \frac{1}{3} = x$

$x =$ ________

13. $\frac{3}{16} + \frac{3}{8} = \frac{9}{y}$

$y =$ ________

14. $\frac{3}{8} + \frac{2}{5} = \frac{w}{40}$

$w =$ ________

Problem Solving

15. A farmer harvested some of his pecan crop on Friday and Saturday. What fraction of the pecan crop was harvested in the two days?

Pecan Harvest	
Day	**Amount**
Friday	$\frac{3}{8}$
Saturday	$\frac{1}{3}$

16. Angel has two chores after school. She rakes leaves for $\frac{3}{4}$ hour and spends $\frac{1}{2}$ hour washing the car. How long does Angel spend on her chores in all?

Brain Builders

17. Processes & Practices 2 **Use Number Sense** Leon raked bags of leaves from his neighbors' yards. Monday he raked $\frac{1}{3}$ of a bag of leaves, Tuesday he raked $\frac{3}{8}$ of a bag, and Wednesday he raked $\frac{1}{4}$ of a bag. If he combines the leaves, will Leon need more than one bag? Explain.

18. Processes & Practices 3 **Which One Doesn't Belong?** Circle the expression that does not belong with the other three. Explain your reasoning.

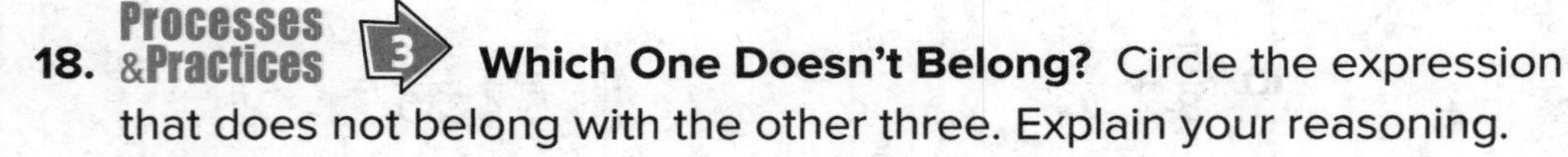

$\frac{5}{6} + \frac{1}{3}$ $\quad$ $\frac{5}{6} + \frac{1}{2}$ $\quad$ $\frac{5}{6} + \frac{2}{6}$ $\quad$ $\frac{5}{6} + \left(\frac{1}{6} + \frac{1}{6}\right)$

Write another example of a problem that would belong.

19. **Building on the Essential Question** How are equivalent fractions used when adding unlike fractions? Include an example to support your answer.

Name

Lesson 5

Add Unlike Fractions

Homework Helper

eHelp

Need help? connectED.mcgraw-hill.com

Find $\frac{1}{6} + \frac{1}{4}$.

Write equivalent, like fractions using the least common denominator, LCD. The LCD of $\frac{1}{6}$ and $\frac{1}{4}$ is 12.

$\frac{1}{6} + \frac{1}{4} = \frac{1 \times 2}{6 \times 2} + \frac{1 \times 3}{4 \times 3}$ Write equivalent fractions using the LCD.

$= \frac{2}{12} + \frac{3}{12}$ Multiply.

$= \frac{2 + 3}{12}$, or $\frac{5}{12}$ Add like fractions.

So, $\frac{1}{6} + \frac{1}{4} = \frac{5}{12}$.

Check The models show that $\frac{1}{6} + \frac{1}{4} = \frac{5}{12}$.

Practice

Add. Write each sum in simplest form.

1. $\frac{5}{8} + \frac{3}{10} =$ ______

2. $\frac{3}{5} + \frac{1}{4} =$ ______

3. $\frac{4}{7} + \frac{1}{8} =$ ______

Problem Solving

4. Tashia ate $\frac{1}{3}$ of a pizza, and Jay ate $\frac{3}{8}$ of the same pizza. What fraction of the pizza was eaten?

5. Basir took a science test on Friday. One-eighth of the questions were multiple choice, and $\frac{3}{4}$ of the questions were true-false questions. What part of the total number of questions were either multiple choice or true-false questions?

Brain Builders

6. **Processes & Practices** 1 **Use Number Sense** Edison delivers $\frac{1}{5}$ of the newspapers in the neighborhood, and Anita delivers $\frac{1}{2}$ of them. After Edison and Anita deliver the newspapers, what fraction of the newspapers remain?

7. Dylan and Sonia are hiking different trails. If Dylan hiked Riverwalk and Mountainview, and Sonia hiked Mountainview and Pine, how many miles did each of them hike? Is it possible to walk two of the trails and walk less than 1 mile? Explain.

Hiking Trails

Trail	Distance (mi)
Riverwalk	$\frac{3}{4}$
Mountainview	$\frac{1}{2}$
Pine	$\frac{3}{5}$

8. **Test Practice** Which expression will have the same sum as $\frac{3}{8} + \frac{1}{4}$?

Ⓐ $\frac{3}{8} + \frac{1}{8}$

Ⓑ $\left(\frac{1}{8} + \frac{1}{8} + \frac{1}{8}\right) + \frac{1}{4}$

Ⓒ $\frac{3}{4} + \frac{1}{4}$

Ⓓ $\left(\frac{1}{8} + \frac{1}{8}\right) + \frac{1}{8}$

Check My Progress

Vocabulary Check 

Write the vocabulary words that describe each set of fractions below.

like fractions **unlike fractions**

1. $\frac{1}{2}$ and $\frac{1}{3}$ ________________
2. $\frac{2}{9}$ and $\frac{2}{9}$ ________________

Concept Check

Round each fraction to 0, $\frac{1}{2}$, or 1. Use a number line if needed.

3. $\frac{9}{16} \approx$ ________
4. $\frac{12}{15} \approx$ ________
5. $\frac{1}{10} \approx$ ________

Add. Write each sum in simplest form.

6. $\frac{5}{9} + \frac{1}{9} =$ ________
7. $\frac{1}{7} + \frac{6}{7} =$ ________
8. $\frac{7}{16} + \frac{3}{16} =$ ________
9. $\frac{1}{6} + \frac{7}{12} =$ ________
10. $\frac{1}{8} + \frac{3}{4} =$ ________
11. $\frac{1}{4} + \frac{3}{8} =$ ________

Subtract. Write each difference in simplest form.

12. $\frac{4}{5} - \frac{1}{5} =$ ________
13. $\frac{11}{20} - \frac{7}{20} =$ ________
14. $\frac{9}{16} - \frac{7}{16} =$ ________

Problem Solving

15. Mike spent $\frac{11}{12}$ hour talking on his cell phone. Brent spent $\frac{7}{12}$ hour talking on his cell phone. How much more time did Mike spend on his cell phone than Brent?

Brain Builders

16. Christie decided to make bracelets for the fifth grade class girls. Two-fifths of the bracelets were finished on Monday, and $\frac{3}{7}$ of the bracelets were finished on Tuesday. If she finishes $\frac{1}{7}$ of the bracelets on Wednesday, will all of the bracelets be finished? Explain.

17. Jamila ate $\frac{3}{8}$ of a salad. Manny ate $\frac{2}{8}$ of the same salad. How much of the salad is left?

18. Test Practice The table shows the fraction of a book Karen read Saturday and Sunday. How much of her book did she read on these two days?

Day	Fraction of Book Read
Saturday	$\frac{1}{3}$
Sunday	$\frac{2}{5}$

Ⓐ $\frac{3}{8}$

Ⓑ $\frac{1}{2}$

Ⓒ $\frac{8}{15}$

Ⓓ $\frac{11}{15}$

Name

Lesson 6 Hands On
Use Models to Subtract Unlike Fractions

ESSENTIAL QUESTION
How can equivalent fractions help me add and subtract fractions?

You can use fraction tiles to subtract fractions with unlike denominators.

Build It

Tools

Akio lives $\frac{4}{5}$ mile from school. Bianca lives $\frac{3}{10}$ mile from school. How much farther from school does Akio live than Bianca?

1. Model each fraction using fraction tiles. Place the $\frac{1}{10}$-tiles below the $\frac{1}{5}$-tiles.

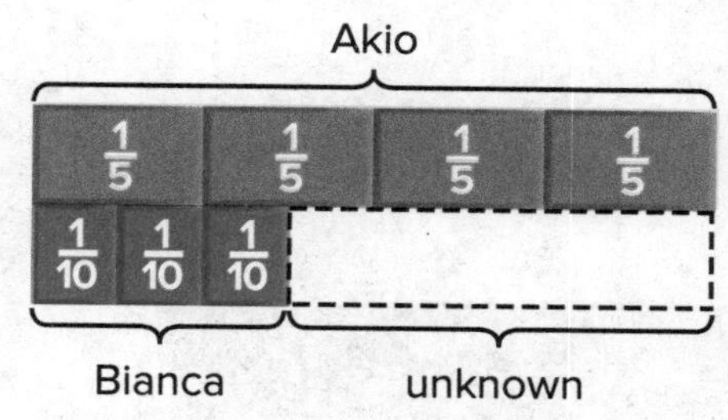

2. Find which fraction tiles will fill in the area of the dotted box.

Try $\frac{1}{3}$-tiles. Do they fill the dotted box? ______

Try $\frac{1}{2}$-tiles. Do they fill the dotted box? ______

How many $\frac{1}{2}$-tiles fill the dotted box? ______

Since $\frac{\square}{\square}$ fills in the area of the dotted box, $\frac{4}{5} - \frac{3}{10} = \frac{\square}{\square}$.

Akio lives $\frac{\square}{\square}$ mile farther from school than Bianca.

Try It

Find $\frac{3}{4} - \frac{1}{6}$.

Model each fraction using fraction tiles. Place the $\frac{1}{6}$-tiles below the $\frac{1}{4}$-tiles.

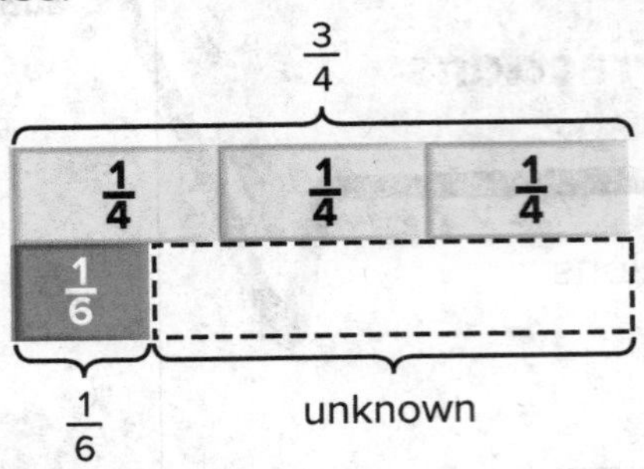

2 Find which fraction tiles will fill in the area of the dotted box.

Try $\frac{1}{3}$-tiles. Do they fill the dotted box? ________

Try $\frac{1}{12}$-tiles. Do they fill the dotted box? ________

How many $\frac{1}{12}$-tiles fill the dotted box? ________

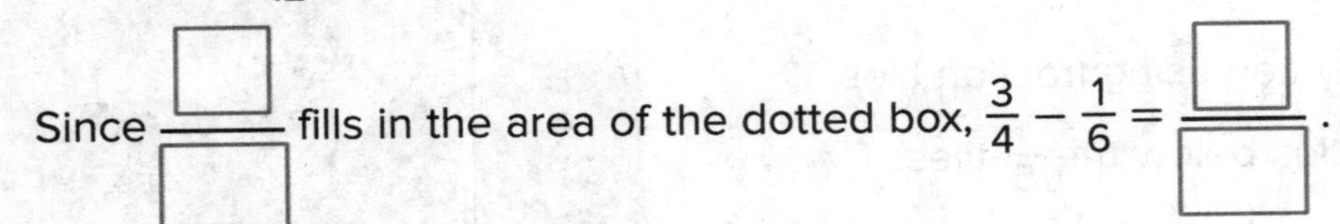

Since $\frac{\square}{\square}$ fills in the area of the dotted box, $\frac{3}{4} - \frac{1}{6} = \frac{\square}{\square}$.

So, $\frac{3}{4} - \frac{1}{6} = \frac{\square}{\square}$.

Talk About It

1. Would any of the other fraction tiles fit inside the dotted box for the first activity? Explain.

____

____

2. **Processes &Practices**  **Stop and Reflect** Describe how you would use fraction tiles to find $\frac{1}{2} - \frac{1}{3}$.

____

____

Name ....

Practice It

Find each difference using fraction tiles. Draw the models.

3. $\frac{2}{3} - \frac{1}{6} =$ ______

4. $\frac{5}{8} - \frac{1}{4} =$ ______

5. $\frac{1}{2} - \frac{1}{6} =$ ______

6. $\frac{3}{5} - \frac{1}{2} =$ ______

7. $\frac{3}{4} - \frac{3}{8} =$ ______

8. $\frac{5}{6} - \frac{1}{4} =$ ______

Apply It

Processes & Practices 5 Use Math Tools Draw fraction tiles to help you solve Exercises 9 and 10.

9. Missy ran $\frac{5}{8}$ mile to warm up for softball practice, while Farrah only ran $\frac{1}{2}$ mile. How much farther did Missy run?

______

10. Pablo used $\frac{1}{2}$ of an oxygen tank while scuba diving. He used another $\frac{1}{6}$ of the tank exploring underwater. How much oxygen is left in the tank?

______

11. **Processes & Practices 4 Model Math** Write a real-world problem that could be represented by the fraction tiles shown.

$\frac{1}{4}$			$\frac{1}{4}$			$\frac{1}{4}$	
$\frac{1}{12}$	$\frac{1}{12}$	$\frac{1}{12}$	$\frac{1}{12}$	$\frac{1}{12}$	$\frac{1}{12}$	$\frac{1}{12}$	$\frac{1}{6}$

______

______

______

Write About It

12. How do fraction tiles help me subtract unlike fractions?

______

______

Name ....

MY Homework

Lesson 6

Hands On: Use Models to Subtract Unlike Fractions

Homework Helper

Need help? connectED.mcgraw-hill.com

Find $\frac{7}{8} - \frac{3}{4}$.

Model each fraction using fraction tiles.

Place the $\frac{1}{4}$-tiles below the $\frac{1}{8}$-tiles.

$\frac{1}{8}$	$\frac{1}{8}$	$\frac{1}{8}$	$\frac{1}{8}$	$\frac{1}{8}$	$\frac{1}{8}$	$\frac{1}{8}$
$\frac{1}{4}$		$\frac{1}{4}$		$\frac{1}{4}$		

Find which fraction will fill in the area of the dotted box.

Try $\frac{1}{3}$-tiles. They do not fill the dotted box.

Try a $\frac{1}{8}$-tile. It does fill the dotted box.

Since $\frac{1}{8}$ fills in the area of the dotted box, $\frac{7}{8} - \frac{3}{4} = \frac{1}{8}$.

Practice

Find each difference using the fraction tiles.

1. $\frac{7}{8} - \frac{1}{2} =$ ________

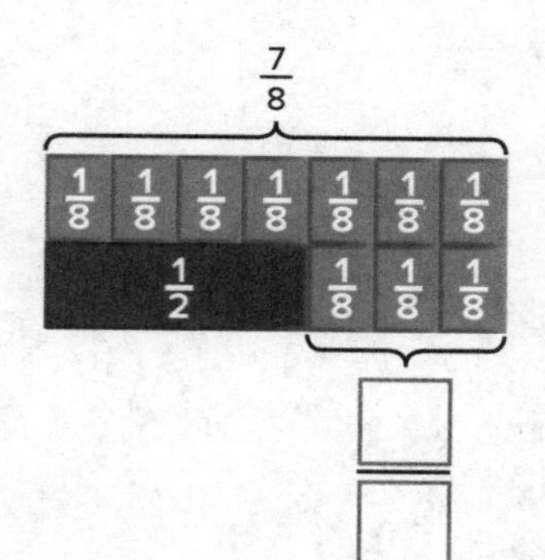

2. $\frac{2}{3} - \frac{1}{4} =$ ________

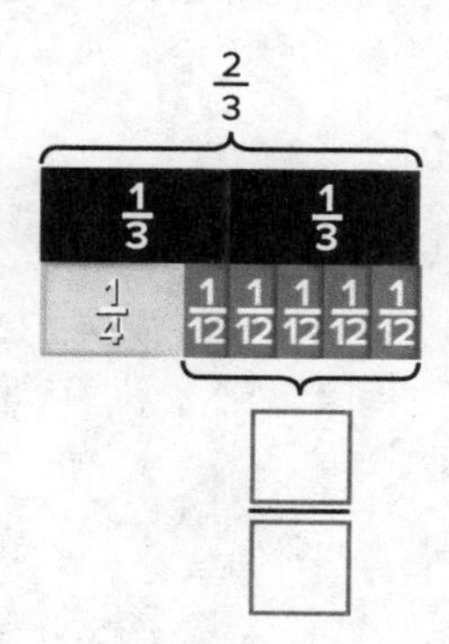

Problem Solving

Processes & Practices 5 **Use Math Tools Draw fraction tiles to help you solve Exercises 3–6.**

3. Noah bought $\frac{1}{2}$ pound of candy to share with his friends. They ate $\frac{3}{8}$ of the candy. How much candy does Noah have left?

____

4. Mr. Corwin gave his students $\frac{3}{4}$ hour to study for a test. After $\frac{1}{3}$ hour, he played a review game for the remaining time. How much time did Mr. Corwin spend playing the review game?

____

5. Mrs. Washer filled the gas tank of her car. She used $\frac{2}{3}$ of a tank of gasoline while driving to the beach. She used another $\frac{1}{6}$ of the tank driving to her hotel. How much gasoline is left in the tank?

____

6. Starting from her hotel, Angie walked $\frac{2}{3}$ mile along the beach in one direction. She turned around and walked $\frac{1}{2}$ mile toward her hotel. How much farther does she need to walk to get to the hotel?

____

Name ....

Lesson 7
Subtract Unlike Fractions

ESSENTIAL QUESTION
How can equivalent fractions help me add and subtract fractions?

Subtracting unlike fractions is similar to adding unlike fractions.

Math in My World

Example 1

A female Cuban tree frog can be up to $\frac{5}{12}$ foot long. A male Cuban tree frog can be up to $\frac{1}{4}$ foot long. How much longer is the female Cuban tree frog than the male?

Find $\frac{5}{12} - \frac{1}{4}$.

Write equivalent, like fractions using the least common denominator, LCD. The LCD of $\frac{5}{12}$ and $\frac{1}{4}$ is 12.

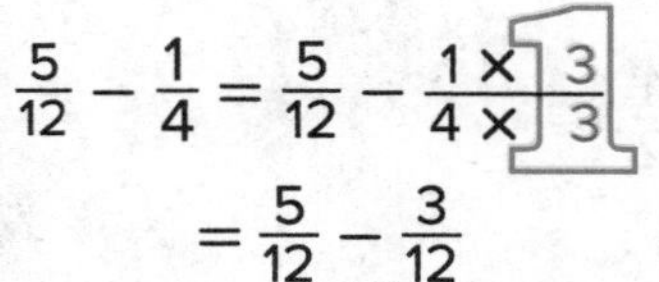

$\frac{5}{12} - \frac{1}{4} = \frac{5}{12} - \frac{1 \times 3}{4 \times 3}$ Write equivalent fractions using the LCD.

$= \frac{5}{12} - \frac{3}{12}$ Multiply.

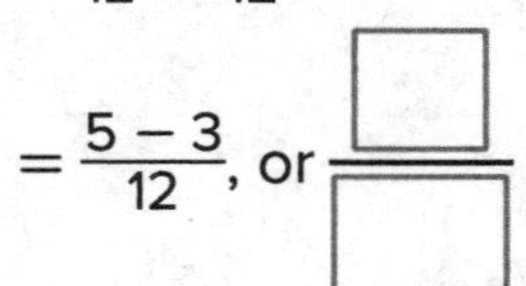

$= \frac{5-3}{12}$, or $\frac{\square}{\square}$ Subtract like fractions.

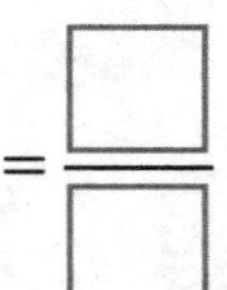

$= \frac{\square}{\square}$ Simplify.

A female Cuban tree frog is $\frac{\square}{\square}$ foot longer than the male.

Check for Reasonableness Use benchmark fractions to check.

Since $\frac{1}{6} < \frac{1}{2}$, your answer is reasonable.

Example 2

Jessie finished $\frac{1}{2}$ of her homework. Lakshani finished $\frac{4}{5}$ of her homework. How much more homework did Lakshani finish than Jessie?

Portion of Homework Completed	
Jessie	$\frac{1}{2}$
Lakshani	$\frac{4}{5}$

Estimate Use benchmark fractions.

$$\frac{4}{5} - \frac{1}{2} \approx 1 - \frac{1}{2} = \frac{\square}{\square}$$

Subtract $\frac{4}{5} - \frac{1}{2}$.

Write equivalent, like fractions using the least common denominator, LCD. The LCD of $\frac{4}{5}$ and $\frac{1}{2}$ is 10.

$\frac{4}{5} - \frac{1}{2} = \frac{4 \times 2}{5 \times 2} - \frac{1 \times 5}{2 \times 5}$ Write equivalent fractions using the LCD.

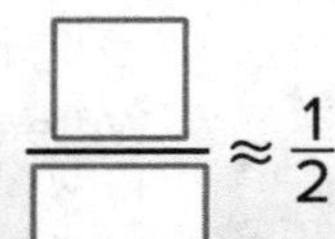

$= \frac{8}{10} - \frac{5}{10}$ Multiply.

$= \frac{8 - 5}{10}$ or $\frac{\square}{\square}$ Subtract like fractions.

Lakshani finished $\frac{\square}{\square}$ more of her homework than Jessie.

Check for Reasonableness Compare to your estimate. $\frac{\square}{\square} \approx \frac{1}{2}$

Guided Practice

1. Subtract. Write in simplest form.

$$\frac{3}{8} - \frac{1}{4} = \frac{\square}{\square}$$

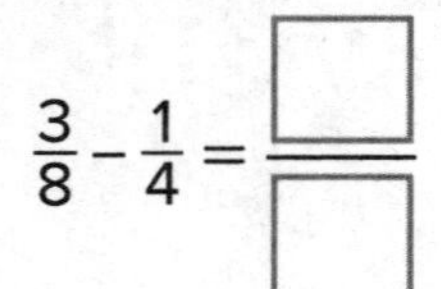

Describe the steps you can use to find $\frac{3}{4} - \frac{1}{12}$.

Name

Independent Practice

Subtract. Write each in simplest form.

2. $\frac{5}{6} - \frac{1}{2} =$ ______

3. $\frac{2}{5} - \frac{1}{4} =$ ______

4. $\frac{4}{5} - \frac{1}{6} =$ ______

5. $\frac{7}{8} - \frac{1}{2} =$ ______

6. $\frac{7}{12} - \frac{1}{3} =$ ______

7. $\frac{5}{6} - \frac{1}{3} =$ ______

8. $\frac{2}{3} - \frac{3}{10} =$ ______

9. $\frac{5}{8} - \frac{1}{2} =$ ______

10. $\frac{4}{5} - \frac{2}{15} =$ ______

Algebra Find the unknown.

11. $\frac{5}{6} - \frac{3}{4} = m$

$m =$ ______

12. $\frac{2}{3} - \frac{3}{5} = \frac{n}{15}$

$n =$ ______

13. $\frac{5}{12} - \frac{1}{6} = p$

$p =$ ______

Problem Solving

14. Angie rides her bicycle $\frac{2}{3}$ mile to school. On Friday, she took a shortcut so that the ride to school was $\frac{1}{9}$ mile shorter. How long was Angie's bicycle ride on Friday?

____

15. Processes &Practices 6 **Be Precise** Ollie used $\frac{1}{2}$ cup of vegetable oil to make brownies. She used another $\frac{1}{3}$ cup of oil to make muffins. How much more oil did she use to make brownies?

____

Brain Builders

16. Danielle poured $\frac{1}{4}$ gallon of water from a $\frac{7}{8}$ gallon bucket. She then poured $\frac{3}{8}$ gallon of the remaining water from the bucket. How much water is left in the bucket?

____

17. Processes &Practices 2 **Use Number Sense** Is finding $\frac{9}{10} - \frac{1}{2}$ the same as finding $\frac{9}{10} - \frac{1}{4} - \frac{1}{4}$? Explain.

____

18. **Building on the Essential Question** Why are equivalent fractions needed when subtracting unlike fractions? Include a model to support your reasoning.

____

____

____

Name ______________________

MY Homework

Lesson 7

Subtract Unlike Fractions

Homework Helper

Need help? connectED.mcgraw-hill.com

Find $\frac{2}{5} - \frac{1}{10}$.

Estimate Use benchmark fractions.

$\frac{2}{5} - \frac{1}{10} \approx \frac{1}{2} - 0 = \frac{1}{2}$

Subtract $\frac{2}{5} - \frac{1}{10}$.

Write equivalent, like fractions using the least common denominator, LCD. The LCD of $\frac{2}{5}$ and $\frac{1}{10}$ is 10.

$\frac{2}{5} - \frac{1}{10} = \frac{2 \times 2}{5 \times 2} - \frac{1}{10}$ Write equivalent fractions using the LCD.

$= \frac{4}{10} - \frac{1}{10}$ Multiply.

$= \frac{4 - 1}{10}$ or $\frac{3}{10}$ Subtract like fractions.

So, $\frac{2}{5} - \frac{1}{10} = \frac{3}{10}$.

Check for Reasonableness Compare to your estimate. $\frac{3}{10} \approx \frac{1}{2}$

Practice

Subtract. Write each in simplest form.

1. $\frac{1}{2} - \frac{1}{4} =$ ________

2. $\frac{7}{8} - \frac{1}{4} =$ ________

3. $\frac{7}{12} - \frac{1}{6} =$ ________

Problem Solving

4. The average rainfall in April and October for Springfield is shown in the table below. How much more rain falls on average in April than in October?

Average Rainfall for Springfield	
Month	**Rainfall (in.)**
April	$\frac{11}{16}$
October	$\frac{3}{8}$

---

5. Trisha helped clean up her neighborhood by picking up plastic. She collected $\frac{3}{4}$ pound of plastic the first day and $\frac{1}{6}$ pound of plastic the second day. How much more trash did she collect the first day than the second day?

Brain Builders

6. Wyatt is hiking a trail that is $\frac{11}{12}$ mile long. After hiking $\frac{1}{4}$ mile, he stops for water. He continues hiking another $\frac{1}{6}$ mile and stops to rest. How much farther must he hike to finish the trail?

---

7. Test Practice The table shows the distance each student ran on Wednesday. How much farther did the student who ran the greatest distance run compared to the student who ran the least distance?

Student	Distance (mi)
Steve	$\frac{1}{6}$
Charlie	$\frac{1}{4}$
Joey	$\frac{2}{3}$

Ⓐ $\frac{1}{12}$ mile

Ⓑ $\frac{5}{12}$ mile

Ⓒ $\frac{1}{2}$ mile

Ⓓ $\frac{5}{6}$ mile

Name ______________________

Lesson 8
Problem-Solving Investigation
STRATEGY: Determine Reasonable Answers

ESSENTIAL QUESTION
How can equivalent fractions help me add and subtract fractions?

Learn the Strategy

Leandra feeds her pet rabbit Bounce the same amount of food each day. Bounce eats three times a day. About how much food does Leandra feed Bounce each day?

Time	Food (cups)
Morning	$\frac{3}{4}$
Afternoon	$\frac{3}{4}$
Evening	$\frac{1}{4}$

1 Understand

What facts do you know?

Leandra feeds the rabbit the same amount every ______.

What do you need to find?

how much food she feeds her rabbit each ______

2 Plan

Use estimation to find a reasonable answer.

3 Solve

Round each fraction to the nearest whole number.

Morning $\frac{3}{4} \rightarrow$ ____

Afternoon $\frac{3}{4} \rightarrow$ ____

Evening $\frac{1}{4} \rightarrow$ ____

In one day, she feeds Bounce about ____ + ____ + ____, or ____ cups of food.

So, Leandra feeds Bounce about ____ cups of food each day.

4 Check

Is my answer reasonable?

The estimate of 2 cups is close to the actual amount of $1\frac{3}{4}$ cups of food.

Practice the Strategy

Jonas needs to determine how much wood to buy at the store to make picture frames. The dimensions for each picture frame are listed in the table. About how many feet of wood does Jonas need to buy to build 5 frames?

Frame	Wood (feet)
Top	$\frac{5}{8}$
Bottom	$\frac{5}{8}$
Left Side	$\frac{3}{4}$
Right Side	$\frac{3}{4}$

Understand

What facts do you know?

____

What do you need to find?

____

Plan

____

3 Solve

Check

Is my answer reasonable?

____

____

____

Name

Apply the Strategy

Determine a reasonable answer to solve each problem.

Bird	Weight (lb)
Flamingo	$9\frac{1}{10}$
Ostrich	$253\frac{1}{2}$

1. Use the table to determine whether 245 pounds, 260 pounds, or 263 pounds is the most reasonable estimate for how much more the ostrich weighs than the flamingo. Explain.

2. Processes & Practices 6 **Explain to a Friend** A grocer sells 12 pounds of apples. Of those, $5\frac{3}{4}$ pounds are green and $3\frac{1}{4}$ pounds are golden. The rest are red. Which is a more reasonable estimate for how many pounds of red apples the grocer sold: 3 pounds or 5 pounds? Explain.

Brain Builders

3. A puzzle book costs $4.25. A novel costs $9.70 more than the puzzle book. Which is the most reasonable estimate for the total cost of 3 puzzle books and 1 novel: $20, $22, or $26? Explain.

4. Hailey has 10 cups of flour. Her muffin recipe uses $1\frac{1}{4}$ cups of flour. Her banana bread recipe uses $2\frac{2}{3}$ cups of flour. Hailey makes 2 batches of muffins. Which is the most reasonable estimate for the total number of banana bread loaves Hailey could make with the remaining flour: 1 loaf, 3 loaves, or 5 loaves?

Review the Strategies

Use any strategy to solve each problem.

- Determine reasonable answers.
- Look for a pattern.
- Solve a simpler problem.
- Find an estimate or exact answer.

5. Use the graph below. Is 21 inches, 24 inches, or 25 inches a more reasonable total amount of rain that fell in May, June, and July?

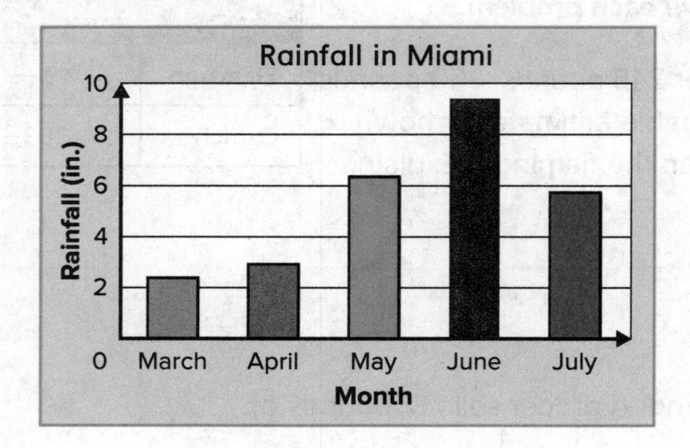

6. Processes & Practices 5 **Use Math Tools** Draw the next figure in the pattern.

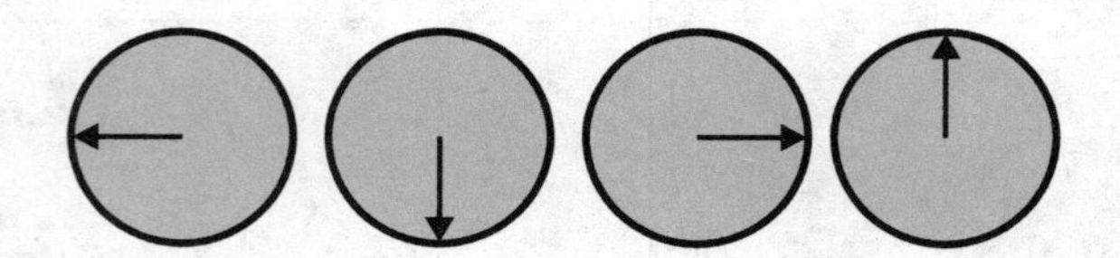

7. A high jumper starts the bar at 48 inches and raises the bar $\frac{1}{2}$ inch after each jump. How high will the bar be after the seventh jump?

8. Ms. Kennedy wants to buy a new interactive whiteboard for her classroom that costs \$989. So far, Ms. Kennedy has collected \$485 in donations, and \$106 in fundraising. About how much more money does she need to buy the interactive whiteboard?

Name ....

Lesson 8

Problem Solving: Determine Reasonable Answers

Homework Helper

Need help? connectED.mcgraw-hill.com

Ms. Montoya makes $2\frac{3}{4}$ pounds of goat cheese in the morning. In the afternoon, she makes $1\frac{1}{5}$ pounds of goat cheese. Is 3 pounds, 4 pounds, or 5 pounds the most reasonable estimate for how much goat cheese Ms. Montoya makes in one day?

1 Understand

What facts do you know?

Ms. Montoya makes $2\frac{3}{4}$ pounds and $1\frac{1}{5}$ pounds of goat cheese in one day.

What do you need to find?

a reasonable estimate of how much goat cheese she makes in one day

2 Plan

Use estimation to find a reasonable answer.

3 Solve

Round each amount of cheese to the nearest whole number.

Morning	**Afternoon**
$2\frac{3}{4} \rightarrow 3$	$1\frac{1}{5} \rightarrow 1$

In one day, she makes about 3 + 1, or 4 pounds of goat cheese.

So, Ms. Montoya makes about 4 pounds of goat cheese in one day.

4 Check

Is my answer reasonable?

Subtract the rounded amount she made in the morning from the estimated total.

$4 - 3 = 1$

So, the answer is reasonable.

Real World

Problem Solving

Processes &Practices 3 **Check for Reasonableness Determine a reasonable answer to solve each problem.**

1. Alyssa needs $7\frac{5}{8}$ inches of ribbon for one project and $4\frac{7}{8}$ inches of ribbon for another project. If she has 11 inches of ribbon, will she have enough to complete both projects? Explain.

2. Josiah has a piece of wood that measures $10\frac{1}{8}$ feet. He wants to make 5 shelves. If each shelf is $1\frac{3}{4}$ feet long, does he have enough to make 5 shelves? Explain.

Brain Builders

3. After school, Philipe spent $1\frac{3}{4}$ hours at baseball practice, $2\frac{1}{4}$ hours on homework, and $\frac{1}{4}$ hour getting ready for bed. About how many hours after school will he be ready for bed? Explain.

4. Kimane and a friend picked apples. Kimane picked $4\frac{2}{3}$ pounds and his friend picked $5\frac{5}{6}$ pounds. About how many more pounds of apples do they need to pick to have about 20 pounds of apples all together? Explain.

Name ....

Lesson 9
Estimate Sums and Differences

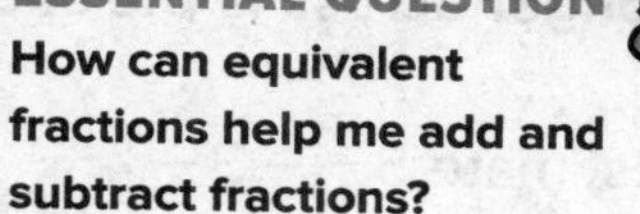

ESSENTIAL QUESTION

How can equivalent fractions help me add and subtract fractions?

Math in My World

Example 1

Lucita and Alexis went scuba diving for $7\frac{1}{3}$ hours. The next day, they went scuba diving for $4\frac{2}{3}$ hours. About how many hours did they spend scuba diving all together?

Estimate $7\frac{1}{3} + 4\frac{2}{3}$.

Round each mixed number to the nearest whole number.

$\frac{1}{3}$ is less than $\frac{1}{2}$.

So, $7\frac{1}{3}$ rounds down to 7.

$\frac{2}{3}$ is greater than $\frac{1}{2}$.

So, $4\frac{2}{3}$ rounds up to 5.

$7\frac{1}{3} + 4\frac{2}{3} \approx$ ______ + ______

$\approx$ ______

So, $7\frac{1}{3} + 4\frac{2}{3} \approx$ ______.

Lucita and Alexis spent about ______ hours scuba diving all together.

Helpful Hint

The ≈ means *about* or *approximately equal to.*

Example 2

One sea anemone is $3\frac{1}{4}$ feet across while a second sea anemone is $5\frac{3}{4}$ feet across. About how much longer across is the second anemone?

Estimate $5\frac{3}{4} - 3\frac{1}{4}$.

Round each mixed number to the nearest whole number.

$\frac{3}{4}$ is greater than $\frac{1}{2}$.

So, $5\frac{3}{4}$ rounds up to 6.

$\frac{1}{4}$ is less than $\frac{1}{2}$.

So, $3\frac{1}{4}$ rounds down to 3.

$5\frac{3}{4} - 3\frac{1}{4} \approx$ ______ − ______

$\approx$ ______

So, $5\frac{3}{4} - 3\frac{1}{4} \approx$ ______.

The second sea anemone is about ______ feet longer across than the first sea anemone.

Guided Practice

Estimate by rounding each mixed number to the nearest whole number.

1. $2\frac{1}{8} + 3\frac{5}{8}$

______

2. $5\frac{2}{9} - 3\frac{2}{3}$

______

3. $3\frac{3}{5} + 1\frac{1}{5}$

______

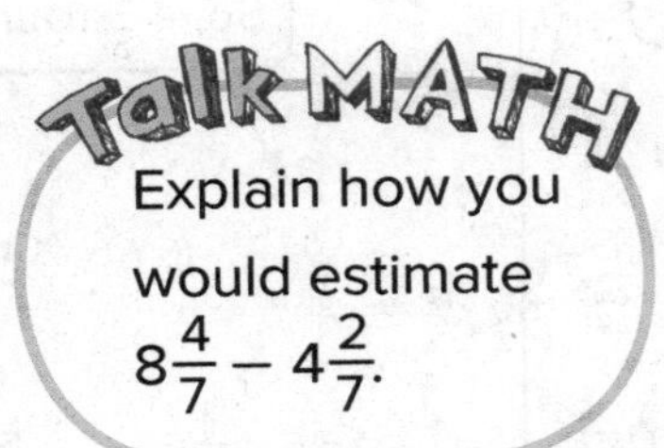

Name

Independent Practice

Estimate by rounding each mixed number to the nearest whole number.

4. $7\frac{2}{3} - 4\frac{1}{4}$

5. $5\frac{1}{3} + 3\frac{7}{9}$

6. $9\frac{1}{7} - 5\frac{6}{7}$

7. $6\frac{7}{10} - 1\frac{1}{5}$

8. $8\frac{11}{12} + 4\frac{1}{3}$

9. $15\frac{3}{7} - 3\frac{4}{7}$

10. $10\frac{2}{7} + 7\frac{5}{7}$

11. $13\frac{4}{11} - 4\frac{1}{4}$

12. $12\frac{7}{10} + 9\frac{3}{5}$

13. $19\frac{3}{7} + \frac{13}{14}$

14. $7\frac{7}{9} - 1\frac{5}{18}$

15. $\frac{9}{16} + 16\frac{5}{8}$

Problem Solving

16. Liam spent $1\frac{3}{4}$ hours playing board games and $2\frac{1}{4}$ hours watching a movie. About how much time did Liam spend all together on these two activities? Round each mixed number to the nearest whole number.

____

17. Processes & Practices 2 **Use Number Sense** Beth walked $10\frac{7}{8}$ miles one week. She walked $2\frac{1}{4}$ fewer miles the following week. About how many miles did she walk the second week? Round each mixed number to the nearest whole number.

____

18. Patrick has played guitar for $3\frac{5}{6}$ years. Gen has played guitar for $6\frac{1}{12}$ years. Estimate how many more years Gen has played guitar than Patrick. Round each mixed number to the nearest whole number.

____

Brain Builders

19. Processes & Practices 3 **Justify Conclusions** Select two mixed numbers whose estimated difference is 1 and whose estimated sum is 5. Justify your selection.

____

____

20. **Building on the Essential Question** How does number sense of fractions help when estimating sums and differences?

____

____

____

Name

MY Homework

Lesson 9

Estimate Sums and Differences

Homework Helper

Need help? connectED.mcgraw-hill.com

Estimate $3\frac{5}{8} - 1\frac{1}{8}$.

Round each mixed number to the nearest whole number.

$\frac{5}{8}$ is greater than $\frac{1}{2}$.
So, $3\frac{5}{8}$ rounds up to 4.

$\frac{1}{8}$ is less than $\frac{1}{2}$.
So, $1\frac{1}{8}$ rounds down to 1.

$$3\frac{5}{8} - 1\frac{1}{8} \approx 4 - 1$$
$$\approx 3$$

So, $3\frac{5}{8} - 1\frac{1}{8}$ is about 3.

Practice

Estimate by rounding each mixed number to the nearest whole number.

1. $2\frac{1}{5} + 6\frac{4}{5}$ ________

2. $6\frac{5}{12} - 2\frac{1}{12}$ ________

3. $11\frac{8}{9} + 4\frac{1}{3}$ ________

4. $7\frac{1}{6} - 5\frac{3}{4}$ ________

5. $\frac{11}{18} + 6\frac{1}{10}$ ________

6. $15\frac{1}{3} - 2\frac{5}{8}$ ________

Problem Solving

For Exercises 7–8, use the pictures shown.

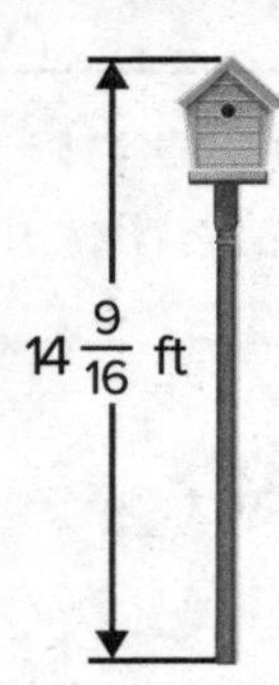

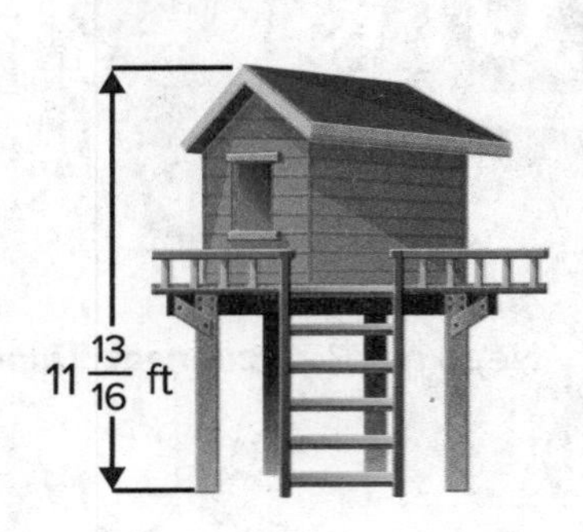

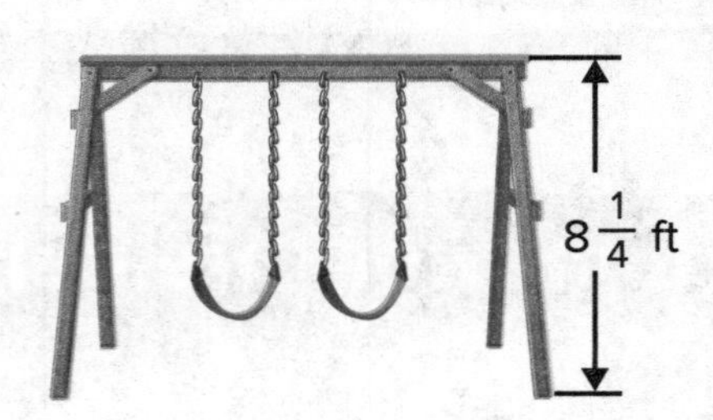

7. About how much taller is the birdhouse than the swing set? Round each mixed number to the nearest whole number.

____

8. About how much taller is the birdhouse than the tree house? Round each mixed number to the nearest whole number.

____

Brain Builders

9. Processes & Practices 1 **Plan Your Solution** Juan is organizing his sports cards in boxes. He has $2\frac{2}{3}$ boxes of baseball cards, $1\frac{5}{6}$ boxes of football cards, and $2\frac{1}{8}$ boxes of basketball cards. Which collections of cards would fit in 4 storage boxes? About how many boxes does Juan need to store all of his cards?

____

____

10. **Test Practice** During week 3, Frank rode his bike about the same amount of time as week 1 and 2 combined. About how many hours did he ride his bike during week 3?

Ⓐ 4 hours

Ⓑ 5 hours

Ⓒ 6 hours

Ⓓ 7 hours

Biking Time	
Week	**Length (h)**
1	$3\frac{1}{4}$
2	$2\frac{5}{6}$

Check My Progress

Vocabulary Check

Use the word bank below to complete each sentence.

benchmark fractions	**denominator**	**least common denominator (LCD)**
mixed number	**numerator**	**unknown**

1. The ____________________ is the bottom number in a fraction.

2. A number that has a whole number part and a fraction part is a ____________________.

3. A missing value in a number sentence or equation is the ____________________.

4. The ____________________ is the top number in a fraction.

Concept Check

Subtract. Write the difference in simplest form.

5. $\frac{7}{10} - \frac{1}{2} =$ ________

6. $\frac{5}{8} - \frac{1}{6} =$ ________

7. $\frac{7}{12} - \frac{1}{3} =$ ________

Estimate by rounding each mixed number to the nearest whole number.

8. $8\frac{9}{15} - \frac{4}{5}$ ________

9. $\frac{9}{16} + 16\frac{5}{8}$ ________

10. $5\frac{6}{11} - 3\frac{4}{5}$ ________

Problem Solving

11. The table shows how many hours Allie swam in two weeks. For about how many hours did she swim all together? Round each mixed number to the nearest whole number.

Day	Length (h)
Week 1	$4\frac{1}{3}$
Week 2	$2\frac{5}{6}$

12. To clean a large painting, you need $3\frac{1}{4}$ ounces of cleaner. A small painting requires only $2\frac{3}{4}$ ounces. You have 8 ounces of cleaner. About how many ounces of cleaner will you have left if you clean a large and a small painting? Round each mixed number to the nearest whole number.

Brain Builders

13. Andrew plays football. On one play, he ran the ball $24\frac{1}{3}$ yards. The following play, he was tackled and lost $3\frac{2}{3}$ yards. The next play, he ran for $5\frac{1}{4}$ yards. The team needs to be about 30 yards down the field after these three plays. Did the team make their 30 yard goal? Explain.

14. Marsha and Tegene are making table decorations for a party. Marsha made $\frac{2}{9}$ of a decoration in half an hour. Tegene made $\frac{2}{3}$ of a decoration in the same amount of time. How much more of a decoration did Tegene make in half an hour?

How much of a decoration can the two girls make in half an hour when working together?

15. **Test Practice** Greta's family drinks $2\frac{1}{8}$ quarts of skim milk, $1\frac{1}{8}$ quarts of whole milk, and $1\frac{5}{6}$ quarts of juice in one week. About how much more milk than juice does Greta's family drink in one week?

Ⓐ $\frac{1}{2}$ quart
Ⓑ 1 quart
Ⓒ 2 quarts
Ⓓ 3 quarts

Name

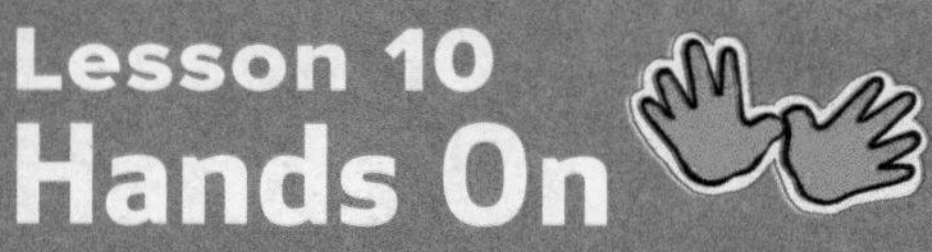

Lesson 10 Hands On

Use Models to Add Mixed Numbers

ESSENTIAL QUESTION
How can equivalent fractions help me add and subtract fractions?

You can use fraction circles to add mixed numbers.

Draw It

Find $2\frac{1}{3} + 1\frac{1}{3}$.

1. Shade and label the fraction circles to represent each mixed number.

$2\frac{1}{3}$ ⟶ ○ ○ ○

$1\frac{1}{3}$ ⟶ ○ ○

2. Combine the whole numbers and fractions.

How many whole fraction circles are there all together? ______

How many thirds are there all together? ______

Add the whole numbers and the fractions.

$$2\frac{1}{3} + 1\frac{1}{3} = 1 + 1 + \frac{1}{3} + 1 + \frac{1}{3}$$

$$= 1 + 1 + 1 + \frac{1}{3} + \frac{1}{3}$$ Group the whole numbers and fractions together.

$$= 3 + \frac{2}{3}$$

$$= 3\frac{2}{3}$$

So, $2\frac{1}{3} + 1\frac{1}{3} = \square\frac{\square}{\square}$

Try It

Find $1\frac{7}{8} + 1\frac{1}{2}$.

1 Shade and label the fraction circles to represent each mixed number. Write $1\frac{1}{2}$ as the equivalent fraction $1\frac{4}{8}$.

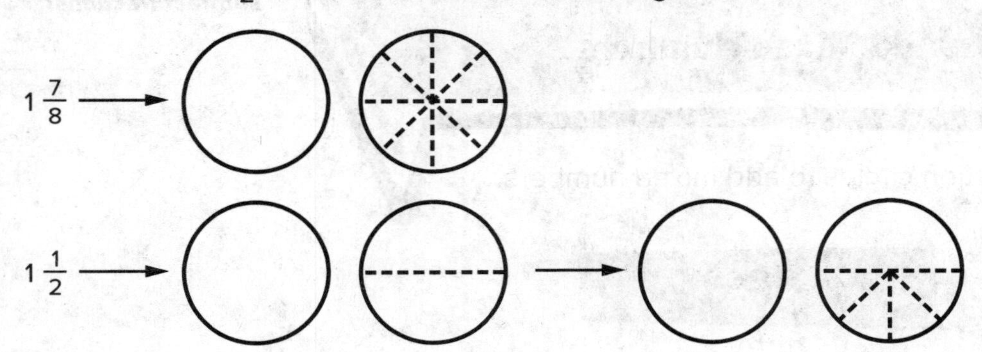

2 Combine the whole numbers and fractions.

How many whole fraction circles are there all together? ______

How many eighths are there all together? ______

Add.

$$1\frac{7}{8} + 1\frac{4}{8} = 1 + \frac{7}{8} + 1 + \frac{4}{8}$$

$$= 1 + 1 + \frac{7}{8} + \frac{4}{8}$$ Group the whole numbers and the fractions together.

$$= 2 + \frac{11}{8}$$

$$= 2 + 1\frac{3}{8}$$ Write $\frac{11}{8}$ as $\frac{8}{8} + \frac{3}{8}$, or $1\frac{3}{8}$.

$$= 3\frac{3}{8}$$

So, $1\frac{7}{8} + 1\frac{1}{2} = \square\frac{\square}{\square}$.

Talk About It

1. **Processes & Practices** 3 **Draw a Conclusion** Refer to Step 1 of the above activity. Explain why it was necessary to write $1\frac{1}{2}$ as $1\frac{4}{8}$.

______

______

Name ....

Practice It

Shade the fraction circles to represent each mixed number. Then find each sum.

2. $1\frac{1}{5} + 2\frac{3}{5} =$ ______

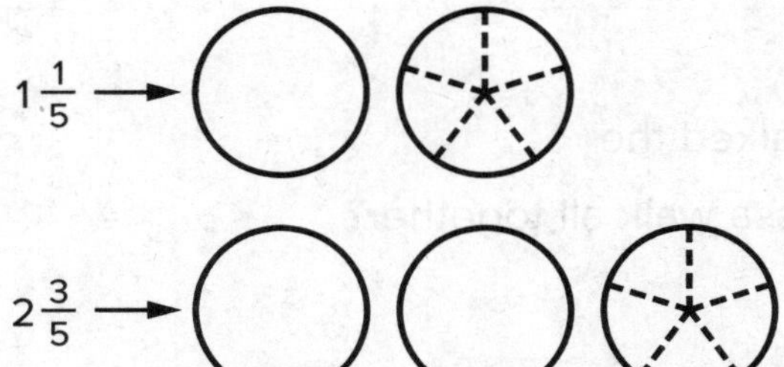

$1\frac{1}{5} \rightarrow$

$2\frac{3}{5} \rightarrow$

3. $2\frac{5}{6} + 1\frac{1}{6} =$ ______

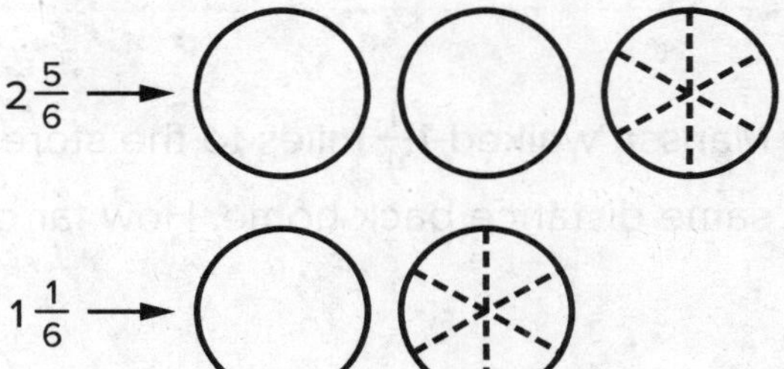

$2\frac{5}{6} \rightarrow$

$1\frac{1}{6} \rightarrow$

4. $2\frac{2}{3} + 1\frac{1}{2} =$ ______

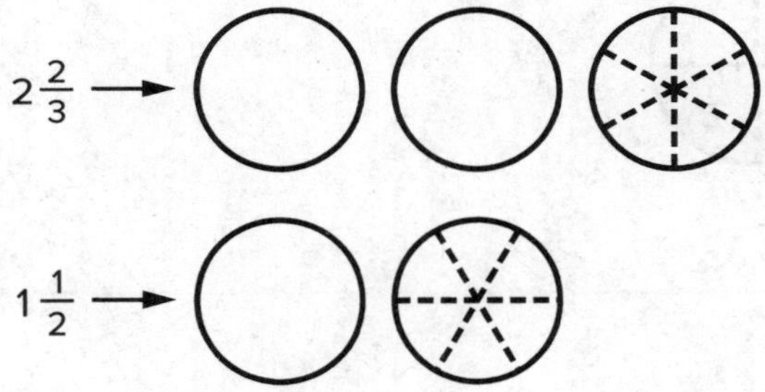

$2\frac{2}{3} \rightarrow$

$1\frac{1}{2} \rightarrow$

5. $2\frac{7}{8} + 1\frac{1}{4} =$ ______

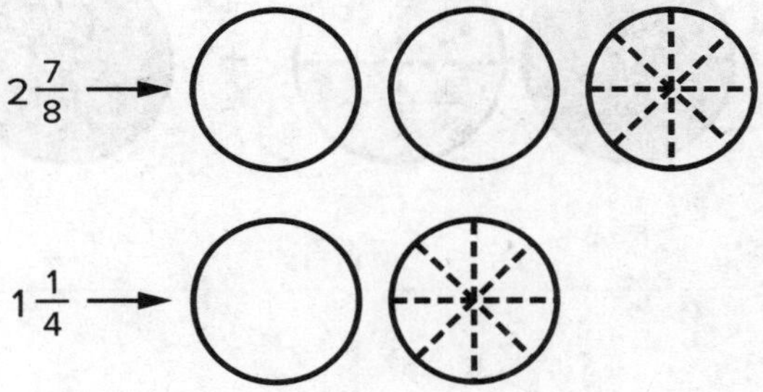

$2\frac{7}{8} \rightarrow$

$1\frac{1}{4} \rightarrow$

6. $2\frac{1}{2} + 2\frac{1}{4} =$ ______

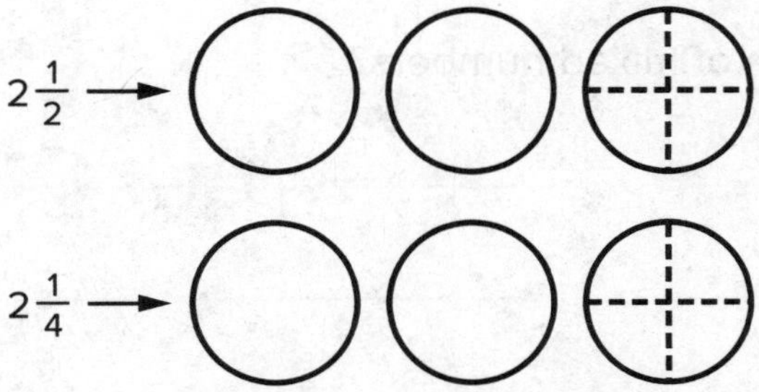

$2\frac{1}{2} \rightarrow$

$2\frac{1}{4} \rightarrow$

7. $2\frac{3}{4} + 2\frac{1}{8} =$ ______

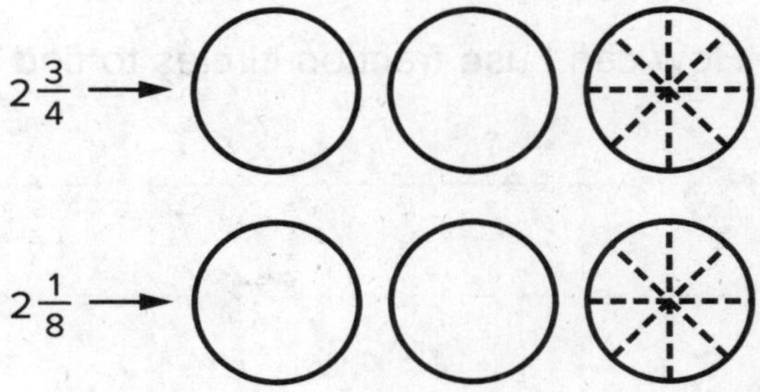

$2\frac{3}{4} \rightarrow$

$2\frac{1}{8} \rightarrow$

Apply It

Processes & Practices 5 **Use Math Tools Use fraction circles to solve Exercises 8 and 9.**

8. Eduardo worked $1\frac{1}{2}$ hours on Monday, $1\frac{1}{2}$ hours on Tuesday, and $2\frac{1}{2}$ hours on Wednesday. How many hours did he work in all?

____

9. Marissa walked $1\frac{1}{4}$ miles to the store. She walked the same distance back home. How far did Marissa walk all together?

____

10. **Processes & Practices 4** **Model Math** Write and solve a real-world problem that could be represented by the fraction circles shown.

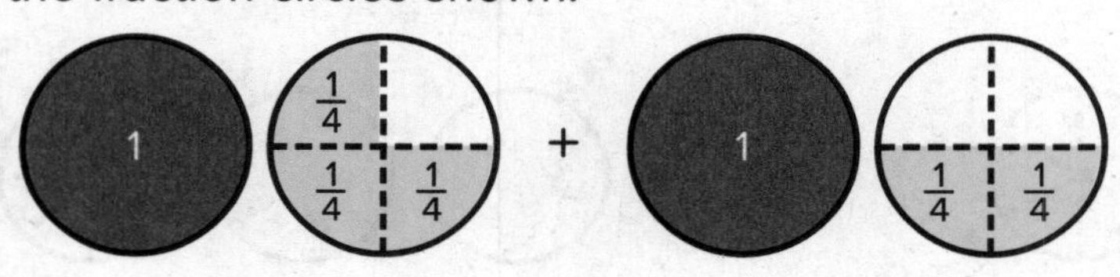

____

____

____

Write About It

11. How can I use fraction circles to find the sum of mixed numbers?

____

____

____

Name

MY Homework

Lesson 10

Hands On: Use Models to Add Mixed Numbers

Homework Helper

Need help? connectED.mcgraw-hill.com

Find $1\frac{3}{5} + 1\frac{4}{5}$.

Shade and label the fraction circles to represent each mixed number.

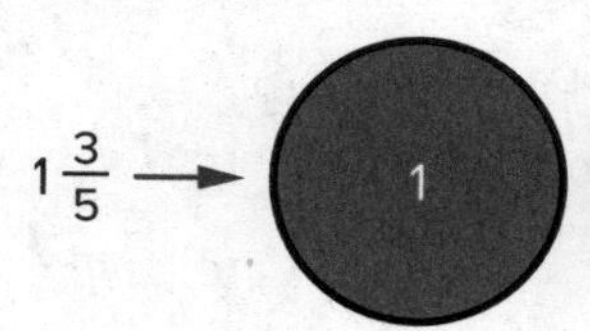

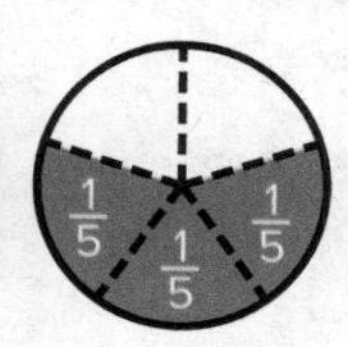

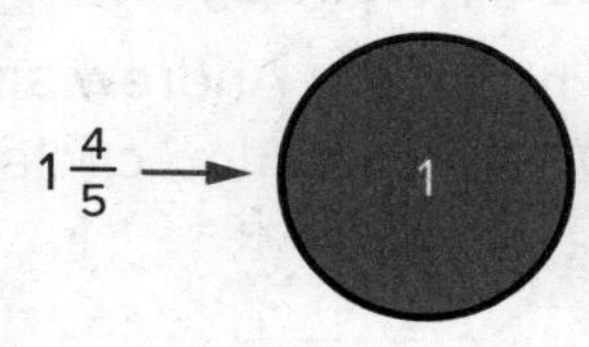

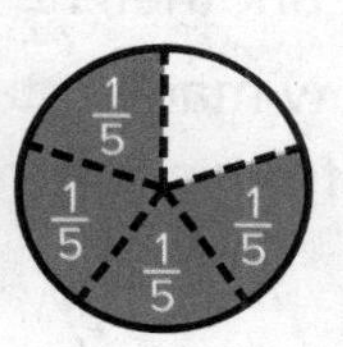

2 Combine the whole numbers and fractions.
Add.

$$\begin{aligned} 1\frac{3}{5} + 1\frac{4}{5} &= 1 + \frac{3}{5} + 1 + \frac{4}{5} \\ &= 1 + 1 + \frac{3}{5} + \frac{4}{5} \\ &= 2 + \frac{7}{5} \\ &= 2 + 1\frac{2}{5} \\ &= 3\frac{2}{5} \end{aligned}$$

Group the whole numbers and the fractions together.

Write $\frac{7}{5}$ as $\frac{5}{5} + \frac{2}{5}$, or $1\frac{2}{5}$.

So, $1\frac{3}{5} + 1\frac{4}{5} = 3\frac{2}{5}$.

Practice

1. Shade the fraction circles to represent each mixed number. Then find the sum.

$1\frac{1}{2} + 1\frac{2}{3} =$ ________

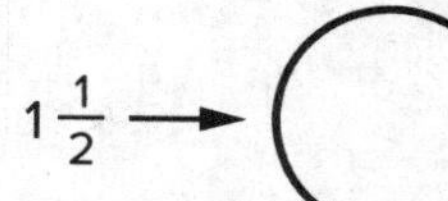

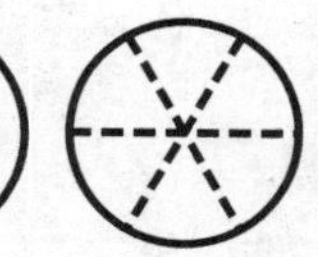

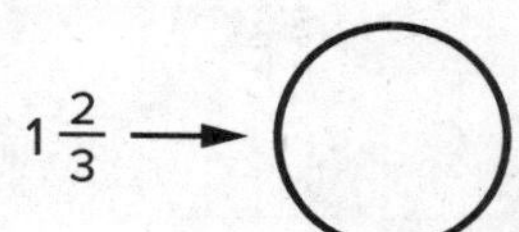

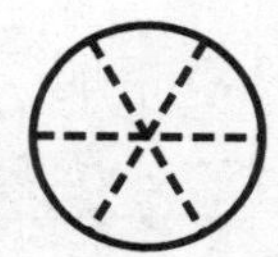

2. Shade the fraction circles to represent each mixed number. Then find the sum.

$1\frac{3}{5} + 2\frac{2}{5} =$ ________

$1\frac{3}{5} \rightarrow$

$2\frac{2}{5} \rightarrow$

Problem Solving

3. Andrew spent $1\frac{1}{6}$ hours studying for his science exam. He spent another $1\frac{1}{2}$ hours studying for his history exam. How many total hours did Andrew spend studying for his exams? Draw fraction circles to solve.

____

4. Kimberly used $1\frac{1}{3}$ cups of sugar to bake a cake and $2\frac{3}{4}$ cups of sugar to make cookies. How many total cups of sugar did she use?

____

5. Processes & Practices 3 **Which One Doesn't Belong?** Find each sum by drawing fraction circles. Then circle the expression that does not belong. Explain.

$1\frac{1}{2} + 2\frac{1}{3}$	$2\frac{2}{3} + 1\frac{1}{6}$
$1\frac{5}{6} + 2\frac{1}{2}$	$1\frac{2}{3} + 2\frac{1}{6}$

____

____

Name

Lesson 11
Add Mixed Numbers

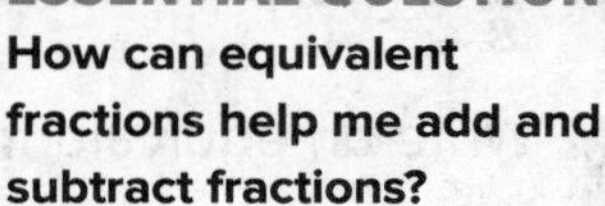

How can equivalent fractions help me add and subtract fractions?

Math in My World

Example 1

A hammerhead shark swam $2\frac{1}{4}$ miles. The next day, it swam $1\frac{1}{4}$ miles. How many miles did it swim all together?

Find $2\frac{1}{4} + 1\frac{1}{4}$.

Estimate $2\frac{1}{4} + 1\frac{1}{4} \approx 2 + 1$, or 3

$2\frac{1}{4} + 1\frac{1}{4} = 1 + 1 + \frac{1}{4} + 1 + \frac{1}{4}$ Write as a sum of wholes and fractions.

$= 1 + 1 + 1 + \frac{1}{4} + \frac{1}{4}$ Group the wholes and the fractions together.

$= 3 + \frac{2}{4}$ $1 + 1 + 1 = 3$ and $\frac{1}{4} + \frac{1}{4} = \frac{2}{4}$

$= \square\frac{\square}{\square}$ Write in simplest form. $\frac{2}{4} = \frac{1}{2}$

So, the hammerhead shark swam 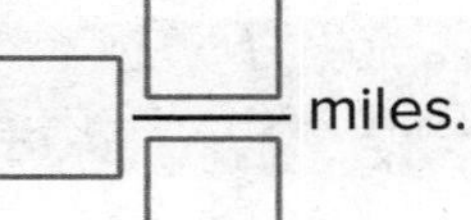 miles.

The models show that $2\frac{1}{4} + 1\frac{1}{4} = 3\frac{1}{2}$.

Check Compare to the estimate, $3\frac{1}{2} \approx 3$. The answer is reasonable.

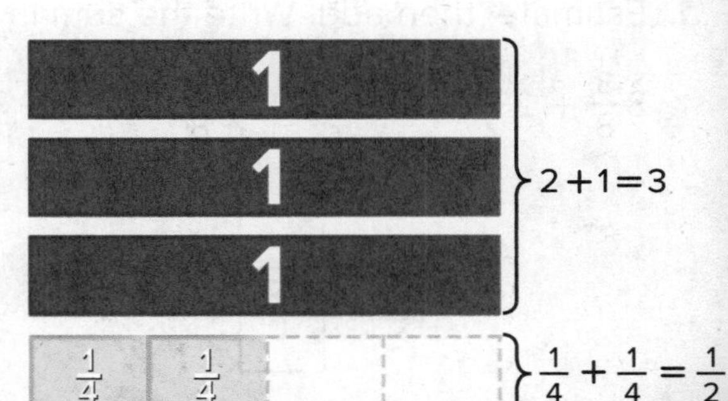

Example 2 

The diagram shows the length of a sea turtle. What is the total length of the sea turtle?

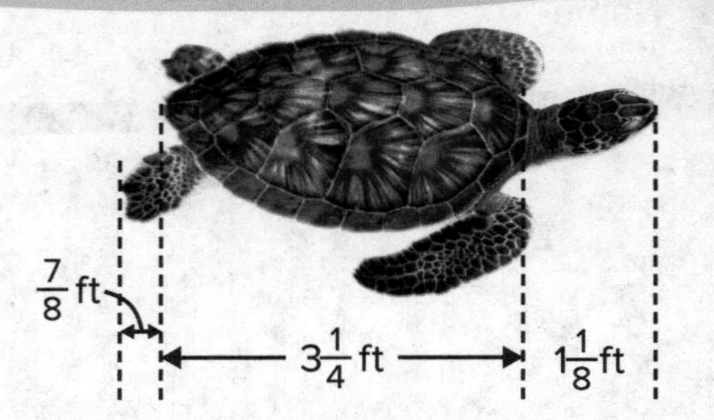

Find $\frac{7}{8} + 3\frac{1}{4} + 1\frac{1}{8}$.

1 Write an equivalent fraction for $3\frac{1}{4}$ so that the fractions all have the same denominator. The LCD is 8.

$3\frac{1}{4} = 3\frac{1 \times 2}{4 \times 2} = 3\frac{2}{8}$ Write an equivalent fraction with a denominator of 8.

2 Add.

$\frac{7}{8} + 3\frac{1}{4} + 1\frac{1}{8} = \frac{7}{8} + 3\frac{2}{8} + 1\frac{1}{8}$ Write $3\frac{1}{4}$ as $3\frac{2}{8}$.

$= 3 + 1 + \frac{7}{8} + \frac{2}{8} + \frac{1}{8}$ Group the wholes and the fractions together.

$= 4 + \frac{10}{8}$ $3 + 1 = 4$ and $\frac{7}{8} + \frac{2}{8} + \frac{1}{8} = \frac{10}{8}$

$= 4 + 1\frac{2}{8}$ Write $\frac{10}{8}$ as $\frac{8}{8} + \frac{2}{8}$, or $1\frac{2}{8}$.

$= ☐\frac{☐}{☐}$ Write in simplest form. $4 + 1 = 5$ and $\frac{2}{8} = \frac{1}{4}$

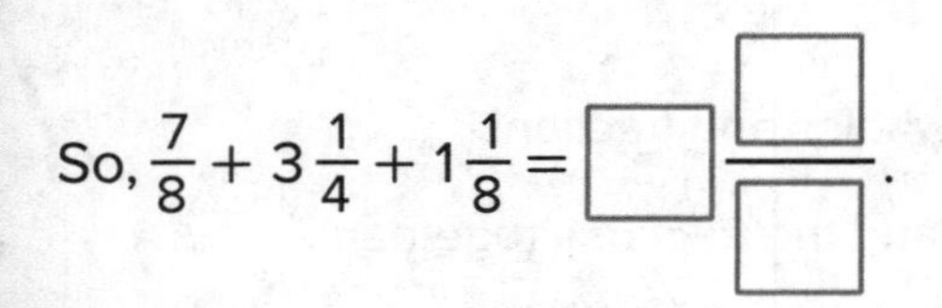

So, $\frac{7}{8} + 3\frac{1}{4} + 1\frac{1}{8} = ☐\frac{☐}{☐}$.

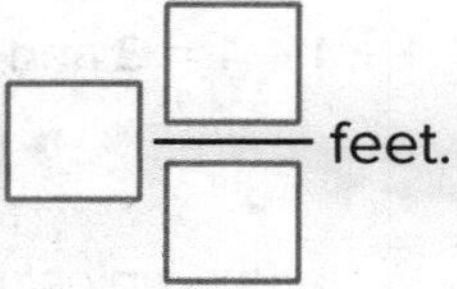

The total length of the sea turtle is $☐\frac{☐}{☐}$ feet.

Guided Practice

1. Estimate, then add. Write the sum in simplest form.

$3\frac{3}{8} + 2\frac{1}{2} = 1 + 1 + 1 + \frac{3}{8} + 1 + 1 + \frac{4}{8}$

$= ☐\frac{☐}{☐}$

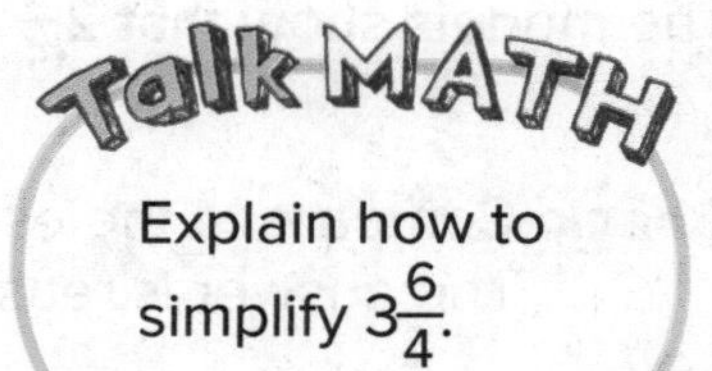

Name ....

Independent Practice

Estimate, then add. Write each sum in simplest form.

2. $4\frac{3}{5} + 3\frac{1}{5} =$ ________

3. $7\frac{4}{11} + 2\frac{6}{11} =$ ________

4. $5\frac{1}{12} + 6\frac{1}{4} =$ ________

5. $8\frac{4}{15} + 3\frac{2}{15} =$ ________

6. $6\frac{1}{9} + 2\frac{1}{3} =$ ________

7. $5\frac{1}{3} + 6\frac{1}{2} =$ ________

8. $3\frac{4}{9} + 4\frac{2}{3}$

9. $6\frac{3}{4} + 3\frac{1}{8}$

10. $4\frac{3}{7} + 7\frac{1}{2}$

Algebra Find each unknown.

11. $9\frac{9}{10} + 7\frac{3}{5} = y$

$y =$ ________

12. $14\frac{19}{20} + 8\frac{1}{4} = k$

$k =$ ________

13. $16\frac{11}{12} + 5\frac{2}{3} = d$

$d =$ ________

Problem Solving

14. Zita made $1\frac{5}{8}$ quarts of punch. Then she made $1\frac{7}{8}$ more quarts. How much punch did she make in all?

______

15. Find *five and two-eighths* plus *three and six-eighths.* Write in words in simplest form.

______

16. **Processes & Practices**  **Make a Plan** Tomás made fruit salad using the recipe. How many cups of fruit are needed all together?

______

Recipe for Fruit Salad	
$3\frac{1}{2}$ c	Bananas
$1\frac{1}{4}$ c	Grapes
$1\frac{3}{4}$ c	Strawberries
$2\frac{1}{4}$ c	Pears

Brain Builders

17. **Processes & Practices** 3 **Find the Error** Antwan is finding $4\frac{1}{5} + 2\frac{3}{5}$. Find and correct his mistake.

$$4\frac{1}{5} + 2\frac{3}{5} = 6\frac{4}{10}$$

______

Explain how Antwan could have known that his answer was incorrect.

______

18. **Building on the Essential Question** How can equivalent fractions help when adding mixed numbers?

______

______

______

Name

MY Homework

Lesson 11

Add Mixed Numbers

Homework Helper

eHelp

Need help? connectED.mcgraw-hill.com

Find $4\frac{3}{8} + 7\frac{1}{4}$.

Write an equivalent fraction for $7\frac{1}{4}$ so that the fractions all have the same denominators. The LCD is 8.

$7\frac{1}{4} = 7\frac{1 \times 2}{4 \times 2} = 7\frac{2}{8}$  Write an equivalent fraction with a denominator of 8.

2 Add.

$4\frac{3}{8} + 7\frac{1}{4} = 4\frac{3}{8} + 7\frac{2}{8}$ Write $7\frac{1}{4}$ as $7\frac{2}{8}$.

$= 4 + 7 + \frac{3}{8} + \frac{2}{8}$ Group the wholes and the fractions together.

$= 11\frac{5}{8}$

So, $4\frac{3}{8} + 7\frac{1}{4} = 11\frac{5}{8}$.

Practice

Estimate, then add. Write each sum in simplest form.

1. $2\frac{1}{10} + 5\frac{7}{10} =$ ________

2. $9\frac{3}{4} + 8\frac{3}{4} =$ ________

3. $3\frac{5}{8} + 6\frac{1}{2} =$ ________

4. $1\frac{1}{12} + 4\frac{5}{12} =$ ________

5. $11\frac{3}{5} + 6\frac{4}{15} =$ ________

6. $9\frac{1}{2} + 12\frac{11}{20} =$ ________

Problem Solving

7. A flower is $9\frac{3}{4}$ inches tall. In one week, it grew $1\frac{1}{8}$ inches. How tall is the flower at the end of the week? Write in simplest form.

8. Find *ten and three-sevenths* plus *eighteen and two-sevenths.* Write in words in simplest form.

Brain Builders

9. Processes & Practices 6 **Explain to a Friend** Connor is filling a 15-gallon wading pool. On his first trip, he carried $3\frac{1}{12}$ gallons of water. He carried $3\frac{5}{6}$ gallons on his second trip and $3\frac{1}{2}$ gallons on his third trip. Suppose he carries two $2\frac{1}{2}$-gallon buckets on his next trip. Will the pool be filled? Explain.

10. **Test Practice** Benjamin had $1\frac{2}{3}$ gallons of fruit punch left after a party. He had $1\frac{3}{4}$ gallons of lemonade left. He had $\frac{2}{3}$ gallon of apple juice left. How many total gallons did he have left?

Ⓐ $4\frac{1}{12}$ gallons

Ⓑ $3\frac{5}{12}$ gallons

Ⓒ $2\frac{5}{12}$ gallons

Ⓓ $2\frac{1}{3}$ gallon

Name ....

Lesson 12
Subtract Mixed Numbers

ESSENTIAL QUESTION
How can equivalent fractions help me add and subtract fractions?

Math in My World

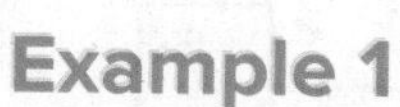

Example 1

One King crab weighs $2\frac{3}{4}$ pounds. A second King crab weighs $1\frac{1}{4}$ pounds. How much more does the one King crab weigh? Use models to find the difference.

Find $2\frac{3}{4} - 1\frac{1}{4}$.

Estimate $3 - 1 =$ ______

Model $2\frac{3}{4}$ using fraction tiles.

Subtract $1\frac{1}{4}$ by crossing out 1 whole and one $\frac{1}{4}$-tile.

1			
1			
$\frac{1}{4}$	$\frac{1}{4}$	$\frac{1}{4}$	

There is one whole and two $\frac{1}{4}$-tiles left,

which is $1\frac{2}{4}$, or $\square\frac{\square}{\square}$.

So, $2\frac{3}{4} - 1\frac{1}{4} = \square\frac{\square}{\square}$.

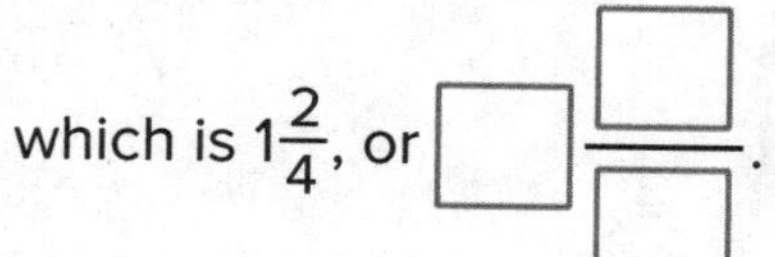

The first King crab weighs $\square\frac{\square}{\square}$ pounds more than the second.

Check for Reasonableness ______ $\approx \square\frac{\square}{\square}$

Example 2

Find $6\frac{11}{16} - 2\frac{5}{8}$.

Estimate $7 - 3 =$ ______

Write an equivalent fraction for $2\frac{5}{8}$ so that the fractions have the same denominator. The LCD is 16.

$2\frac{5}{8} = 2\frac{5 \times 2}{8 \times 2} \rightarrow \square\frac{\square}{\square}$

2 Subtract the wholes. Then subtract the fractions.

$6\frac{11}{16} - 2\frac{5}{8}$

Subtract the wholes.
$6 - 2 = 4$
Subtract the fractions.
$\frac{11}{16} - \frac{10}{16} = \frac{1}{16}$

$\rightarrow 6\frac{11}{16} - 2\frac{\square}{\square} = \square\frac{\square}{\square}$

So, $6\frac{11}{16} - 2\frac{5}{8} = \square\frac{\square}{\square}$

Check for Reasonableness ______ $\approx \square\frac{\square}{\square}$

Guided Practice

Estimate, then subtract. Write each difference in simplest form.

1. $4\frac{2}{3} - 2\frac{1}{3}$

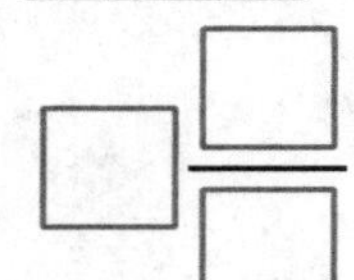

2. $5\frac{4}{5} - 3\frac{2}{5}$

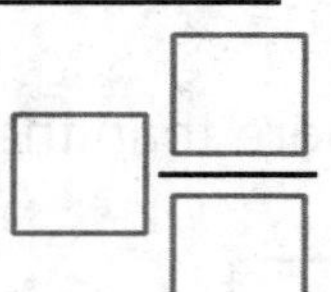

Talk MATH

Describe the steps you would take to find $3\frac{5}{8} - 2\frac{3}{8}$.

Name

Independent Practice

Estimate, then subtract. Write each difference in simplest form.

3. $5\frac{3}{4} - 2\frac{1}{2}$

4. $6\frac{5}{7} - 3\frac{3}{7}$

5. $7\frac{8}{9} - 5\frac{1}{3}$

6. $15\frac{11}{12} - 4\frac{1}{3}$

7. $13\frac{9}{10} - 4\frac{2}{5}$

8. $12\frac{5}{6} - 7\frac{1}{3}$

9. $8\frac{3}{8} - 2\frac{1}{4} =$ ________

10. $7\frac{7}{8} - 4\frac{1}{2} =$ ________

11. $12\frac{7}{10} - 7\frac{2}{5} =$ ________

Processes & Practices 

Use Algebra **Find each unknown.**

12. $11\frac{11}{12} - 2\frac{1}{12} = x$

$x =$ ________

13. $14\frac{9}{14} - 5\frac{2}{7} = c$

$c =$ ________

14. $18\frac{11}{15} - 9\frac{2}{5} = n$

$n =$ ________

Problem Solving

15. The length of Mr. Cho's garden is $8\frac{5}{6}$ feet. Find the width of Mr. Cho's garden if it is $3\frac{1}{6}$ feet less than the length.

____

16. Timberly spent $3\frac{4}{5}$ hours and Misty spent $2\frac{1}{10}$ hours at gymnastics practice over the weekend. How many more hours did Timberly spend than Misty at gymnastics practice?

____

Brain Builders

17. Processes & Practices 1 **Make Sense of Problems** Warner lives $9\frac{1}{4}$ blocks away from the ocean. Shelly lives $12\frac{7}{8}$ blocks away from the ocean. If Shelly moves $1\frac{3}{4}$ blocks closer to the ocean, how many more blocks will she live away from the ocean than Warner?

____

Use a model to illustrate the solution.

18. Processes & Practices 4 **Model Math** Write and solve a real-world problem involving the subtraction of two mixed numbers whose difference is less than $2\frac{1}{2}$.

____

____

____

19. **Building on the Essential Question** How can number sense help me to know if I have subtracted two mixed numbers correctly?

____

____

Name

MY Homework

Lesson 12

Subtract Mixed Numbers

Homework Helper

Need help? connectED.mcgraw-hill.com

Find $8\frac{1}{2} - 3\frac{1}{6}$.

Estimate $9 - 3 = 6$

1 Write an equivalent fraction for $8\frac{1}{2}$ so that the fractions have the same denominator. The LCD is 6.

$$8\frac{1}{2} = 8\frac{1 \times 3}{2 \times 3} \rightarrow 8\frac{3}{6}$$

2 Subtract the wholes. Then subtract the fractions.

$$\begin{array}{rcr} 8\frac{1}{2} & \rightarrow & 8\frac{3}{6} \\ -\ 3\frac{1}{6} & \rightarrow & -\ 3\frac{1}{6} \\ \hline & & 5\frac{2}{6} \text{ or } 5\frac{1}{3} \end{array}$$

Subtract the wholes.
$8 - 3 = 5$
Subtract the fractions.
$\frac{3}{6} - \frac{1}{6} = \frac{2}{6}$ or $\frac{1}{3}$

So, $8\frac{1}{2} - 3\frac{1}{6} = 5\frac{2}{6}$ or $5\frac{1}{3}$.

Check for Reasonableness $6 \approx 5\frac{1}{3}$

Practice

Estimate, then subtract. Write each difference in simplest form.

1. $\begin{array}{r} 6\frac{5}{8} \\ -\ 2\frac{3}{8} \\ \hline \end{array}$

2. $\begin{array}{r} 9\frac{3}{4} \\ -\ 1\frac{1}{3} \\ \hline \end{array}$

3. $\begin{array}{r} 4\frac{5}{6} \\ -\ 4\frac{1}{3} \\ \hline \end{array}$

Problem Solving

4. Mrs. Gabel bought $7\frac{5}{6}$ gallons of punch for the class party. The students drank $4\frac{1}{2}$ gallons of punch. How much punch was left at the end of the party? Write in simplest form.

5. Bella is $10\frac{5}{12}$ years old. Franco is $12\frac{7}{12}$ years old. What is the difference in their ages? Write in simplest form.

Brain Builders

6. During week 1 of a recycling drive, the fifth grade class recycled $9\frac{2}{3}$ pounds of glass and $12\frac{3}{4}$ pounds of newspaper. During week 2, the fifth grade class recycled $8\frac{2}{3}$ pounds of glass and 13 pounds of newspaper. How many more pounds of newspaper than glass did the class recycle?

7. Processes &Practices 2 **Use Number Sense** A snack mix recipe calls for $5\frac{3}{4}$ cups of cereal and $3\frac{5}{12}$ cups less of raisins. How many total cups of snack mix does the recipe make? Write in simplest form.

8. **Test Practice** What is the difference between the two weights?

$4\frac{1}{2}$ oz

$3\frac{1}{8}$ oz

Ⓐ $\frac{1}{2}$ ounce

Ⓑ $\frac{7}{8}$ ounce

Ⓒ $1\frac{3}{8}$ ounces

Ⓓ $1\frac{7}{8}$ ounces

Name ....

Lesson 13
Subtract with Renaming

ESSENTIAL QUESTION
How can equivalent fractions help me add and subtract fractions?

Sometimes the fraction in the first mixed number is less than the fraction in the second mixed number. In this case, rename the first mixed number.

Math in My World

Example 1

A black sea cucumber has an average length of 2 feet. A spotted sea cucumber has an average length of $1\frac{1}{3}$ feet. How much longer is the average black sea cucumber than the average spotted sea cucumber?

Find $2 - 1\frac{1}{3}$.

Estimate $2 - 1 =$ ______

You cannot subtract $\frac{1}{3}$ from 0 thirds.

Rename 2 as $1\frac{3}{3}$ to show more thirds.

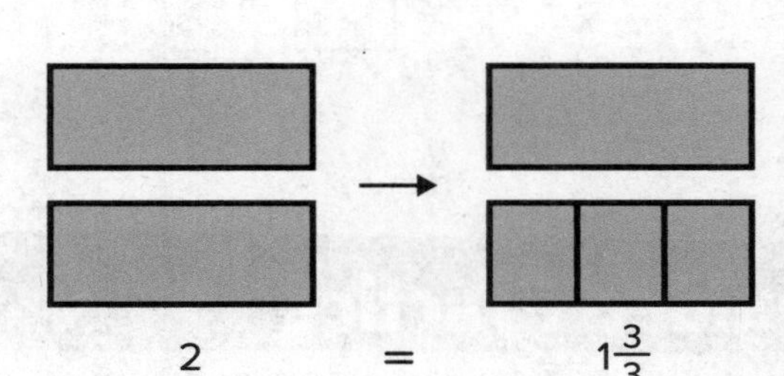

$$\begin{array}{r} 2 \\ -\,1\frac{1}{3} \\ \hline \end{array} \rightarrow \begin{array}{r} 1\frac{3}{3} \\ -\,1\frac{1}{3} \\ \hline \frac{\square}{\square} \end{array}$$

Rename 2 as $1\frac{3}{3}$

Subtract the wholes.
$1 - 1 = 0$
Subtract the fractions.
$\frac{3}{3} - \frac{1}{3} = \frac{2}{3}$

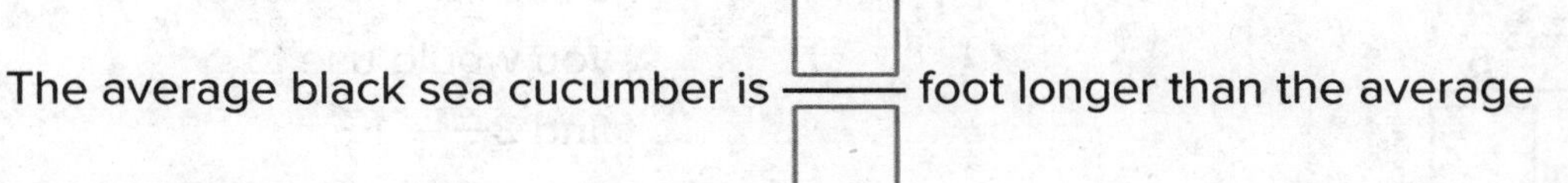

The average black sea cucumber is $\frac{\square}{\square}$ foot longer than the average spotted sea cucumber.

Check for Reasonableness $1 \approx \frac{2}{3}$

Example 2

Find $4\frac{1}{4} - 2\frac{5}{8}$.

Write equivalent fractions. The LCD is 8.

$$4\frac{1}{4} - 2\frac{5}{8} = 4\frac{1 \times 2}{4 \times 2} - 2\frac{5}{8}$$

$$= 4\frac{2}{8} - 2\frac{5}{8}$$

Helpful Hint

You can always check by using fraction tiles or drawing a picture.

You cannot subtract $\frac{5}{8}$ from $\frac{2}{8}$. So, rename $4\frac{2}{8}$ to show more eighths.

$4\frac{2}{8} \rightarrow$ $\square\frac{\square}{\square}$ Rename $4\frac{2}{8}$ as $3\frac{10}{8}$.

$-2\frac{5}{8} \rightarrow$ $-2\frac{5}{8}$

$\square\frac{\square}{\square}$ Subtract.

So, $4\frac{1}{4} - 2\frac{5}{8} = \square\frac{\square}{\square}$.

Guided Practice

1. Estimate, then subtract. Write each difference in simplest form.

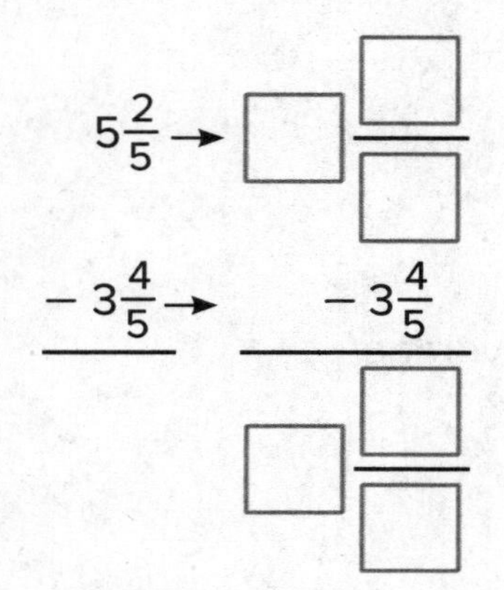

$5\frac{2}{5} \rightarrow \square\frac{\square}{\square}$

$-3\frac{4}{5} \rightarrow -3\frac{4}{5}$

$\square\frac{\square}{\square}$

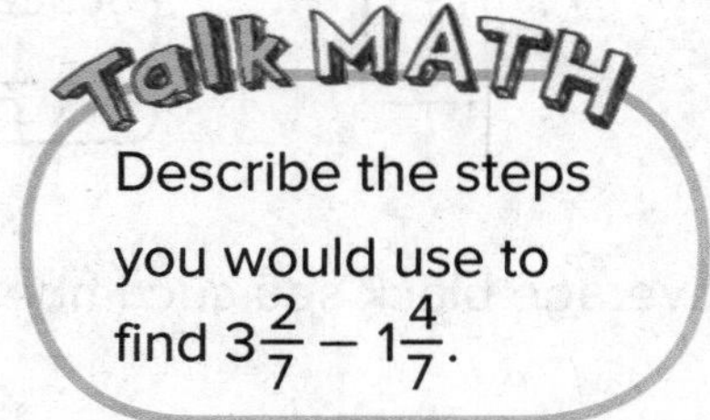

Describe the steps you would use to find $3\frac{2}{7} - 1\frac{4}{7}$.

Name

Independent Practice

Estimate, then subtract. Write each difference in simplest form.

2. $4\frac{3}{8} - 1\frac{5}{8}$

3. $3\frac{1}{6} - 1\frac{1}{3}$

4. $5\frac{1}{4} - 4\frac{1}{2}$

5. $7\frac{1}{2} - 3\frac{4}{5}$

6. $4 - 1\frac{1}{8}$

7. $12 - 5\frac{1}{6}$

8. $7\frac{2}{7} - 6\frac{4}{7} =$ ______

9. $9\frac{3}{10} - 5\frac{7}{10} =$ ______

10. $10\frac{1}{3} - 3\frac{2}{3} =$ ______

11. $18 - 9\frac{1}{4} =$ ______

12. $13 - 4\frac{1}{3} =$ ______

13. $5\frac{1}{4} - 1\frac{1}{2} =$ ______

Problem Solving

Use the table for Exercises 14 and 15. The table shows the average lengths of some insects in the United States.

Insect	Length (in.)
Monarch Butterfly	$3\frac{1}{2}$
Walking Stick	4
Grasshopper	$1\frac{3}{4}$
Bumble Bee	$\frac{5}{8}$

14. Processes & Practices 2 **Reason** Is the difference in length between a Monarch butterfly and a bumble bee greater or less than the difference in length between a walking stick and grasshopper? Explain your reasoning.

15. Find the difference in length between a walking stick and a bumble bee.

Brain Builders

16. Arlo had 5 gallons of paint. He used $2\frac{1}{8}$ gallons for one project and $1\frac{3}{4}$ gallons for another project. How much paint does he have left?

17. Processes & Practices 1 **Plan Your Solution** It takes Danny $9\frac{1}{4}$ minutes longer to walk to the park than to run. It takes Danny $3\frac{2}{3}$ minutes less time to ride his bike to the park than to run. It takes Danny 15 minutes to walk to the park. How much time does it take Danny to ride his bike to the park?

18. **Building on the Essential Question** How are equivalent fractions used in renaming when subtracting? Include an example.

Name ....

MY Homework

Lesson 13

Subtract with Renaming

Homework Helper

Need help? connectED.mcgraw-hill.com

Find $2 - 1\frac{1}{4}$.

Estimate $2 - 1 = 1$

You cannot subtract $\frac{1}{4}$ from 0 fourths.

Rename 2 as $1\frac{4}{4}$ to show more fourths.

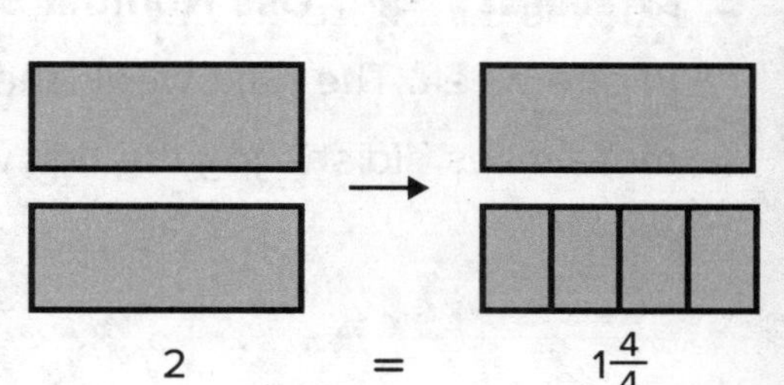

$$\begin{array}{r} 2 \\ -\,1\frac{1}{4} \\ \hline \end{array} \rightarrow \begin{array}{r} 1\frac{4}{4} \\ -\,1\frac{1}{4} \\ \hline \frac{3}{4} \end{array}$$

Rename 2 as $1\frac{4}{4}$.

Subtract the wholes.
$1 - 1 = 0$
Subtract the fractions.
$\frac{4}{4} - \frac{1}{4} = \frac{3}{4}$

So, $2 - 1\frac{1}{4} = \frac{3}{4}$.

Check for Reasonableness $1 \approx \frac{3}{4}$

Practice

Estimate, then subtract. Write each difference in simplest form.

1. $\begin{array}{r} 2\frac{1}{8} \\ -\,1\frac{7}{8} \\ \hline \end{array}$

2. $\begin{array}{r} 12\frac{1}{4} \\ -\,5\frac{2}{3} \\ \hline \end{array}$

3. $8\frac{1}{6} - 3\frac{5}{6} =$ ______

Problem Solving

4. **Processes & Practices 1** **Make a Plan** Sherman's backpack weighs $6\frac{1}{4}$ pounds. Brie's backpack weighs $5\frac{3}{4}$ pounds. How much heavier is Sherman's backpack than Brie's backpack?

____

5. **Processes & Practices 2** **Use Number Sense** Veronica jogged $10\frac{3}{16}$ miles in one week. The next week she jogged $8\frac{7}{16}$ miles. How many more miles did she jog the first week?

____

6. **Processes & Practices 6** **Be Precise** Careta swam $4\frac{3}{8}$ miles in the morning and $2\frac{6}{8}$ miles in the afternoon. Joey swam $2\frac{5}{8}$ miles in the morning and 3 miles in the afternoon. How many more miles did Careta swim than Joey?

____

7. **Test Practice** Ross has $6\frac{1}{3}$ yards of material. He bought $2\frac{1}{6}$ more yards. Then he used $6\frac{5}{6}$ yards. How many yards of material does he have left?

Ⓐ $1\frac{2}{3}$ yards

Ⓑ $2\frac{1}{3}$ yards

Ⓒ $3\frac{2}{3}$ yards

Ⓓ $8\frac{1}{2}$ yards

Review

Chapter 9

Add and Subtract Fractions

Vocabulary Check

Write each of the following on the lines to make a true sentence.

equivalent fractions **least common denominator** **like fractions**

mixed number **unlike fractions**

1. Fractions that have the same denominator, such as $\frac{1}{3}$ and $\frac{2}{3}$, are

________________________.

2. Fractions that have the same value, such as $\frac{1}{4}$ and $\frac{2}{8}$, are

________________________.

3. The ________________________ of $\frac{5}{6}$ and $\frac{7}{12}$ is 12.

4. A number that has a whole number part and a fraction part is called a ________________________.

5. An example of two ________________________ are $\frac{1}{8}$ and $\frac{3}{4}$.

Concept Check

Round each fraction to 0, $\frac{1}{2}$, or 1.

6. $\frac{4}{7} \approx$ ________

7. $\frac{9}{10} \approx$ ________

8. $\frac{2}{9} \approx$ ________

Add. Write each sum in simplest form.

9. $\frac{5}{9} + \frac{1}{9} =$ ______

10. $\frac{6}{7} + \frac{1}{7} =$ ______

11. $\frac{5}{8} + \frac{1}{8} =$ ______

12. $\frac{3}{5} + \frac{1}{10} =$ ______

13. $\frac{1}{2} + \frac{1}{8} =$ ______

14. $\frac{2}{7} + \frac{5}{14} =$ ______

15. $6\frac{3}{4} + 2\frac{2}{4} =$ ______

16. $1\frac{1}{12} + 3\frac{2}{12} =$ ______

17. $12\frac{1}{3} + 6\frac{2}{6} =$ ______

Estimate, then subtract. Write each difference in simplest form.

18. $\frac{7}{16} - \frac{3}{16} =$ ______

19. $\frac{11}{12} - \frac{7}{12} =$ ______

20. $\frac{4}{5} - \frac{2}{5} =$ ______

21. $\frac{8}{9} - \frac{5}{6} =$ ______

22. $\frac{11}{12} - \frac{7}{8} =$ ______

23. $\frac{7}{10} - \frac{1}{5} =$ ______

Problem Solving

24. Russ is putting his vacation photographs in an album that is $12\frac{1}{8}$ inches long and $10\frac{1}{4}$ inches wide. Should he trim the edges of the photographs to 12 inches long and 10 inches wide or to $12\frac{1}{2}$ inches long and $10\frac{1}{2}$ inches wide?

25. Steve watched television for $\frac{3}{4}$ hour on Monday and $\frac{5}{6}$ hour on Tuesday. How much longer did he watch television on Tuesday than on Monday?

Brain Builders

26. When Ricki walks to school on the sidewalk, she walks $\frac{7}{10}$ mile. She then takes a shortcut across the field, which is $\frac{1}{4}$ mile long. How long is Ricki's route to and from school?

27. Test Practice Peta was swimming with stingrays. The first stingray she swam with was $5\frac{1}{4}$ feet long. The second one she swam with was $4\frac{3}{4}$ feet long. The third stingray was $5\frac{1}{8}$ feet long. How much longer was the longest stingray than the shortest stingray?

Ⓐ $\frac{3}{8}$ foot

Ⓑ $\frac{1}{2}$ foot

Ⓒ $1\frac{1}{4}$ feet

Ⓓ $1\frac{1}{2}$ feet

Reflect

Chapter 9

Answering the ESSENTIAL QUESTION

Use what you learned about fraction operations to complete the graphic organizer.

Write an Example

Real-World Example

ESSENTIAL QUESTION

How can equivalent fractions help me add and subtract fractions?

Estimate

Vocabulary

Now reflect on the ESSENTIAL QUESTION Write your answer below.

Name

Date

Score

Performance Task

Community Pick Up

The community is organizing into two teams that will pick up trash alongside roads, parks, and creeks. They will spend a week on each project.

Show all your work to receive full credit.

Part A

In Week 1, the two teams are concentrating on picking up trash alongside roads. The table shows how many miles of road each team will clean on the given day.

Day	Team Cougars	Team Wolves
Monday	$\frac{1}{3}$ mile	$\frac{1}{3}$ mile
Tuesday	$\frac{3}{4}$ mile	$\frac{2}{3}$ mile
Wednesday	$\frac{2}{3}$ mile	$\frac{1}{2}$ mile
Thursday	$\frac{1}{3}$ mile	$\frac{3}{4}$ mile
Friday	$\frac{3}{4}$ mile	$\frac{2}{3}$ mile

Which team cleaned more miles of road? Explain.

Part B

In Week 2, the two teams are concentrating on picking up trash in the parks. They are going to split into smaller teams and have a combined goal of reaching 50 parks. On Monday, $12\frac{1}{3}$ of the parks were cleaned. On Tuesday, $8\frac{1}{4}$ of the parks were cleaned. On Wednesday, $11\frac{1}{3}$ of the parks were cleaned. On Thursday, $10\frac{1}{4}$ of the parks were cleaned. How many parks are left to clean on Friday in order to meet their goal? Explain.

Part C

In Week 3, the two teams concentrate on picking up the trash around creeks. The Cougars predict that they can fill $110\frac{3}{4}$ trash bags, and the Wolves predict that they can fill $128\frac{2}{3}$ trash bags. At the end of the week the Cougars actually filled $110\frac{1}{6}$ bags, and the Wolves actually filled $128\frac{1}{4}$ bags. Which team was closer to their prediction? Explain.

Chapter

10 Multiply and Divide Fractions

ESSENTIAL QUESTION

What strategies can be used to multiply and divide fractions?

In My Kitchen

Watch

Watch a video!

Name ....

MY Chapter Project

I'm Game

1. In your group, discuss the different types of games you know (*Examples:* card games, spinner games, trivia games). Make a list of these games on a separate sheet of paper.
2. Discuss with your group what type of game to create. Remember, the game should use the concepts you are learning in this chapter.
3. As a group, write the title, rules, and how to win on a separate sheet of paper. Work together to draw sketches of any boards, cards, or other items you'll use for your game.
4. Collect all the materials you need and work together to create your game. Make sure everyone in your group participates!
5. Add new content to your game to incorporate new concepts you learn throughout the chapter.
6. At the end of the chapter, have a Game Day where you play other groups' games and other groups play your game.

Name

Estimate. Use rounding. Show how you estimated.

1. $8\frac{5}{6} + 7\frac{1}{5} \approx$ ______

2. $17\frac{1}{4} - 3\frac{7}{8} \approx$ ______

3. Tyra filled the two water coolers shown with flavored sports drink. About how many gallons of sports drink does she have all together?

______

Estimate. Use rounding or compatible numbers. Show how you estimated.

4. $196 \times 12 \approx$ ______

5. $79 \times 31 \approx$ ______

6. $304 \div 6 \approx$ ______

7. $558 \div 8 \approx$ ______

Add or subtract. Write in simplest form.

8. $\frac{1}{6} + \frac{1}{6} =$ ______

9. $\frac{3}{4} + \frac{2}{5} =$ ______

10. $\frac{8}{9} - \frac{5}{9} =$ ______

11. $\frac{7}{12} - \frac{1}{12} =$ ______

How Did I Do?

Shade the boxes to show the problems you answered correctly.

1	2	3	4	5	6	7	8	9	10	11

Name

MY Math Words

Vocab

Review Vocabulary

decimal point	denominator	digit
divide	equivalent	greatest common factor (GCF)
least common multiple (LCM)	mixed numbers	multiply
number line		

Making Connections

Use the Venn diagram to categorize the review vocabulary.

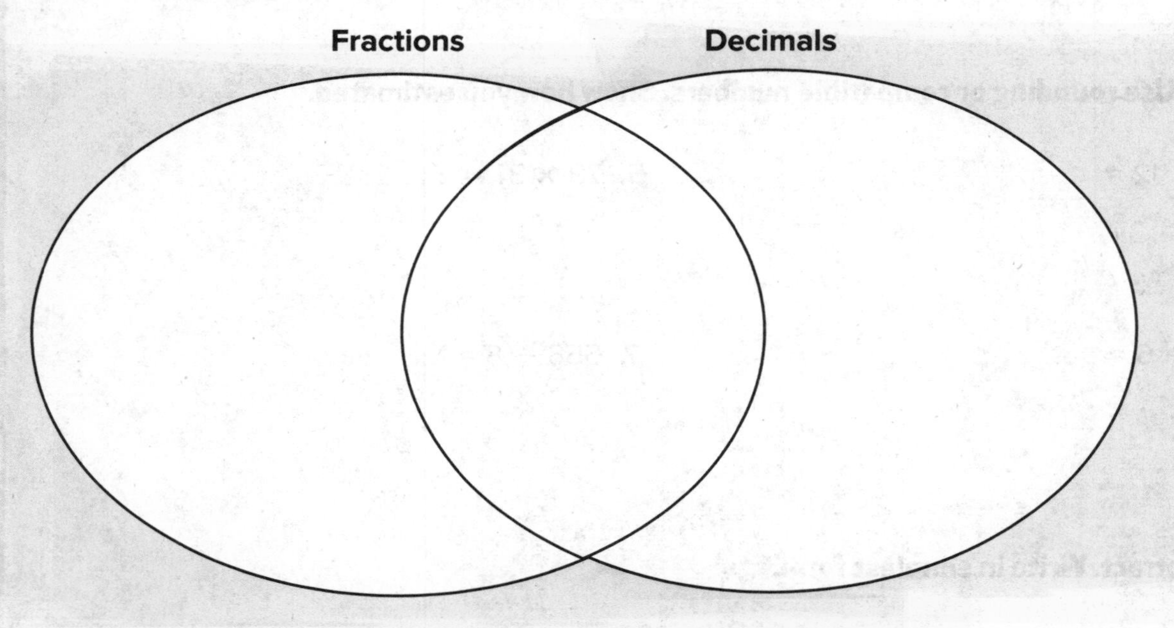

How is adding and subtracting fractions similar to adding and subtracting whole numbers?

MY Vocabulary Cards

Processes & Practices

Lesson 10-8

scaling

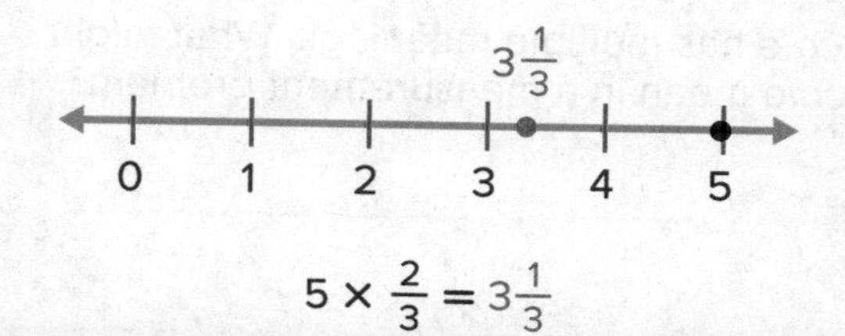

$5 \times \frac{2}{3} = 3\frac{1}{3}$

Lesson 10-9

unit fraction

$\frac{1}{3}$ ← numerator = 1

Ideas for Use

- During the year, create a separate stack of cards for key math verbs, such as *scaling.* They will help you in problem solving.
- Use the blank cards to write review vocabulary cards.
- Group like ideas you find throughout the chapter, such as the relationship between multiplication and division. Use the blank cards to write notes about these concepts.

A fraction with a numerator of 1.

The Latin root *fract* means "break." How does this relate to the meaning of *fraction?*

The process of resizing a number when it is multiplied by a fraction that is greater than or less than 1.

Scale has multiple meanings. What might *scale* mean in a measurement problem?

MY Foldable

FOLDABLES® Follow the steps on the back to make your Foldable.

$\frac{3}{4}$ cup sugar

$\frac{1}{3}$ cup butter

1 egg

$\frac{1}{2}$ teaspoon vanilla

$1\frac{1}{3}$ cups flour

$1\frac{1}{4}$ teaspoons baking powder

$\frac{1}{2}$ teaspoon salt

$\frac{3}{4}$ cup milk

_______ cups sugar

_______ cup butter

_______ eggs

_______ teaspoon vanilla

_______ cups flour

_______ teaspoons baking powder

_______ teaspoon salt

_______ cups milk

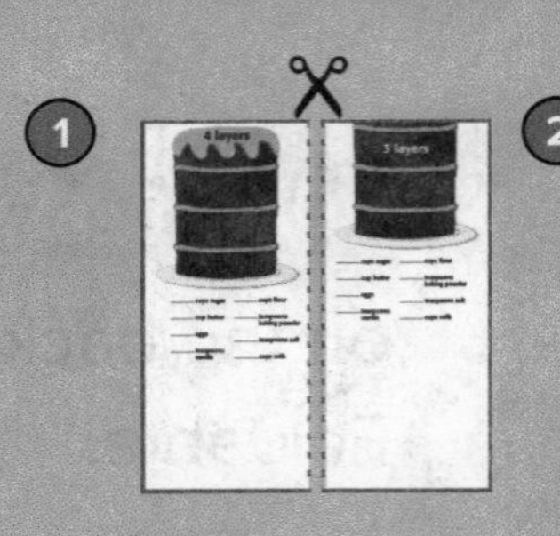

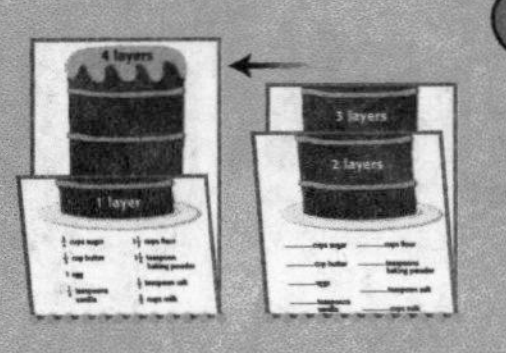

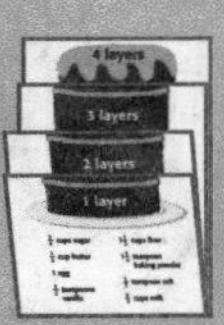

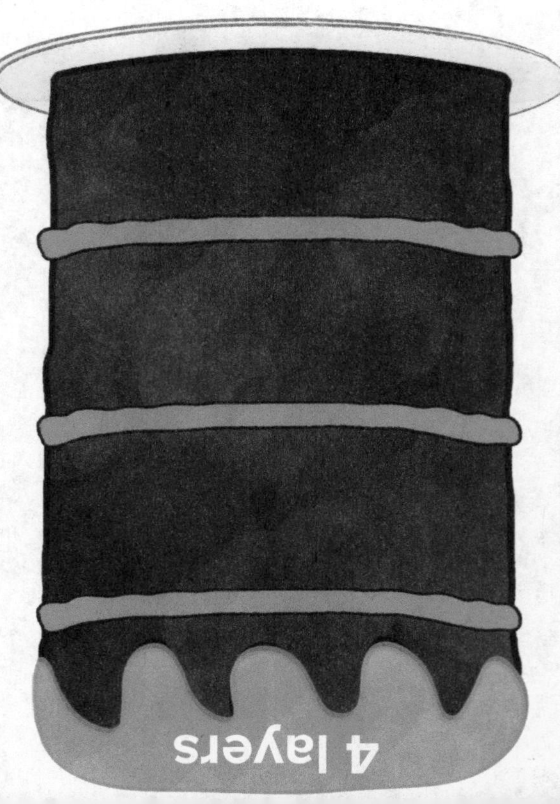

- cups sugar
- cups butter
- eggs
- teaspoons vanilla
- cups flour
- teaspoons baking powder
- teaspoons salt
- cups milk

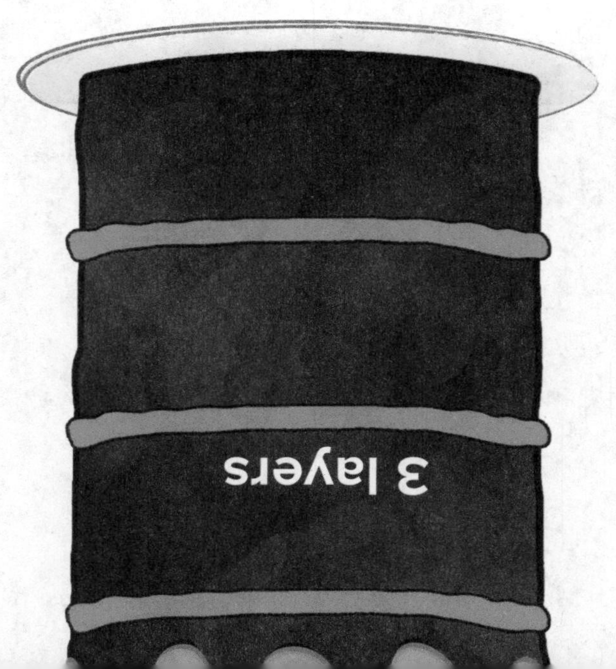

- cups sugar
- cup butter
- eggs
- teaspoons vanilla
- cups flour
- teaspoons baking powder
- teaspoons salt
- cups milk

Name

Lesson 1 Hands On 

Part of a Number

ESSENTIAL QUESTION
What strategies can be used to multiply and divide fractions?

You can use bar diagrams to find parts of a number.

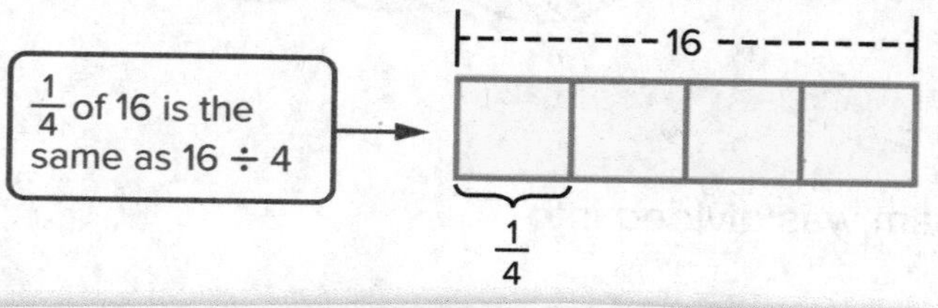

Draw It

Lelah threw 16 pitches in the first inning of a softball game. Of the pitches she threw, $\frac{3}{4}$ of them were strikes. How many strikes did she throw in the first inning?

Find $\frac{3}{4}$ of 16.

 The bar diagram represents the number of pitches she threw.

16 pitches

4 pitches	4 pitches	4 pitches	4 pitches

$\frac{3}{4}$

 Since the denominator is 4, the bar diagram was divided into

_______ equal sections.

Each section of the bar represents _______ pitches.

 Use the diagram to determine $\frac{3}{4}$ of 16.

4 + 4 + 4 = _______

$\frac{3}{4}$ of 16 is _______.

So, Lelah threw _______ strikes.

Try It

Find $\frac{1}{3}$ of 15 using a bar diagram.

Label the bar diagram that represents 15.

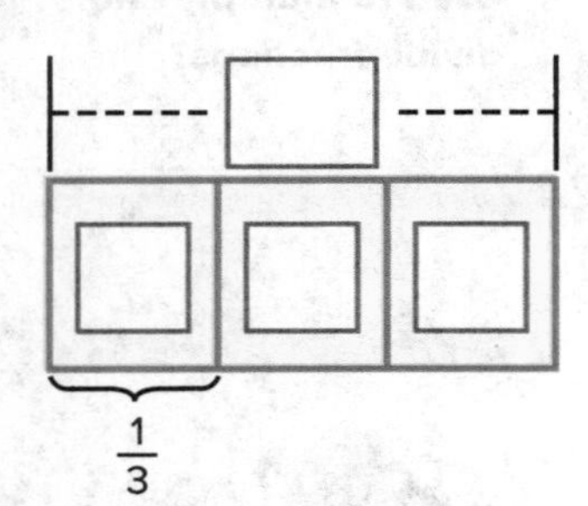

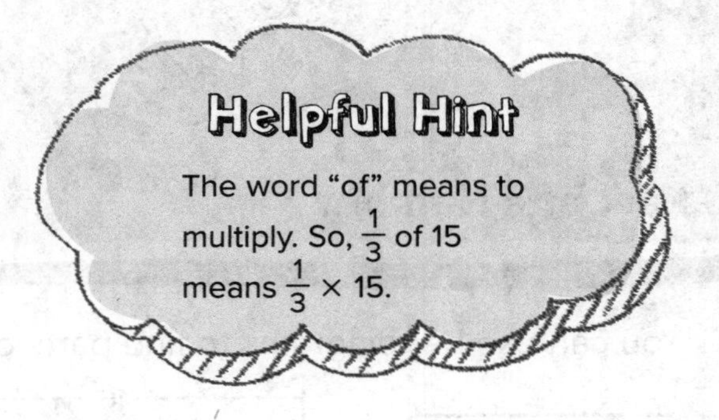

2 Since the denominator is 3, the bar diagram was divided into

_______ equal sections.

Each section of the bar represents _______.

3 Use the diagram to determine $\frac{1}{3}$ of 15.

$\frac{1}{3}$ of 15 is the same as $15 \div 3$, which is the same as $\frac{1}{3} \times 15$.

What is $\frac{1}{3}$ of 15? _______

So, $\frac{1}{3}$ of 15 = _______.

Talk About It

1. **Processes & Practices 6** **Explain to a Friend** Explain why $\frac{1}{4}$ of 16 is the same as $16 \div 4$.

____

____

2. Explain why $\frac{3}{4}$ of 16 is the same as $3 \times 16 \div 4$.

____

____

Name ....

Practice It

Draw a bar diagram to find each product.

3. $12 \times \frac{1}{2} =$ ______

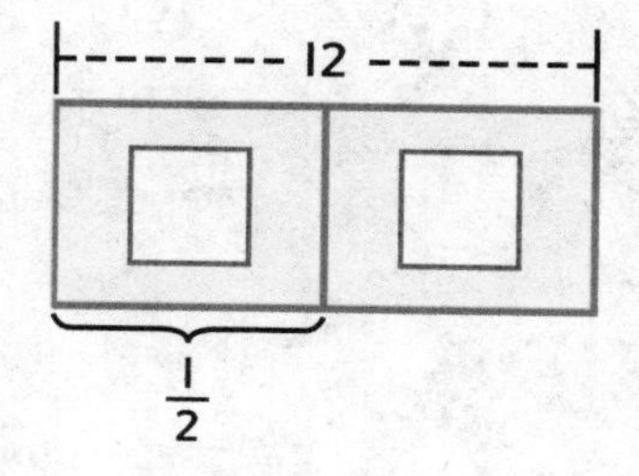

4. $\frac{2}{3}$ of $15 =$ ______

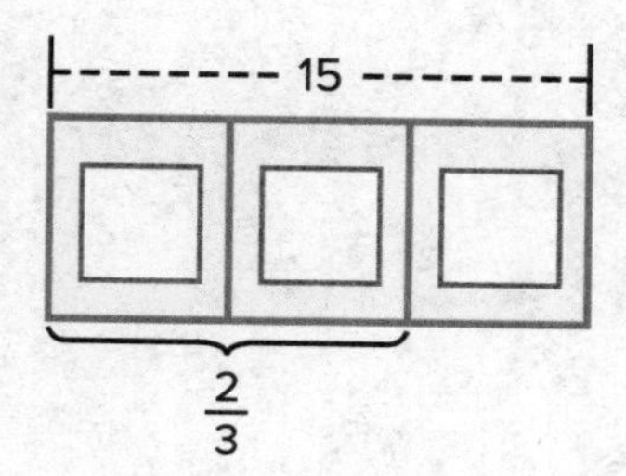

5. $\frac{2}{3}$ of $18 =$ ______

6. $9 \times \frac{1}{3} =$ ______

7. $8 \times \frac{1}{4} =$ ______

8. $\frac{1}{2}$ of $16 =$ ______

9. $25 \times \frac{2}{5} =$ ______

10. $24 \times \frac{3}{4} =$ ______

Apply It

Draw a bar diagram to help solve Exercises 11 and 12.

11. Leon used plant fertilizer on $\frac{4}{7}$ of his potted flowers. If he has 28 potted flowers, on how many did he use plant fertilizer?

12. Processes & Practices 5 **Use Math Tools** Jeremy washed $\frac{3}{8}$ of the plates from dinner. If 16 plates were used, how many plates did Jeremy wash?

13. Processes & Practices 4 **Model Math** Write a real-world problem that could represent the bar diagram shown.

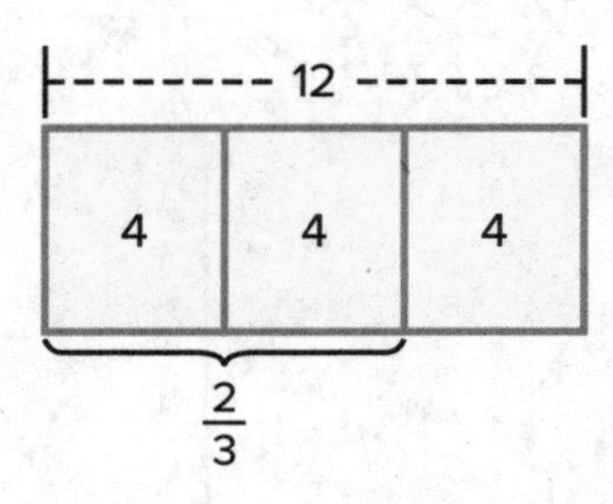

Write About It

14. How can I use models to find part of a number?

Name ....

Lesson 1

Hands On: Part of a Number

Homework Helper

Need help? connectED.mcgraw-hill.com

Find $\frac{3}{4}$ of 40 using a bar diagram.

1. The bar diagram represents 40.

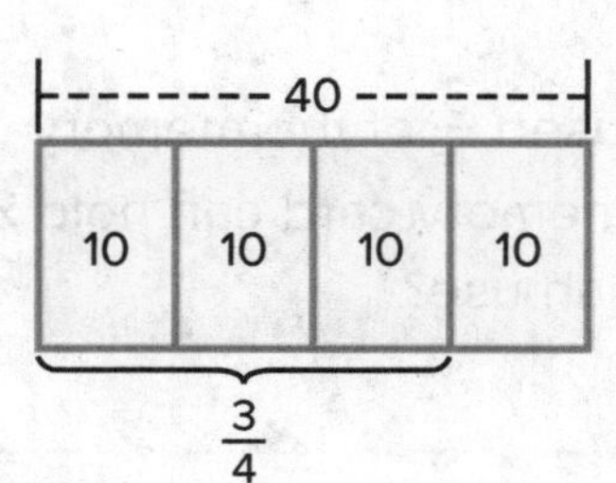

2. Since the denominator is 4, the bar diagram is divided into 4 equal sections.

Each section of the bar represents 10.

3. Use the diagram to determine $\frac{3}{4}$ of 40.

Since $\frac{1}{4}$ of 40 is the same as $40 \div 4$, or 10, then $\frac{3}{4}$ of 40 is the same as $3 \times 40 \div 4$, or 30.

So, $\frac{3}{4}$ of $40 = 30$.

Practice

Draw a bar diagram to find each product.

1. $\frac{2}{3}$ of $36 =$ ________

2. $35 \times \frac{3}{5} =$ ________

Problem Solving

Draw a bar diagram to help solve Exercises 3–5.

3. Hope used $\frac{1}{3}$ of the flour in the container to make cookies. If the container holds 12 cups of flour, how many cups did Hope use?

____

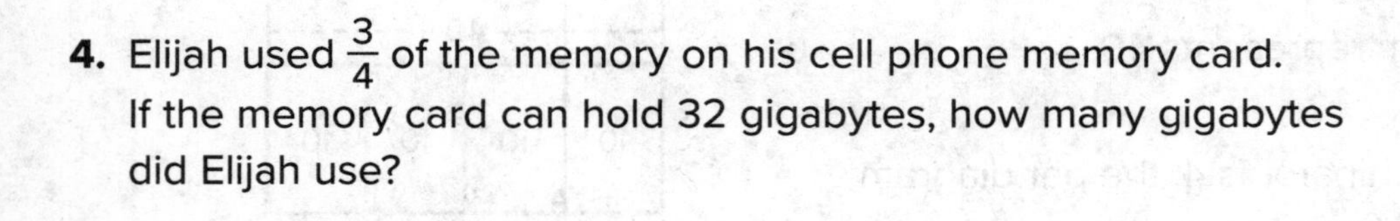

4. Elijah used $\frac{3}{4}$ of the memory on his cell phone memory card. If the memory card can hold 32 gigabytes, how many gigabytes did Elijah use?

____

5. **Processes & Practices 4** **Model Math** Jeremy used $\frac{5}{6}$ of a loaf of bread throughout the week. If there were 24 slices of bread, how many slices did Jeremy use?

____

6. Write a real-world problem that could represent the bar diagram shown.

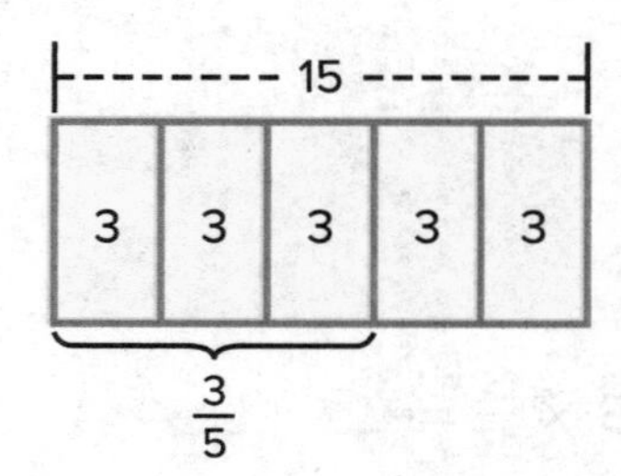

____

____

____

____

Name ....

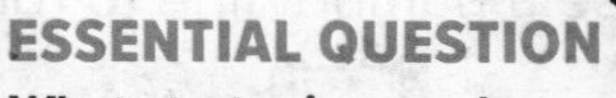

Lesson 2
Estimate Products of Fractions

ESSENTIAL QUESTION
What strategies can be used to multiply and divide fractions?

You have already estimated products of decimals using rounding and compatible numbers. You can use the same strategies to estimate the products of fractions.

Math in My World

Example 1

About $\frac{1}{3}$ of the containers of yogurt in a box are strawberry. If the box contains 17 containers, about how many containers are strawberry?

Estimate $\frac{1}{3} \times 17$.

Use compatible numbers.

$\frac{1}{3} \times 17 \approx \frac{1}{3} \times$ ______ ← What number is close to 17 and also compatible with 3?

Complete the bar diagram to find the product of $\frac{1}{3} \times$ ______.

How many yogurt containers are represented in each section of the bar diagram? ______

So, $\frac{1}{3} \times 18 =$ ______.

About ______ yogurt containers are strawberry.

Example 2

Estimate $\frac{2}{5} \times \frac{6}{7}$.

Use a number line to round each fraction to 0, $\frac{1}{2}$, or 1.

$\frac{2}{5} \times \frac{6}{7} \approx \frac{\square}{\square} \times 1$

$\approx \frac{\square}{\square}$ Identity Property

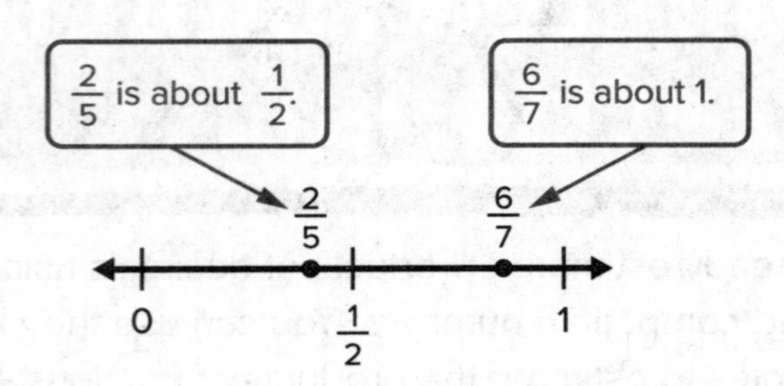

So, $\frac{2}{5} \times \frac{6}{7}$ is about $\frac{\square}{\square}$.

Example 3

Estimate $7\frac{2}{7} \times 3\frac{7}{8}$.

Round each mixed number to the nearest whole number.

Round $7\frac{2}{7}$ down to ______ and round $3\frac{7}{8}$ up to ______.

$7\frac{2}{7} \times 3\frac{7}{8} \approx$ ______ $\times$ ______

$\approx$ ______

So, $7\frac{2}{7} \times 3\frac{7}{8}$ is about ______.

Explain how you would estimate the product of $\frac{4}{5} \times \frac{5}{6}$.

Guided Practice

1. Estimate the product of $\frac{1}{2} \times 25$.

   Use compatible numbers.

   $\frac{1}{2} \times 25 \approx \frac{1}{2} \times$ ______ Think: What is half of 24?

   $\approx$ ______

   So, $\frac{1}{2} \times 25$ is about ______.

Name

Independent Practice

Estimate each product. Draw a bar diagram if necessary.

2. $\frac{2}{3} \times 13$

3. $\frac{1}{3} \times 20$

4. $\frac{1}{2} \times 33$

5. $17 \times \frac{1}{4}$

6. $\frac{7}{8} \times \frac{1}{9}$

7. $\frac{3}{5} \times \frac{8}{9}$

8. $\frac{1}{6} \times \frac{5}{7}$

9. $\frac{1}{4} \times \frac{8}{9}$

10. $2\frac{2}{3} \times 3\frac{1}{6}$

11. $6\frac{4}{5} \times 5\frac{7}{8}$

12. $10\frac{1}{7} \times 4\frac{4}{5}$

13. $2\frac{6}{7} \times 6\frac{2}{9}$

Problem Solving

14. A cup of chocolate chips weighs about 9 ounces. A recipe calls for $3\frac{3}{4}$ cups of chocolate chips. About how many ounces of chocolate chips are needed?

15. Processes & Practices 5 **Use Math Tools** Isabella sent out 22 invitations to her birthday party. If about $\frac{1}{4}$ of the invitations are for her school friends, about how many classmates did she invite? Draw a bar diagram to help you solve.

Brain Builders

16. Processes & Practices 1 **Make a Plan** Write a real-world problem involving the multiplication of two mixed numbers whose product is about 14. Explain how you determined your mixed numbers. Then solve the problem.

17. **Building on the Essential Question** Explain when estimation would not be the best method for solving a problem. Give an example.

Name

MY Homework

Lesson 2

Estimate Products of Fractions

Homework Helper

Need help? connectED.mcgraw-hill.com

Find the estimated area of the rug shown.

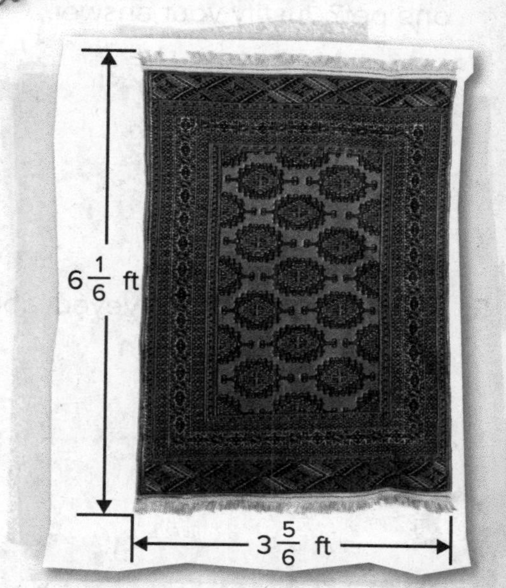

Estimate $6\frac{1}{6} \times 3\frac{5}{6}$.

Round each mixed number to the nearest whole number.

Round $6\frac{1}{6}$ down to 6. Round $3\frac{5}{6}$ up to 4.

$6\frac{1}{6} \times 3\frac{5}{6} \approx 6 \times 4$

≈ 24

So, the area of the floor rug is about 24 square feet.

Practice

Estimate each product. Draw a bar diagram if necessary.

1. $\frac{2}{3} \times 26$ ____________

2. $\frac{7}{8} \times \frac{5}{6}$ ____________

3. $5\frac{1}{5} \times 8\frac{5}{6}$ ____________

Real World

Problem Solving

Use the table to answer Exercises 4 and 5.

Students with Pets	
Number of Pets	**Fraction of Students with Pets**
0	$\frac{1}{20}$
1	$\frac{1}{2}$
2 or more	$\frac{9}{20}$

4. Processes & Practices 3 **Justify Conclusions** The table shows the results of a class survey about pets. Suppose 53 students were surveyed. About how many students have one pet? Justify your answer.

5. If 100 students are surveyed, about how many students own 2 or more pets? Explain.

Brain Builders

6. A rectangular floor measures $10\frac{1}{8}$ feet by $13\frac{3}{4}$ feet. If Mary buys 150 square feet of carpet, will she have enough carpet to cover the floor? Explain.

7. **Test Practice** A cup of water holds about 8 ounces. If a recipe calls for $\frac{1}{3}$ cup of water, which is the best estimate of the number of ounces that are needed? Draw a bar diagram if necessary.

Ⓐ 1 ounce
Ⓑ 3 ounces
Ⓒ 6 ounces
Ⓓ 8 ounces

Name

Lesson 3
Hands On
Model Fraction Multiplication

ESSENTIAL QUESTION
What strategies can be used to multiply and divide fractions?

You can multiply a whole number by a fraction using repeated addition.

Draw It

Find $\frac{1}{3} \times 3$. Use repeated addition.

1. Since the denominator is 3, each model is divided into ______ equal sections.

2. Shade $\frac{1}{3}$ of each model.

How many sections of each model are shaded? ______

Add. How many total sections of the models are shaded? ______

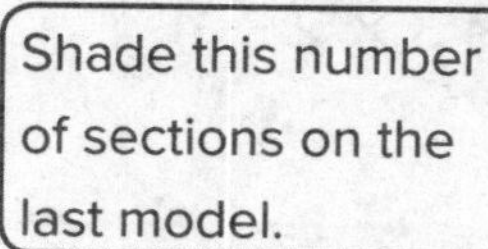

3. The last model shows the product of $\frac{1}{3} \times 3$.

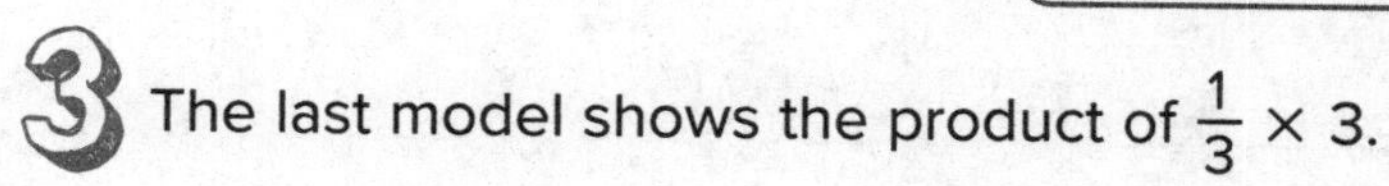

What is $\frac{3}{3}$ equal to? ______

So, $\frac{1}{3} \times 3 = 3 \div 3$, or ______.

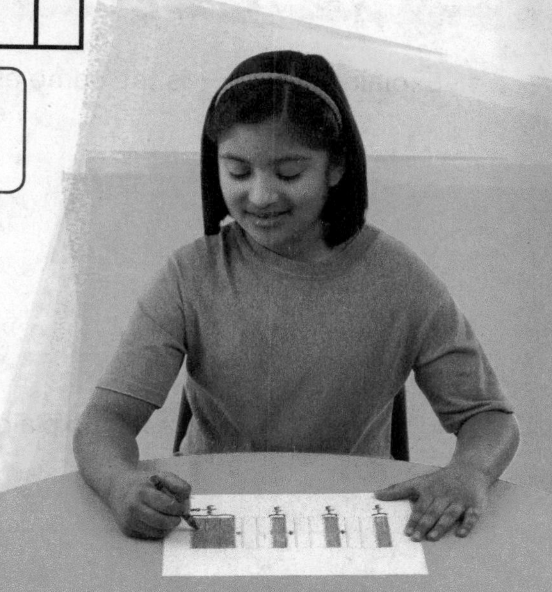

Try It

Find the unknown in $2 \times \frac{3}{4} = \blacksquare$. Use models. Write in simplest form.

1 Divide each model below into ________ equal sections since the denominator is 4.

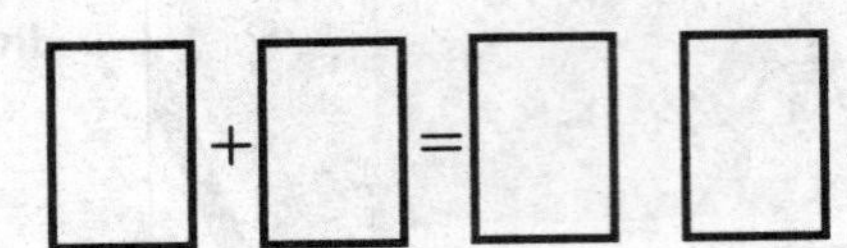

2 Shade $\frac{3}{4}$ of each of the first two models.

How many sections of each model are shaded? ________

Add. How many total sections of the models are shaded? ________

3 The last two models show the product of $2 \times \frac{3}{4}$.

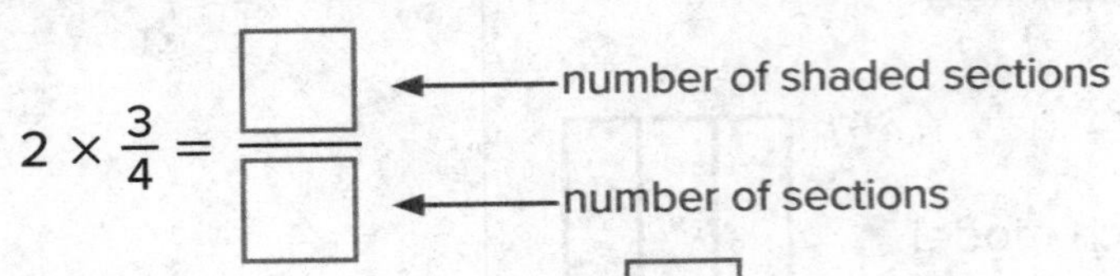

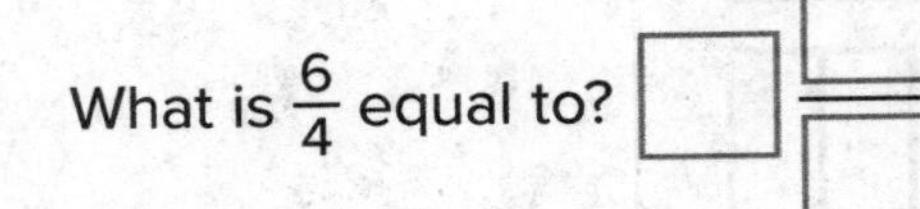

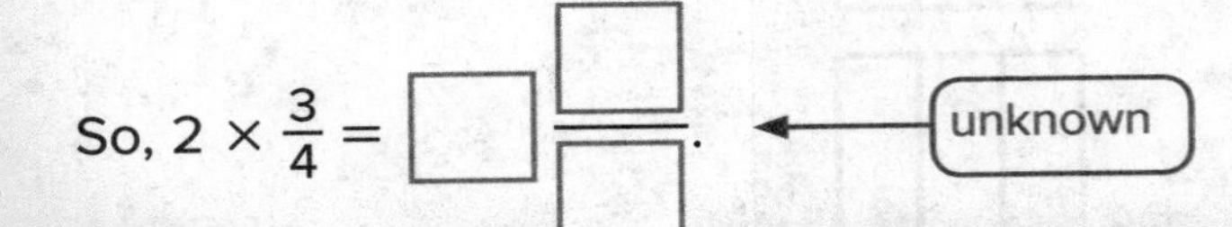

Helpful Hint: $2 \times \frac{3}{4}$ is the same as $3 \times 2 \div 4$.

Talk About It

1. Explain why $2 \times \frac{3}{4}$ is the same as $3 \times 2 \div 4$.

____

____

____

2. Processes & Practices 2 **Reason** Explain how you could find $2 \times \frac{3}{4}$ without using models.

____

____

Name ________________

Practice It

Shade the models to find each product. Write in simplest form.

3. $4 \times \frac{1}{2} =$ ________

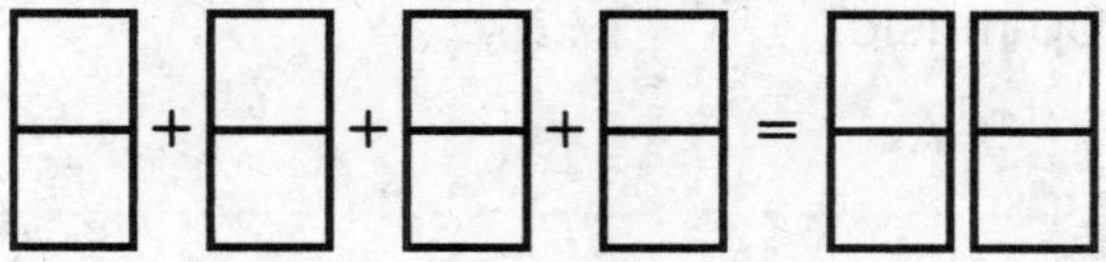

+ + + =

4. $4 \times \frac{1}{3} =$ ________

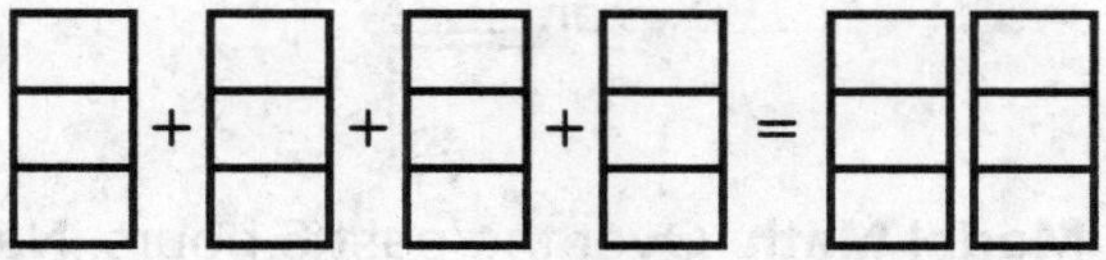

+ + + =

5. $\frac{2}{3} \times 5 =$ ________

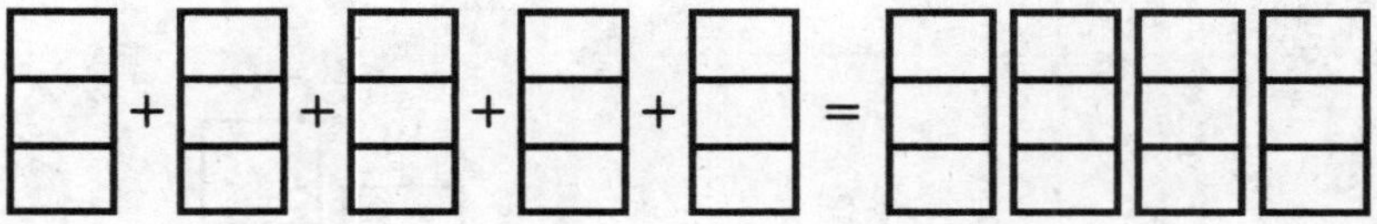

+ + + + =

6. $3 \times \frac{2}{3} =$ ________

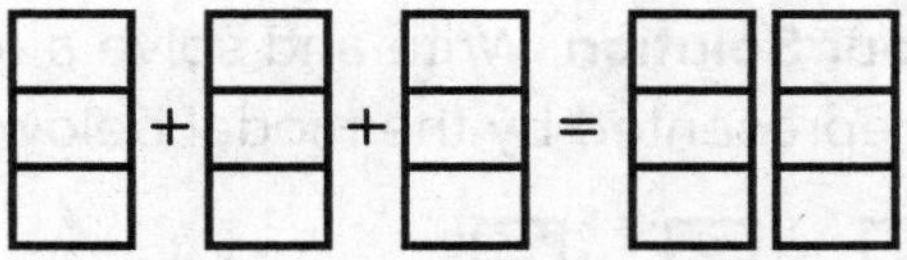

+ + =

Algebra **Find each unknown. Shade the models to find each product. Write in simplest form.**

7. $\frac{1}{5} \times 6 = ■$

■ = ________

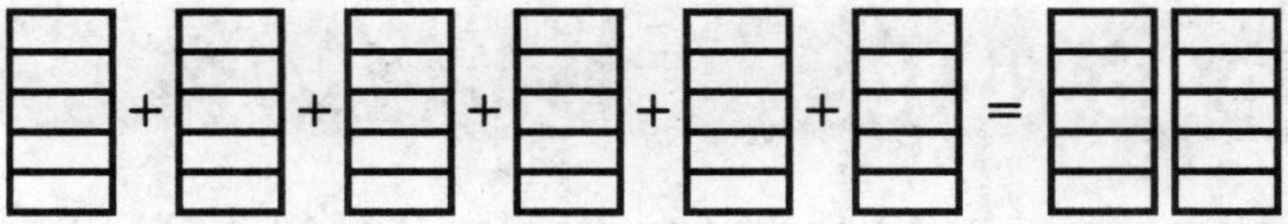

+ + + + + =

8. $\frac{2}{5} \times 6 = ■$

■ = ________

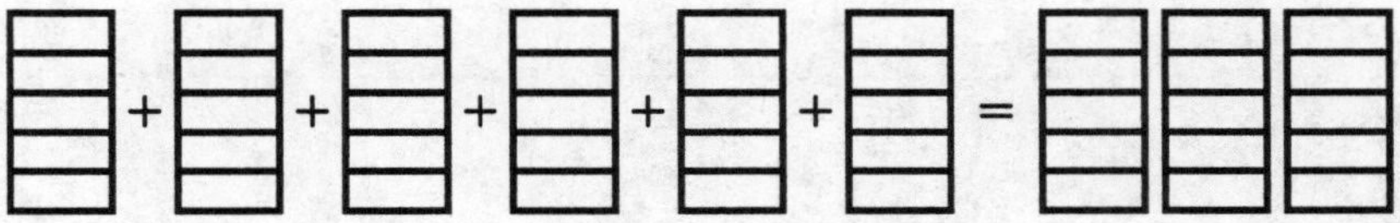

+ + + + + =

Apply It

Algebra Use models to help you solve Exercises 9 and 10. Then complete the equation.

9. Brandon saved 2 gigabytes of music on his MP3 player. Of the music saved, $\frac{1}{4}$ is hip hop. What fraction of a gigabyte did Brandon use for his hip hop music?

____

Equation: $2 \times \frac{1}{4} = 2 \div$ ______, or $\frac{\square}{\square}$

10. Processes &Practices 4 **Model Math** Over the past 6 hours, Natalie spent $\frac{1}{5}$ of each hour kneading bread. How much time did she spend kneading bread in all?

____

Equation: $6 \times \frac{1}{5} = 6 \div$ ______, or $\square\frac{\square}{\square}$

11. Processes &Practices 1 **Plan Your Solution** Write and solve a real-world problem that could be represented by the model below.

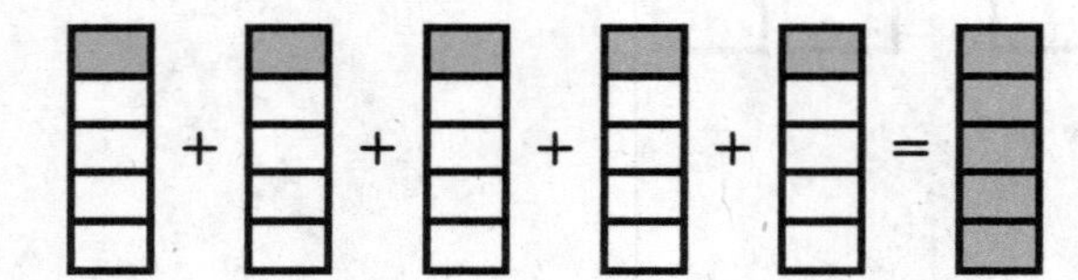

____

____

Write About It

12. Explain why $\frac{2}{3} \times 12$ can be written as $2 \times 12 \div 3$.

____

____

Name

MY Homework

Lesson 3

Hands On: Model Fraction Multiplication

Homework Helper

Need help? connectED.mcgraw-hill.com

Find $5 \times \frac{2}{5}$ using models. Write in simplest form.

1. Each model below is divided into 5 equal sections since the denominator is 5.

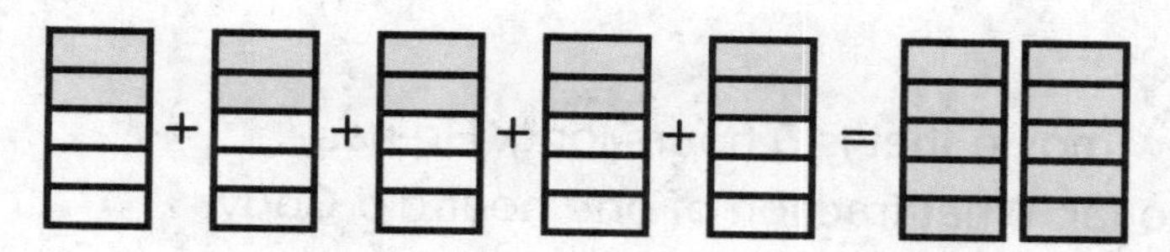

2. Of each model, $\frac{2}{5}$ is shaded. There are 2 sections of each model that are shaded.
Add. A total of 10 sections of the models are shaded.

3. The last two models show the product of $5 \times \frac{2}{5}$.

$5 \times \frac{2}{5} = \frac{10}{5}$ ← number of shaded sections / ← number of sections

$= 2$ Simplify.

So, $5 \times \frac{2}{5} = 2$.

Practice

Shade the models to find each product. Write in simplest form.

1. $4 \times \frac{2}{3} =$ ________

2. $\frac{3}{4} \times 2 =$ ________

Problem Solving

  **Model Math** Use models to help you solve Exercises 3–5. Then complete the equation.

3. Melinda spent 4 hours reviewing for her midterm exams. She spent $\frac{1}{4}$ of the time studying for social studies. How many hours did she spend on social studies?

____

Equation: $4 \times \frac{1}{4} = 4 \div$ ______, or $\square$

4. Cody wants to watch a new movie that is 3 hours long. He has watched $\frac{1}{6}$ of the movie so far. What fraction of one hour did Cody spend watching the new movie?

____

Equation: $3 \times \frac{1}{6} = 3 \div$ ______, or $\frac{\square}{\square}$

5. The distance from Paula's house to school is 5 miles. There are railroad tracks one-fourth of the distance from her house on the way to school. How far are the railroad tracks from Paula's house?

____

Equation: $5 \times \frac{1}{4} = 5 \div$ ______, or $\square\frac{\square}{\square}$

Name ....

Lesson 4
Multiply Whole Numbers and Fractions

ESSENTIAL QUESTION
What strategies can be used to multiply and divide fractions?

You can write a whole number as a fraction.

$$12 \longrightarrow \frac{12}{1}$$

Math in My World

Example 1

Wild parrots spend $\frac{1}{6}$ of the day looking for food. How many hours a day does a parrot spend looking for food?

Find $\frac{1}{6} \times 24$. ← There are 24 hours in a day.

$\frac{1}{6} \times 24 = \frac{1}{6} \times \frac{24}{1}$ — Write 24 as a fraction.

$= \frac{1 \times 24}{6 \times 1}$ ← Multiply the numerators. ← Multiply the denominators.

$= \frac{24}{6}$ or ______ — Simplify.

So, a wild parrot spends ______ hours a day looking for food.

Check $\frac{1}{6} \times 24 = 1 \times 24 \div 6$, or ______

Key Concept Multiply Fractions

To multiply a whole number by a fraction, write the whole number as a fraction. Then multiply the numerators and multiply the denominators.

Example 2

Find the unknown in $2 \times \frac{4}{5} = \blacksquare$.

Estimate $2 \times 1 =$ ______

$2 \times \frac{4}{5} = \frac{2}{1} \times \frac{4}{5}$ Write 2 as a fraction.

$= \frac{2 \times 4}{1 \times 5}$ ← Multiply the numerators. ← Multiply the denominators.

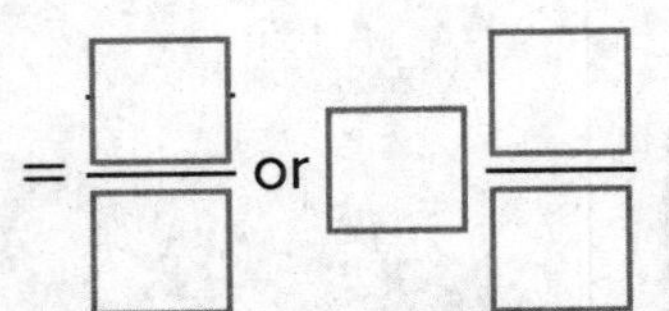

Simplify.

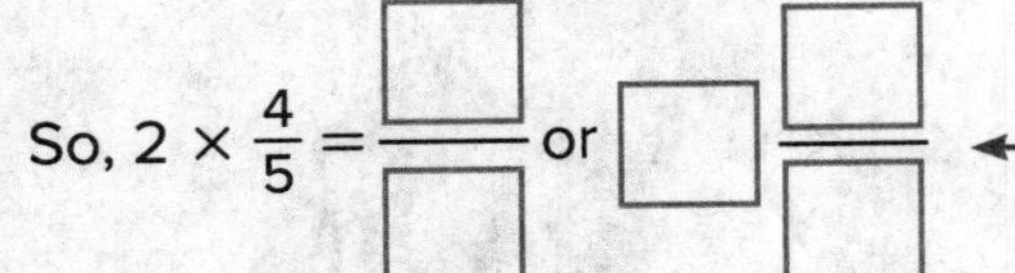

Check

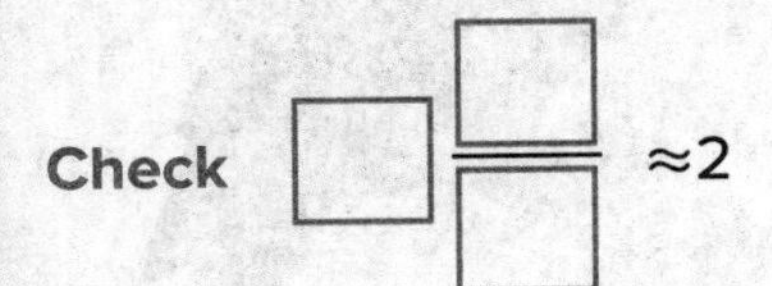

Talk MATH

Explain how you could find the product of 50 and $\frac{2}{5}$ mentally.

Guided Practice

1. Find $\frac{1}{2} \times 4$. Write in simplest form.

$\frac{1}{2} \times 4 = \frac{1}{2} \times \frac{4}{1}$ Write 4 as $\frac{4}{1}$.

$= \frac{1 \times 4}{2 \times 1}$ Multiply.

$= \frac{\square}{\square}$ or ______ Simplify.

So, $\frac{1}{2} \times 4 =$ ______.

Name ....

Independent Practice

Multiply. Write in simplest form.

2. $\frac{1}{3} \times 12 =$ ______

3. $\frac{1}{4} \times 20 =$ ______

4. $\frac{5}{6} \times 18 =$ ______

5. $\frac{1}{5} \times 7 =$ ______

6. $\frac{2}{3} \times 14 =$ ______

7. $\frac{2}{5} \times 11 =$ ______

8. $12 \times \frac{1}{6} =$ ______

9. $13 \times \frac{2}{13} =$ ______

10. $24 \times \frac{3}{4} =$ ______

Processes & Practices 2 **Use Algebra Find the unknown in each equation. Write in simplest form.**

11. $5 \times \frac{1}{4} = ■$

■ = ______

12. $15 \times \frac{7}{10} = ■$

■ = ______

13. $32 \times \frac{5}{6} = ■$

■ = ______

Problem Solving

14. Arleta is making nacho cheese dip for a party. She needs to make 5 batches. How much salsa will she need?

____

15. Maria ate $\frac{1}{4}$ of a pizza. If there were 20 slices of pizza, how many slices did Maria eat?

____

Brain Builders

16. Processes & Practices 4 **Model Math** Andrew spends $\frac{3}{5}$ hour every day feeding his pets. How much time does Andrew spend feeding his pets each week? Write an expression to represent the situation. Then solve.

____

17. Processes & Practices 3 **Which One Doesn't Belong?** Circle the expression that does not belong with the other three. Explain how you could change the expression so that it does belong with the other three.

$\frac{1}{2} \times 12$	$18 \times \frac{1}{3}$	$\frac{1}{4} \times 20$	$\frac{1}{6} \times 36$

____

18. **Building on the Essential Question** Can all whole numbers be written as fractions? Explain.

____

____

Name

MY Homework

Lesson 4

Multiply Whole Numbers and Fractions

Homework Helper

Need help? connectED.mcgraw-hill.com

Leon was given 10 days to finish his art project. He used $\frac{2}{3}$ of his time to paint the project. For how many days did Leon paint his project?

Estimate $\frac{2}{3} \times 9 = 6$

$\frac{2}{3} \times 10 = \frac{2}{3} \times \frac{10}{1}$ Write 10 as a fraction.

$= \frac{2 \times 10}{3 \times 1}$ ← Multiply the numerators. ← Multiply the denominators.

$= \frac{20}{3}$ or $6\frac{2}{3}$ Simplify.

So, Leon spent $6\frac{2}{3}$ days painting his project.

Check $\frac{2}{3} \times 10 = 2 \times 10 \div 3$, or $6\frac{2}{3}$

Practice

Multiply. Write in simplest form.

1. $\frac{2}{3} \times 12 =$ ______

2. $\frac{3}{10} \times 8 =$ ______

3. $8 \times \frac{1}{5} =$ ______

4. $13 \times \frac{1}{2} =$ ______

5. $20 \times \frac{3}{5} =$ ______

6. $\frac{3}{10} \times 7 =$ ______

Real World Problem Solving

7. The length of a popcorn machine is $\frac{3}{4}$ of its height. The height is 24 inches. What is the length of the machine?

____

8. Quinten is making bread and wants to triple the recipe. The recipe calls for $\frac{2}{3}$ cup of sugar. How much sugar will he need?

____

Brain Builders

9. Joaquin has $24. He used $\frac{5}{8}$ of his money to buy a pair of jeans. How much money does Joaquin have left?

____

10. Processes & Practices 1 **Plan Your Solution** The length of a rectangle is $\frac{1}{3}$ of its width. The width of the rectangle is 9 feet. What is the area of the rectangle?

____

11. **Test Practice** Rico is making punch for 18 people. How much punch should Rico make if each person will drink $\frac{1}{6}$ gallon of punch?

Ⓐ 2 gallons　　Ⓒ 4 gallons

Ⓑ 3 gallons　　Ⓓ 5 gallons

Check My Progress

Vocabulary Check

Draw lines to match each word with its correct description or meaning.

1. fraction	• a number that is multiplied to form a product
2. factor	• the result of a multiplication problem
3. product	• a number representing part of a whole or part of a set

Concept Check

Estimate each product. Draw a bar diagram if necessary.

4. $\frac{1}{5} \times 11$ ____________

5. $\frac{3}{5} \times \frac{10}{11}$ ____________

6. $4\frac{7}{8} \times 2\frac{1}{12}$ ____________

7. Shade the model to find the product. Write in simplest form.

$2 \times \frac{3}{5} =$ ____________

Multiply. Write in simplest form.

8. $\frac{2}{5} \times 16 =$ ________

9. $18 \times \frac{1}{6} =$ ________

10. $6 \times \frac{3}{5} =$ ________

Problem Solving

11. Diane sent 25 invitations to her birthday party. If $\frac{1}{6}$ of the invitations were for her family, about how many family members did she invite?

12. A recipe calls for $\frac{2}{3}$ cup of brown sugar. Marilyn is making 5 batches. How many cups of brown sugar should Marilyn use?

13. There are 24 students in Mr. Sampson's math class. One-third of the students received a B on a recent test. The remaining students received an A on the test. How many students received an A on the test?

14. Jacinta is working on a jigsaw puzzle that has 403 pieces. She has completed $\frac{1}{4}$ of the puzzle. About how many pieces does Jacinta have left to place in the puzzle? Explain.

15. Test Practice George has a collection of 16 postcards from the states shown in the table. How many postcards does he have from California?

Ⓐ 12 postcards

Ⓑ 10 postcards

Ⓒ 8 postcards

Ⓓ 6 postcards

Postcards	
State	**Fraction**
Florida	$\frac{1}{4}$
California	$\frac{3}{8}$
Texas	$\frac{1}{8}$
New York	$\frac{1}{4}$

Name

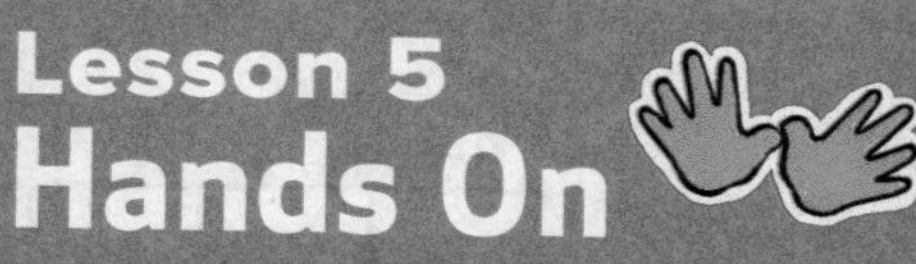

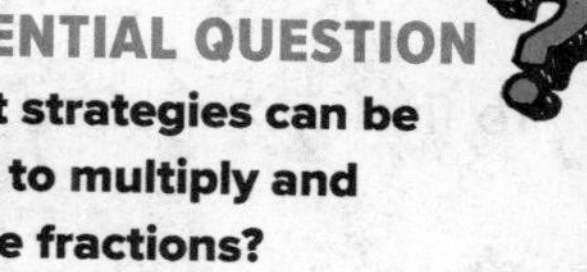

Draw It

Find $\frac{1}{3} \times \frac{1}{4}$. Write in simplest form.

To find $\frac{1}{3} \times \frac{1}{4}$, find the area of a $\frac{1}{3}$- by $\frac{1}{4}$-unit rectangle.

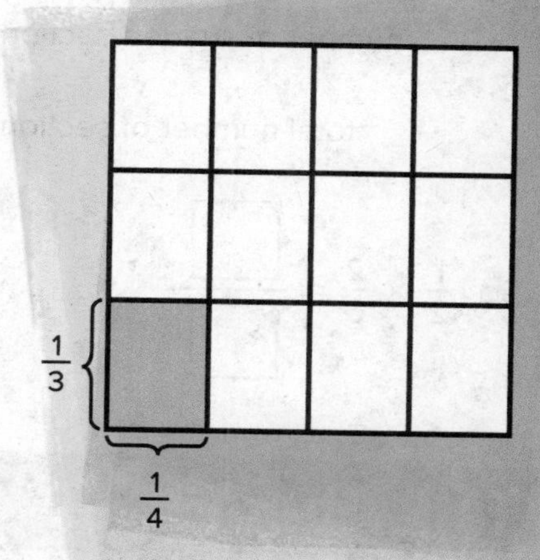

1. Divide the square into ________ equal rows since the denominator of the first fraction is 3.

2. Divide the square into ________ equal columns since the denominator of the second fraction is 4.

3. Shade the portion of the model where $\frac{1}{3}$ and $\frac{1}{4}$ intersect.

   How many sections of the model are shaded? ________

4. Write a fraction that compares the number of shaded sections to the total number of sections.

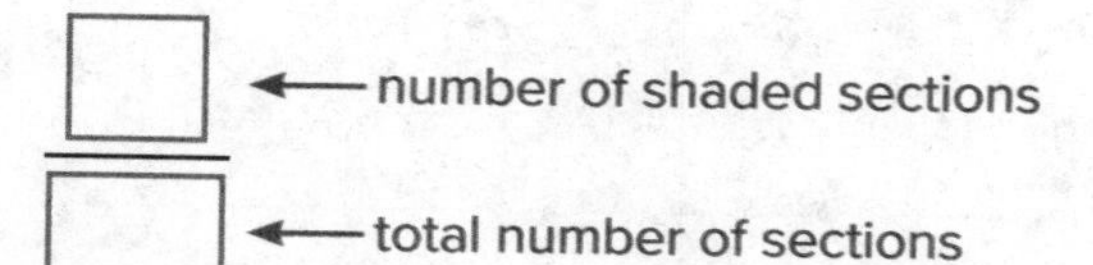

So, $\frac{1}{3} \times \frac{1}{4} = \frac{\square}{\square}$.

Try It

Find $\frac{1}{2} \times \frac{2}{3}$. Write in simplest form.

To find $\frac{1}{2} \times \frac{2}{3}$, find the area of a $\frac{1}{2}$- by $\frac{2}{3}$-unit rectangle.

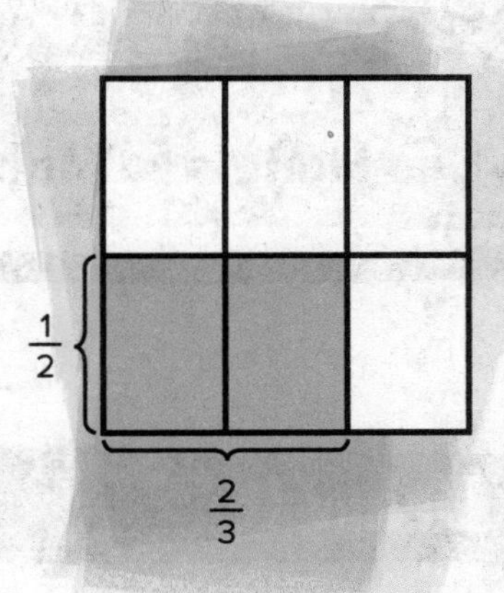

1. Divide the square into ________ equal rows since the denominator of the first fraction is 2.

2. Divide the square into ________ equal columns since the denominator of the second fraction is 3.

3. Shade the portion of the model where $\frac{1}{2}$ and $\frac{2}{3}$ intersect.

   How many sections of the model are shaded? ________

4. Write a fraction that compares the number of shaded sections to the total number of sections. Simplify the fraction.

   number of shaded sections → $\frac{\square}{\square} = \frac{\square}{\square}$ } Simplify.
   total number of sections →

So, $\frac{1}{2} \times \frac{2}{3} = \frac{\square}{\square}$.

Talk About It

1. In the second activity, how do the numerators of the fractions $\frac{1}{2}$ and $\frac{2}{3}$ relate to the total number of shaded sections, 2?

   ____

2. In the second activity, how do the denominators of the fractions relate to the total number of sections, 6?

   ____

3. **Processes & Practices** 2 **Stop and Reflect** Write a rule you can use to multiply fractions without using models.

   ____

Name

Practice It

Processes & Practices  **Use Math Tools** Shade the models to find each product. Write in simplest form.

4. $\frac{1}{6} \times \frac{2}{3} =$ ________

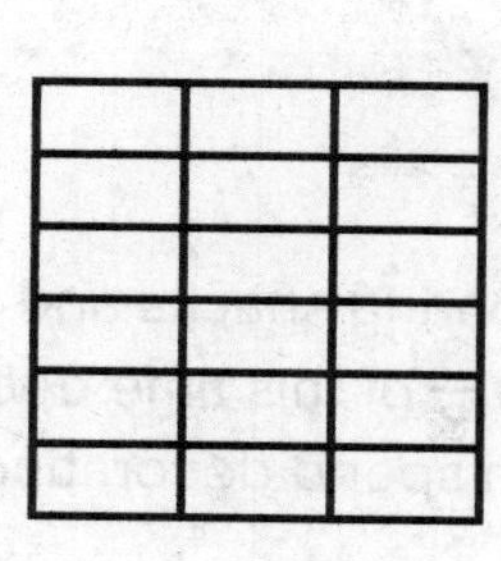

5. $\frac{3}{5} \times \frac{1}{3} =$ ________

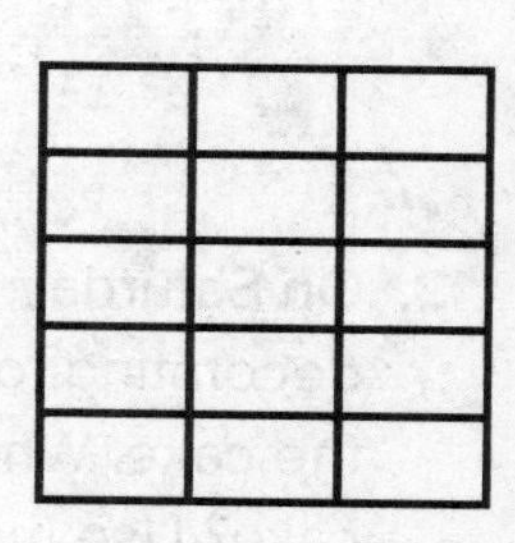

6. $\frac{3}{4} \times \frac{1}{6} =$ ________

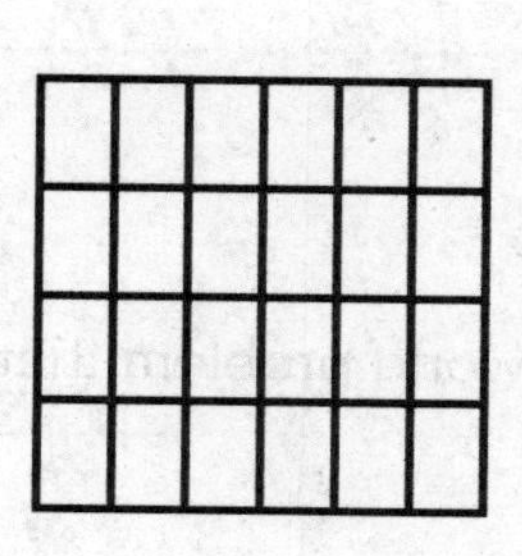

7. $\frac{3}{5} \times \frac{3}{4} =$ ________

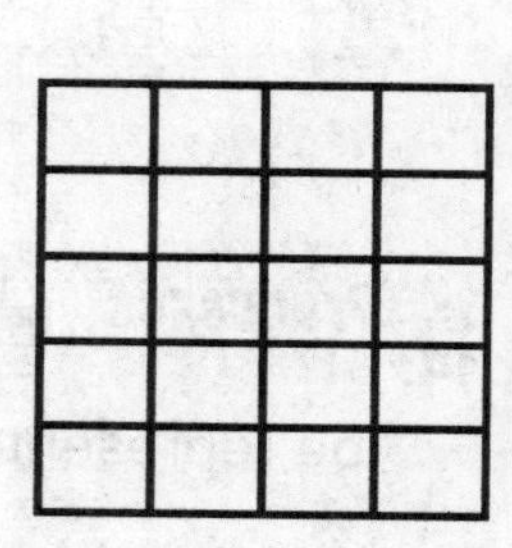

8. $\frac{3}{5} \times \frac{1}{5} =$ ________

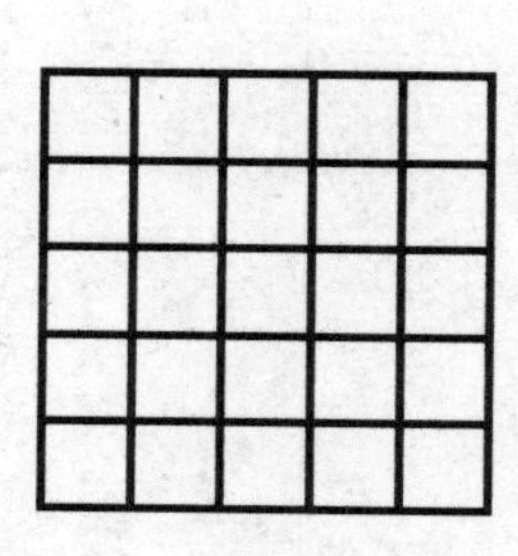

9. $\frac{1}{2} \times \frac{3}{5} =$ ________

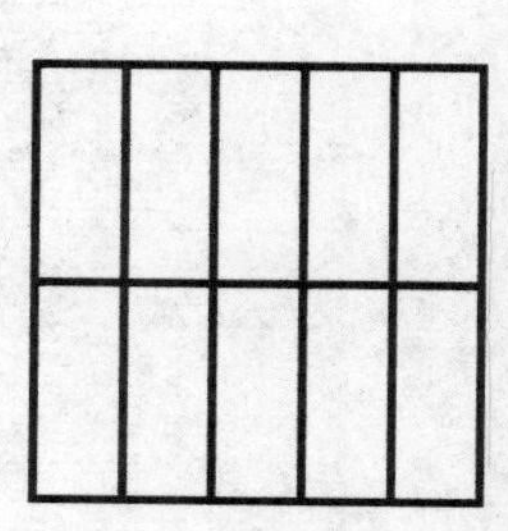

10. $\frac{3}{4} \times \frac{2}{5} =$ ________

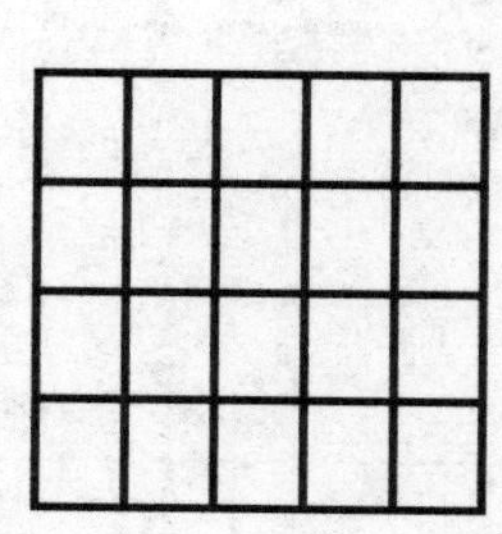

11. $\frac{2}{3} \times \frac{2}{3} =$ ________

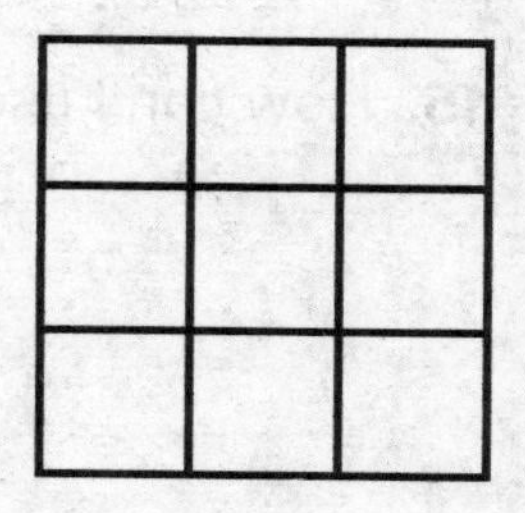

Real World

Apply It

12. Processes & Practices 5 **Use Math Tools** Ruby biked a trail that is $\frac{3}{5}$ mile each way. After biking $\frac{1}{4}$ of the trail, she stopped to rest. What fraction of a mile did Ruby bike before she stopped to rest? Use models to help you solve.

13. On Saturday, Tom spent $\frac{1}{3}$ of the day preparing snacks and decorating for a birthday party. Tom spent $\frac{3}{8}$ of this time decorating the cake. What fraction of the day did Tom spend decorating the cake? Use models to help you solve.

14. Processes & Practices 4 **Model Math** Write a real-world problem that could be represented by the model. Then solve.

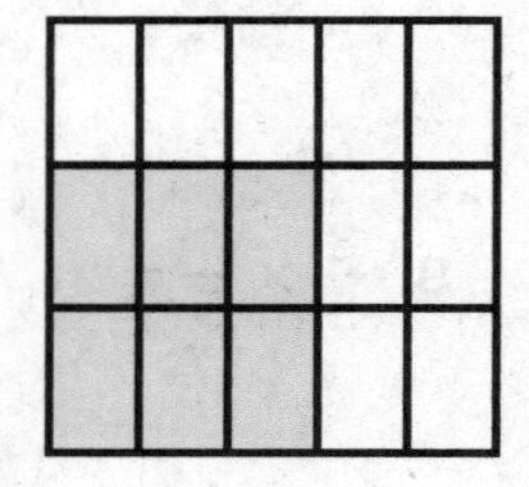

Write About It

15. How can I use models to multiply fractions?

Name ____________________

MY Homework

Lesson 5

Hands On: Use Models to Multiply Fractions

Homework Helper

Need help? connectED.mcgraw-hill.com

Find $\frac{2}{5} \times \frac{1}{2}$. Write in simplest form.

To find $\frac{2}{5} \times \frac{1}{2}$, find the area of a $\frac{2}{5}$- by $\frac{1}{2}$-unit rectangle.

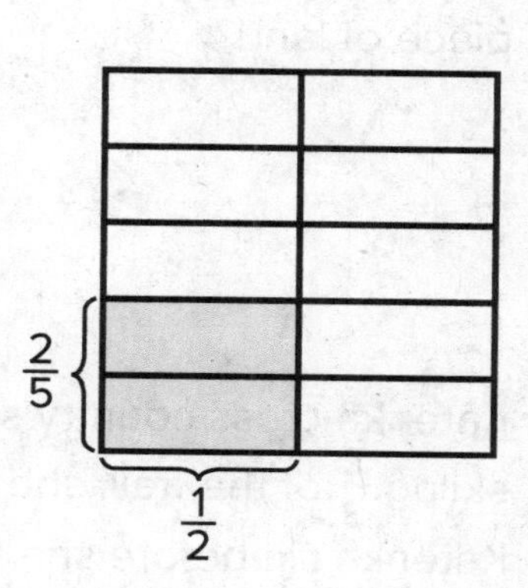

1. The square is divided into 5 equal rows since the denominator of the first fraction is 5.

2. The square is divided into 2 equal columns since the denominator of the second fraction is 2.

3. The portion of the model where $\frac{2}{5}$ and $\frac{1}{2}$ intersect is shaded. There are 2 sections of the model that are shaded.

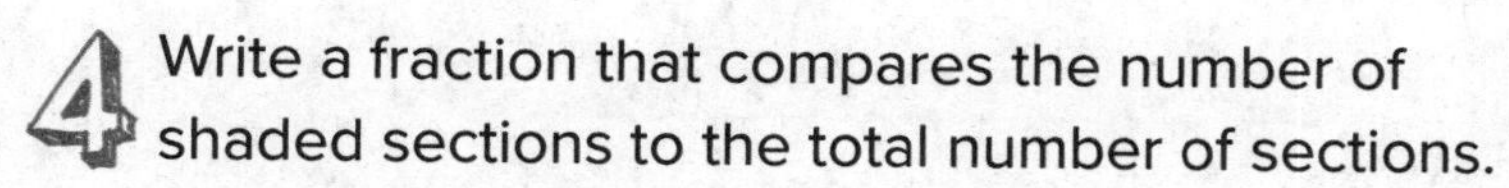

4. Write a fraction that compares the number of shaded sections to the total number of sections.

number of shaded sections → $\frac{2}{10} = \frac{1}{5}$ } Simplify.
total number of sections →

So, $\frac{2}{5} \times \frac{1}{2} = \frac{2}{10}$ or $\frac{1}{5}$.

Practice

Shade the models to find each product. Write in simplest form.

1. $\frac{1}{2} \times \frac{1}{2} =$ ______

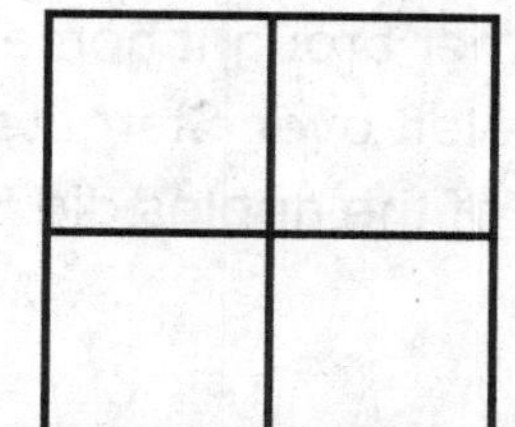

2. $\frac{4}{5} \times \frac{1}{2} =$ ______

Problem Solving

Draw models to help you solve Exercises 3–7.

3. Charlotte spent $\frac{1}{2}$ of the day shopping at the mall. She spent $\frac{1}{4}$ of this time trying on jeans. What fraction of the day did Charlotte spend trying on jeans?

______

4. Ms. Hendricks is buying a rectangular piece of land that is $\frac{2}{5}$ mile on one side and $\frac{5}{6}$ mile on an adjacent side. What is the area of the piece of land?

______

5. Katenka cross country skied a trail that was $\frac{3}{4}$ mile each way. After skiing $\frac{2}{3}$ of the trail, she turned around. What fraction of a mile did Katenka ski before she turned around?

______

6. Processes & Practices 5 **Use Math Tools** Merriam spent $\frac{3}{4}$ of her allowance at the mall. Of the money spent at the mall, $\frac{1}{3}$ was spent on new earrings. What part of her allowance did Merriam spend on earrings?

______

7. Lexi ate some of the apples that her mother brought home from the farmer's fair. One-half of the apples were left over. Starr ate $\frac{1}{6}$ of the apples that were left over. What fraction of the apples did Starr eat?

______

Name ....

Lesson 6
Multiply Fractions

ESSENTIAL QUESTION
What strategies can be used to multiply and divide fractions?

Multiplying fractions is similar to multiplying a whole number and a fraction. To multiply fractions, multiply the numerators and multiply the denominators.

$$\frac{3}{5} \times \frac{1}{2} = \frac{3 \times 1}{5 \times 2}$$

Math in My World

Example 1

Franco ate some pizza. Two-thirds of a pizza is left. Diendra ate $\frac{3}{4}$ of the leftover pizza. How much of a whole pizza did Diendra eat?

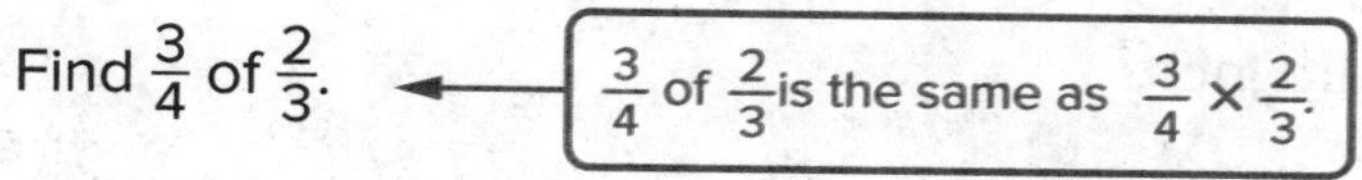

Find $\frac{3}{4}$ of $\frac{2}{3}$. ← $\frac{3}{4}$ of $\frac{2}{3}$ is the same as $\frac{3}{4} \times \frac{2}{3}$.

One Way **Multiply first. Then simplify.**

$\frac{3}{4} \times \frac{2}{3} = \frac{3 \times 2}{4 \times 3}$

$= \frac{6}{12}$ Multiply.

$= \frac{6 \div 6}{12 \div 6} = \frac{\square}{\square}$ Simplify.

So, Diendra ate $\frac{\square}{\square}$ of the whole pizza.

Another Way **Simplify first.**

$\frac{3}{4} \times \frac{2}{3} = \frac{\overset{1}{\cancel{3}} \times \overset{1}{\cancel{2}}}{\underset{2}{\cancel{4}} \times \underset{1}{\cancel{3}}}$

Simplify.
Divide 3 and 3 by 3.
Divide 4 and 2 by 2.

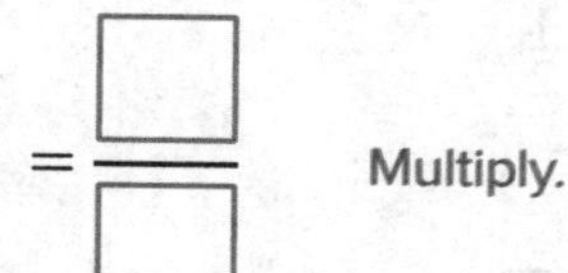

$= \frac{\square}{\square}$ Multiply.

To find the area of a rectangle, multiply the length and the width.

Example 2

Find the area of the rectangle with a length of $\frac{3}{4}$ inch and a width of $\frac{5}{9}$ inch.

One Way **Tile the rectangle as part of a unit square.**

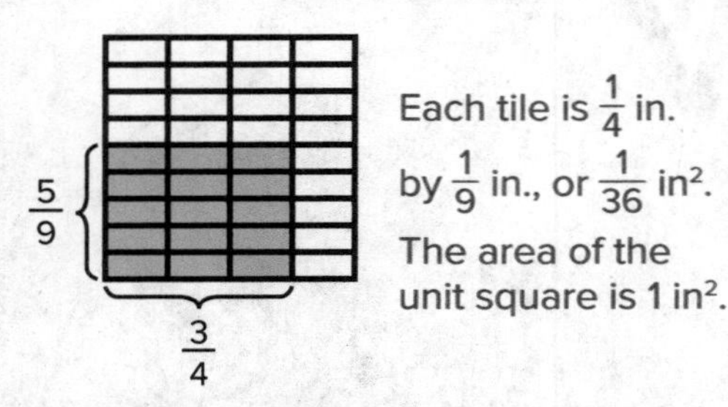

Each tile is $\frac{1}{4}$ in. by $\frac{1}{9}$ in., or $\frac{1}{36}$ in^2. The area of the unit square is 1 in^2.

Count the shaded tiles.
There are 15 tiles shaded.

$15 \times \frac{1}{36} = \frac{15}{36}$

$= \frac{15 \div 3}{36 \div 3} = \frac{\square}{\square}$ Simplify.

Another Way **Multiply the side lengths.**

$\frac{3}{4} \times \frac{5}{9} = \frac{3 \times 5}{4 \times 9}$

$= \frac{15}{36}$ Multiply.

$= \frac{15 \div 3}{36 \div 3} = \frac{\square}{\square}$ Simplify.

So, the area of the rectangle is $\frac{\square}{\square}$ square inch.

Guided Practice

1. Find $\frac{1}{7} \times \frac{1}{2}$. Write in simplest form.

$\frac{1}{7} \times \frac{1}{2} = \frac{1 \times 1}{7 \times 2}$

$= \frac{\square}{\square}$

So, $\frac{1}{7} \times \frac{1}{2} = \frac{\square}{\square}$.

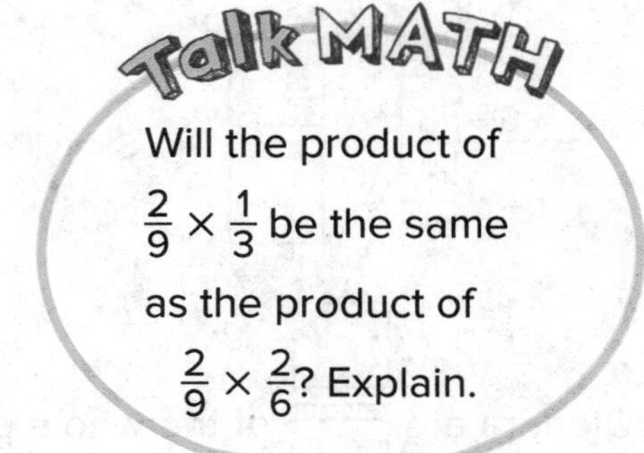

Name

Independent Practice

Shade the models to find each product. Write in simplest form.

2. $\frac{5}{6} \times \frac{3}{4} =$ ________

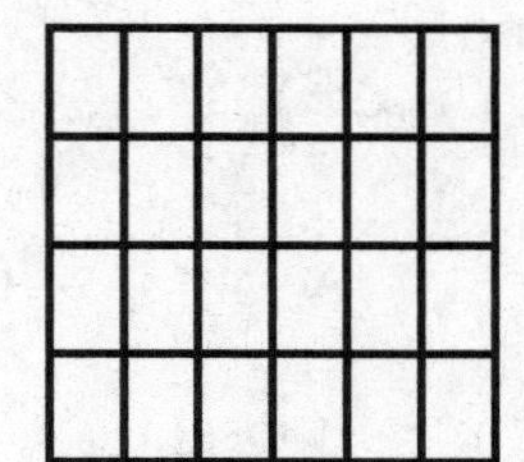

3. $\frac{3}{10} \times \frac{5}{6} =$ ________

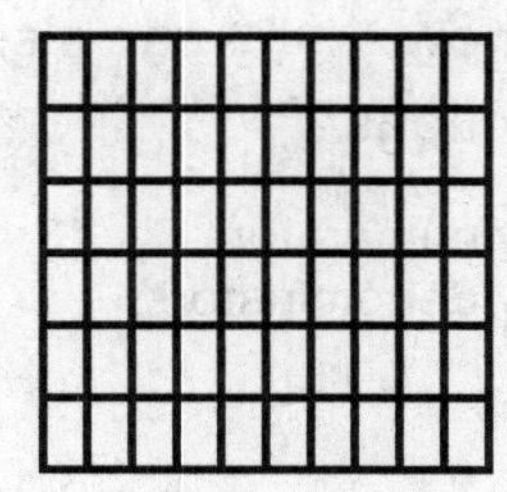

4. $\frac{3}{4} \times \frac{2}{7} =$ ________

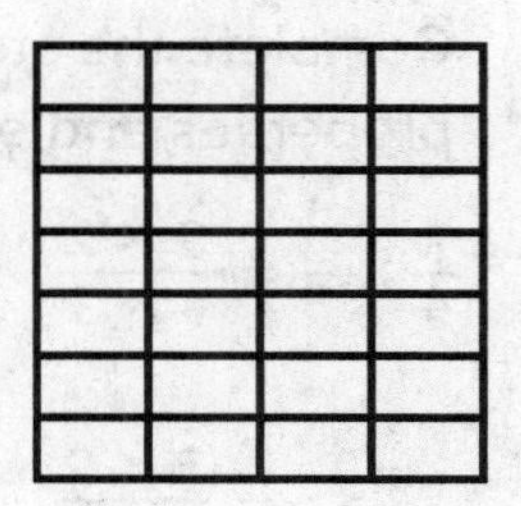

Multiply. Write in simplest form.

5. $\frac{3}{5} \times \frac{5}{6} =$ ________

6. $\frac{3}{4} \times \frac{5}{12} =$ ________

7. $\frac{2}{9} \times \frac{1}{3} =$ ________

8. $\frac{3}{4} \times \frac{2}{5} =$ ________

9. $\frac{1}{3} \times \frac{2}{5} =$ ________

10. $\frac{3}{5} \times \frac{5}{7} =$ ________

Algebra Find the unknown in each equation. Write in simplest form.

11. $\frac{6}{7} \times \frac{3}{4} = x$

$x =$ ________

12. $\frac{2}{3} \times \frac{1}{4} = m$

$m =$ ________

13. $\frac{5}{6} \times \frac{2}{3} = p$

$p =$ ________

Problem Solving

14. Algebra Find the area of the rectangle shown. Write an equation.

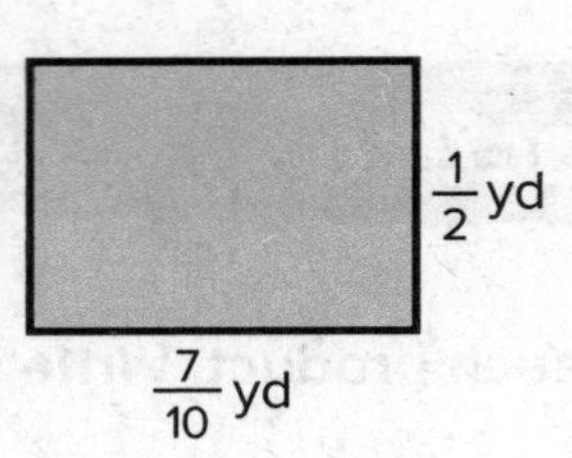

____

15. **Processes & Practices 7** **Identify Structure** Refer to Example 1. Complete the steps to show how $\frac{3}{4} \times \frac{2}{3}$ can also be found using properties and equivalent fractions.

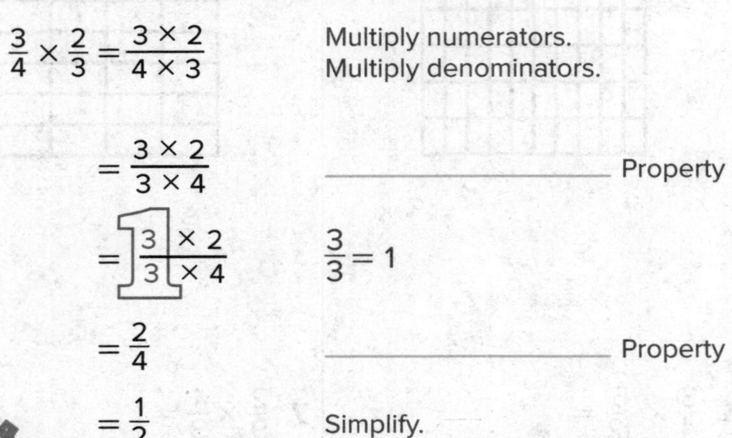

$\frac{3}{4} \times \frac{2}{3} = \frac{3 \times 2}{4 \times 3}$ Multiply numerators. Multiply denominators.

$= \frac{3 \times 2}{3 \times 4}$ ______________ Property

$= \frac{3 \times 2}{3 \times 4}$ $\frac{3}{3} = 1$

$= \frac{2}{4}$ ______________ Property

$= \frac{1}{2}$ Simplify.

Brain Builders

16. **Processes & Practices 4** **Model Math** Write a real-world problem that describes the model.

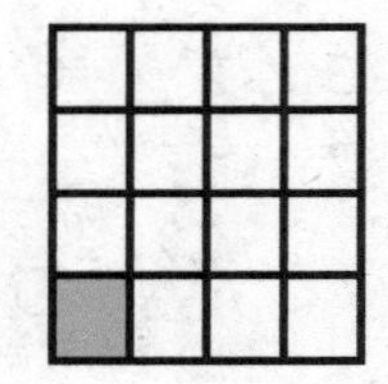

____

____

____

17. **Building on the Essential Question** How can I multiply fractions? Include an example and use a model to find the solution.

____

____

____

Name

Lesson 6

Multiply Fractions

Homework Helper

Need help? connectED.mcgraw-hill.com

Two-thirds of the vegetables that Michael planted were green. Two-fifths of the green vegetables were peppers. What fraction of the vegetable garden is green peppers? Write in simplest form.

Find $\frac{2}{3} \times \frac{2}{5}$.

$$\frac{2}{3} \times \frac{2}{5} = \frac{2 \times 2}{3 \times 5}.$$

$$= \frac{4}{15}$$ Multiply.

So, $\frac{4}{15}$ of the garden is green peppers.

Practice

Multiply. Write in simplest form.

1. $\frac{4}{9} \times \frac{3}{8} =$ ______

2. $\frac{3}{4} \times \frac{5}{8} =$ ______

3. $\frac{1}{8} \times \frac{3}{4} =$ ______

4. $\frac{3}{5} \times \frac{1}{3} =$ ______

5. $\frac{5}{7} \times \frac{1}{4} =$ ______

6. $\frac{7}{8} \times \frac{2}{9} =$ ______

Problem Solving

Algebra **Write an equation to solve Exercises 7 and 8.**

7. Carrie ate some of the oranges that her mother brought home from the grocery store. Of the oranges her mother bought, $\frac{1}{4}$ were left over. Shyla, Carrie's sister, ate $\frac{2}{5}$ of the oranges that were left over. What fraction of the oranges did Shyla eat?

____

8. Find the area of the rectangle shown.

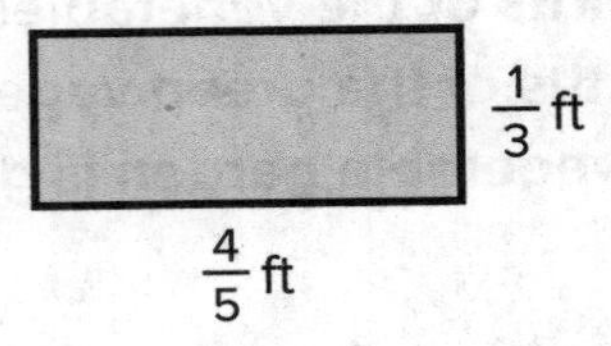

____

Brain Builders

9. **Processes & Practices 7** **Identify Structure** Complete the steps to show how $\frac{4}{5} \times \frac{5}{6}$ can also be found using properties and equivalent fractions.

$\frac{4}{5} \times \frac{5}{6} = \frac{4 \times 5}{5 \times 6}$ Multiply numerators. Multiply denominators.

$= \frac{5 \times 4}{5 \times 6}$ ______________ Property

$= \frac{5 \times 4}{5 \times 6}$ $\frac{5}{5} = 1$

$= \frac{4}{6}$ ______________ Property

$= \frac{2}{3}$ Simplify.

Write another example that would use the same properties for each step.

____

10. **Test Practice** James made flash cards for $\frac{5}{6}$ of the vocabulary words. Lauretta made $\frac{1}{2}$ of the amount of cards that James made. What fraction of the vocabulary words did Lauretta make cards for?

Ⓐ $\frac{5}{12}$ Ⓑ $\frac{1}{2}$ Ⓒ $\frac{7}{12}$ Ⓓ $\frac{3}{4}$

Name

Lesson 7 Multiply Mixed Numbers

ESSENTIAL QUESTION

What strategies can be used to multiply and divide fractions?

Math in My World

Example 1

A blueberry muffin recipe calls for $\frac{1}{2}$ cup of blueberries. A blueberry pie recipe calls for $3\frac{1}{2}$ times more blueberries. How many cups of blueberries are needed to make the pie?

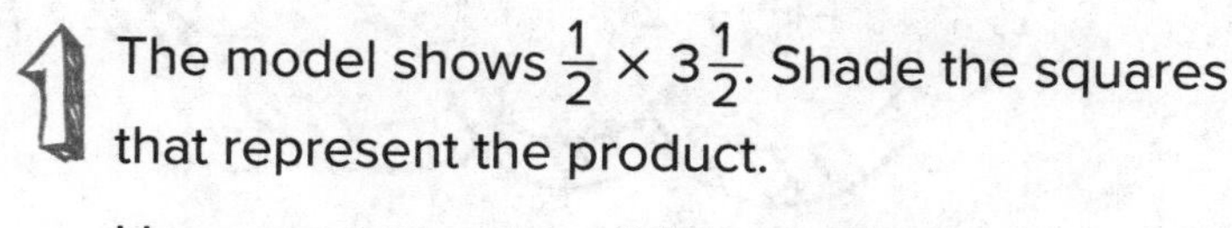

1. The model shows $\frac{1}{2} \times 3\frac{1}{2}$. Shade the squares that represent the product.

   How many squares did you shade? ______

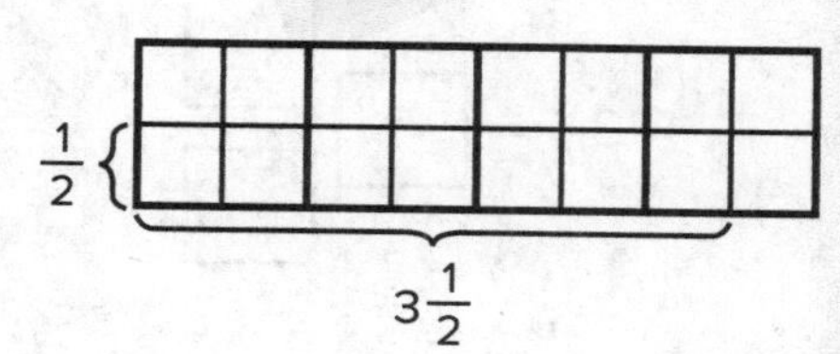

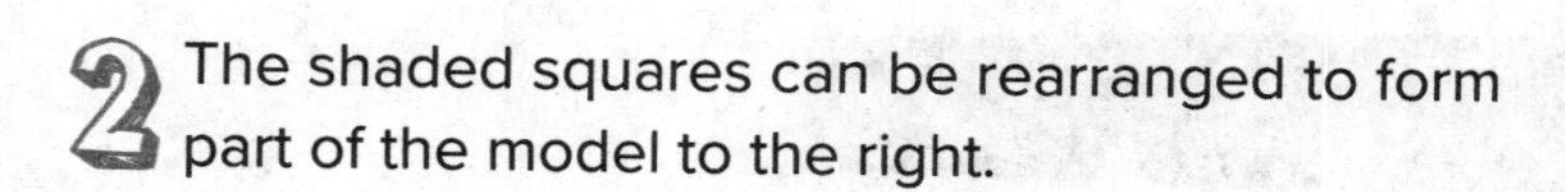

2. The shaded squares can be rearranged to form part of the model to the right.

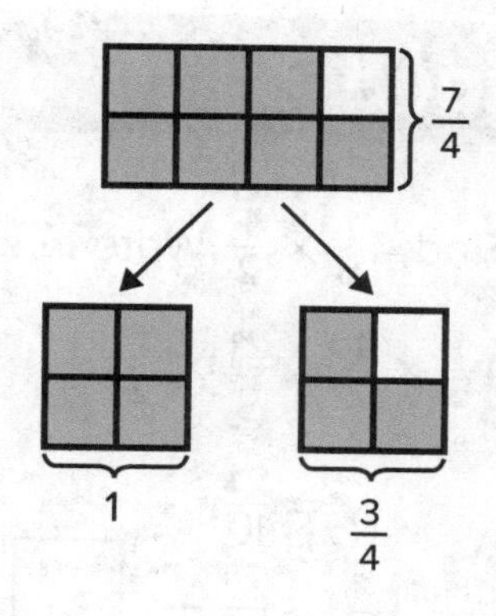

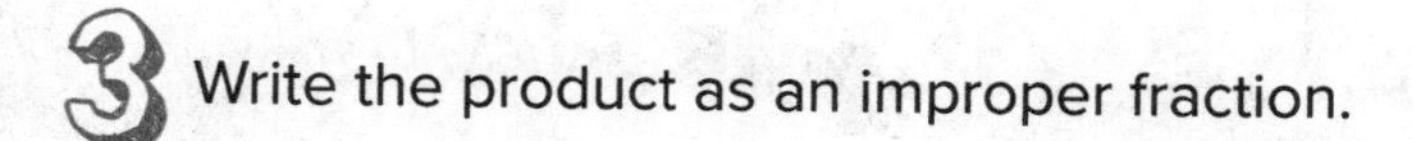

3. Write the product as an improper fraction.

   $\frac{\square}{\square}$ ← number of shaded sections / ← number of squares in each section of the model

   As a mixed number, this is $1\frac{3}{4}$.

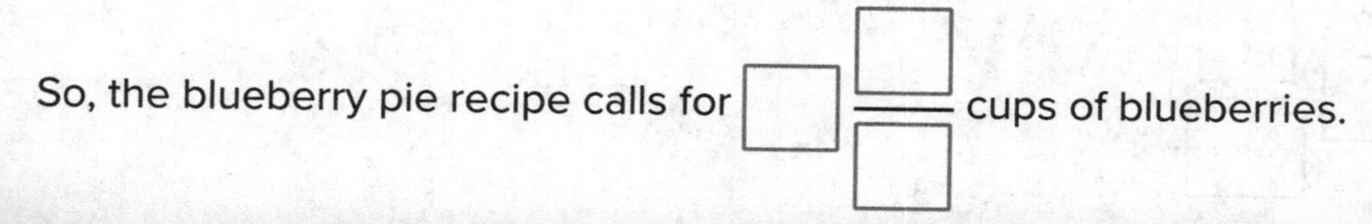

So, the blueberry pie recipe calls for $\square\frac{\square}{\square}$ cups of blueberries.

Key Concept Multiply Mixed Numbers

To multiply mixed numbers, write the mixed numbers as improper fractions. Then multiply as with fractions.

Example 2 

Find the unknown in $1\frac{1}{2} \times 3\frac{3}{4} = $ ■.

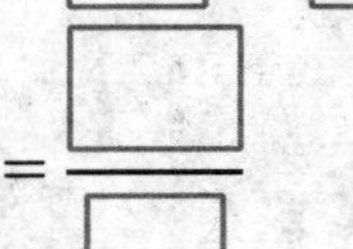

$1\frac{1}{2} \times 3\frac{3}{4} = \frac{\square}{\square} \times \frac{\square}{\square}$ Write each mixed number as an improper fraction.

$= \frac{\square}{\square}$ Multiply.

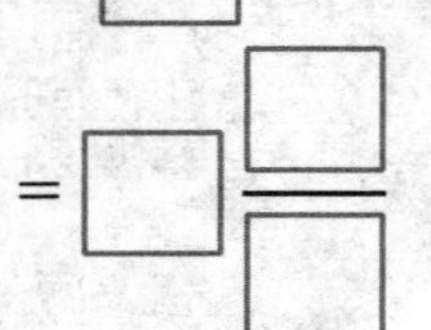

$= \square\frac{\square}{\square}$ Simplify.

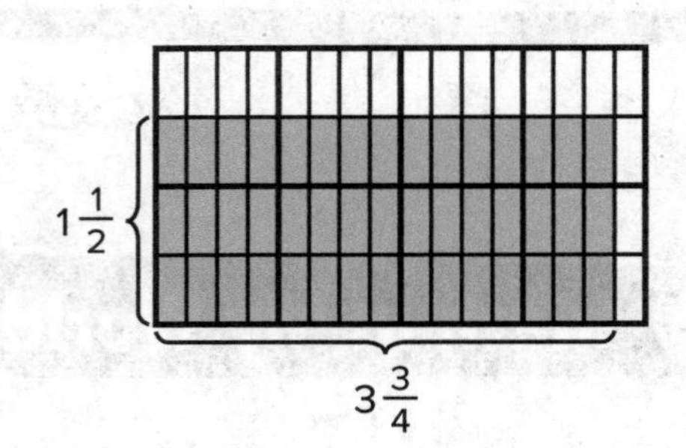

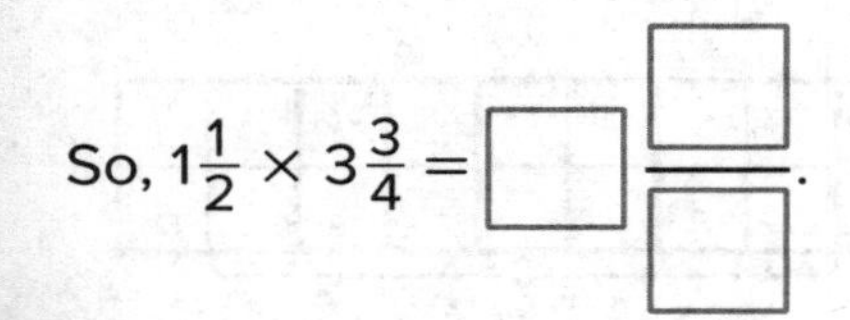

So, $1\frac{1}{2} \times 3\frac{3}{4} = \square\frac{\square}{\square}$.

Helpful Hint

To write a mixed number as an improper fraction, multiply the denominator by the whole number and add the numerator. Keep the original denominator.

$1\frac{1}{2} \rightarrow 2 \times 1 + 1 = \frac{3}{2}$

Guided Practice

1. Find $4\frac{1}{5} \times \frac{1}{2}$. Write in simplest form.

$4\frac{1}{5} \times \frac{1}{2} = \frac{21}{5} \times \frac{1}{2}$

$= \frac{21}{10}$

$= \square\frac{\square}{\square}$

So, $4\frac{1}{5} \times \frac{1}{2} = \square\frac{\square}{\square}$.

Explain how to find the product of two mixed numbers.

Name

Independent Practice

Multiply. Write in simplest form.

2. $1\frac{1}{3} \times \frac{2}{3} =$ ______

3. $\frac{2}{5} \times 2\frac{3}{4} =$ ______

4. $3\frac{3}{5} \times \frac{1}{4} =$ ______

5. $2\frac{1}{2} \times 4\frac{1}{5} =$ ______

6. $1\frac{1}{3} \times 3\frac{2}{3} =$ ______

7. $4\frac{2}{5} \times 1\frac{3}{4} =$ ______

8. $2\frac{4}{5} \times 6\frac{1}{8} =$ ______

9. $1\frac{5}{6} \times 4\frac{1}{3} =$ ______

10. $2\frac{3}{5} \times 3\frac{7}{8} =$ ______

Algebra **Write a multiplication equation represented by each model. Shade the product on the model.**

11.

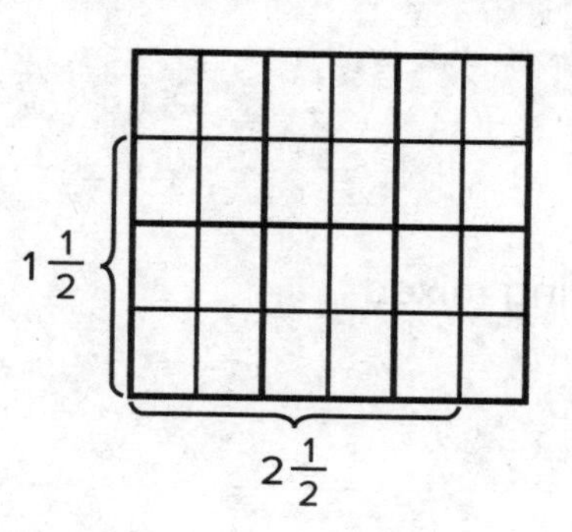

______

12.

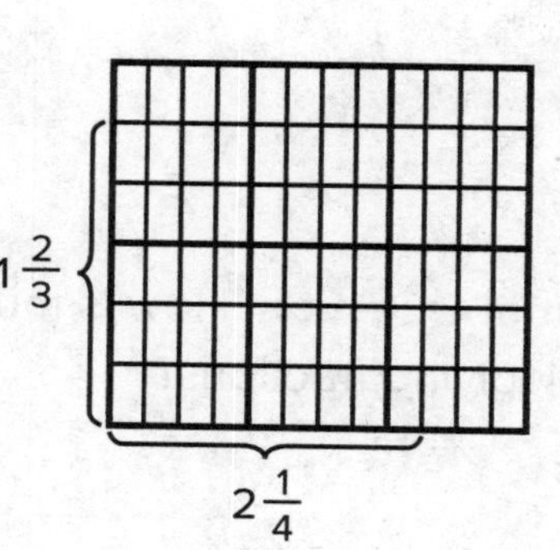

______

13.

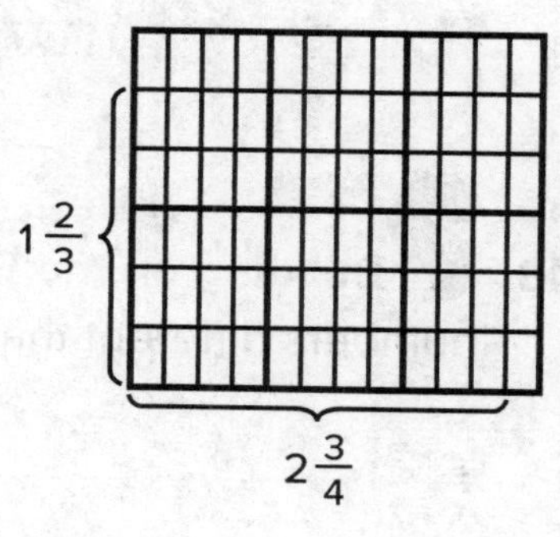

______

Problem Solving

14. Processes & Practices 4 **Model Math** Alexandra made a rectangular quilt that measured $3\frac{1}{4}$ feet in length by $2\frac{3}{4}$ feet in width. To find the area, multiply the length and width. What is the area of the quilt in square feet? Write an equation to solve.

15. Laney bought $1\frac{2}{3}$ pounds of grapes. She also bought bananas that were $2\frac{1}{4}$ times the weight of the grapes. How much did the bananas weigh?

Brain Builders

16. Mrs. Plesich has a $17\frac{3}{4}$ ounce bag of chocolate chips. She will use $1\frac{1}{4}$ ounces of the chocolate chips for baking the cake. Then she will use $\frac{1}{4}$ of the remaining bag for decorating a cake. How many ounces will she use for decorating?

17. Processes & Practices 2 **Use Number Sense** Write and solve a real-world problem that involves finding the product of a fraction and a mixed number.

18. **Building on the Essential Question** How is multiplying mixed numbers different than multiplying fractions?

Name

Lesson 7

Multiply Mixed Numbers

Homework Helper

Need help? connectED.mcgraw-hill.com

A swimming pool has two diving boards. The shorter diving board is $2\frac{1}{4}$ yards high. The taller diving board is $1\frac{1}{3}$ the height of the shorter board. How tall is the taller diving board?

Find $2\frac{1}{4} \times 1\frac{1}{3}$.

$2\frac{1}{4} \times 1\frac{1}{3} = \frac{9}{4} \times \frac{4}{3}$ Write each mixed number as an improper fraction.

$= \frac{36}{12}$ Multiply.

$= 3$ Simplify.

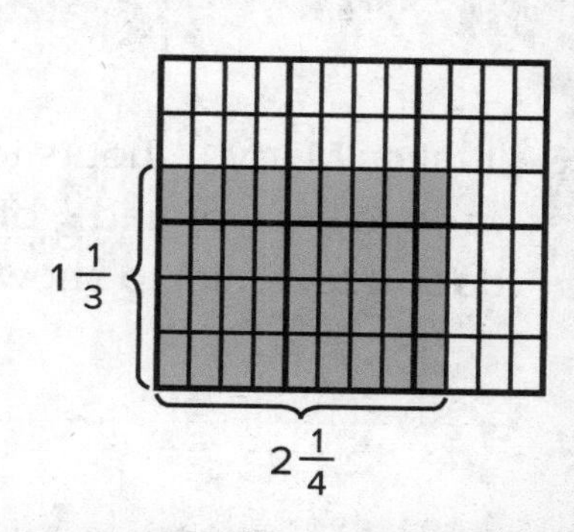

So, the taller diving board is 3 yards high.

Practice

Multiply. Write in simplest form.

1. $5\frac{1}{3} \times 1\frac{1}{4} =$ ______

2. $1\frac{2}{5} \times 3\frac{1}{6} =$ ______

3. $7\frac{1}{8} \times 2\frac{5}{6} =$ ______

Problem Solving

4. The table shows some ingredients in lasagna. If you make three times the recipe, how many cups of cheese are needed?

Tomato Sauce	Chopped Onion	Cheese
$3\frac{1}{2}$ cups	$\frac{1}{4}$ cup	$2\frac{2}{3}$ cups

5. Karyn purchased a square picture frame. Each side measures $1\frac{1}{4}$ feet. What is the area of the picture frame in square feet?

Brain Builders

6. It takes Marty $1\frac{1}{4}$ hours to get ready for school. If $\frac{1}{5}$ of that time is used to shower and $\frac{1}{3}$ of that time is used to eat breakfast, how long does it take him to shower and eat breakfast?

7. Processes & Practices 2 **Use Algebra** Kallisto built a rectangular sign that measured $2\frac{3}{4}$ feet in length. The width of the sign is $1\frac{1}{4}$ feet shorter than the length. To find the area, multiply the length and width. What is the area of the sign in square feet? Write an equation to solve.

8. **Test Practice** Antoinette bought $2\frac{2}{3}$ pounds of grapes and $1\frac{1}{5}$ pounds of apples. If she bought bananas that weighed $1\frac{1}{4}$ times as much as the grapes, how much did the bananas weigh?

Ⓐ $2\frac{1}{6}$ pounds

Ⓑ $3\frac{1}{3}$ pounds

Ⓒ $3\frac{1}{4}$ pounds

Ⓓ $3\frac{1}{2}$ pounds

Name

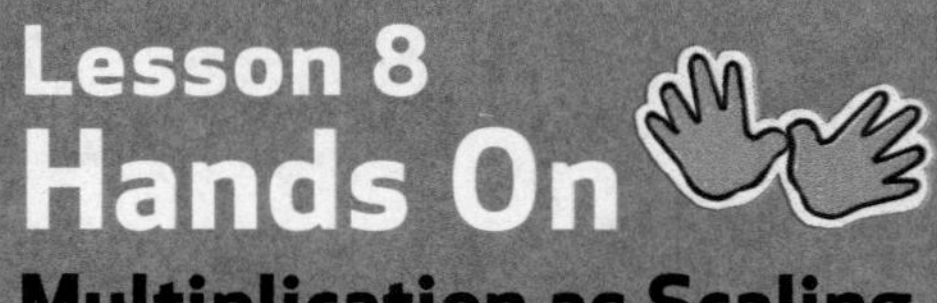

Lesson 8
Hands On
Multiplication as Scaling

ESSENTIAL QUESTION
What strategies can be used to multiply and divide fractions?

Scaling is the process of resizing a number when you multiply by a fraction that is greater than or less than 1.

Draw It

Multiply the number 2 by three fractions greater than 1.

1. Multiply the number 2 by three different fractions greater than 1, such as $1\frac{1}{5}$, $1\frac{1}{2}$, and $1\frac{3}{4}$.

$2 \times 1\frac{1}{5} = \square\frac{\square}{\square}$ $2 \times 1\frac{1}{2} = \square$ $2 \times 1\frac{3}{4} = \square\frac{\square}{\square}$

2. Plot 2 on the number line. Then plot the products on the number line.

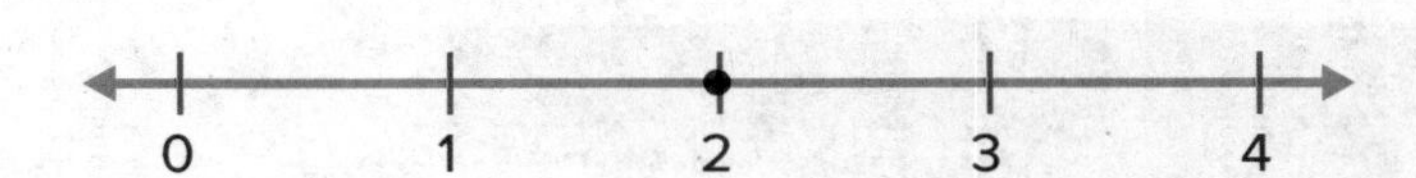

3. Compare the products. Circle whether the products are greater than, less than, or equal to 2.

greater than 2 less than 2 equal to 2

Multiplying a number by a fraction greater than one results

in a product that is ____________ than the number.

Try It

Multiply the number 2 by three fractions less than 1.

1. Multiply 2 by three different fractions less than 1 such as $\frac{1}{4}$, $\frac{1}{2}$, and $\frac{5}{8}$.

$2 \times \frac{1}{4} = \frac{\square}{\square}$ $2 \times \frac{1}{2} = \square$ $2 \times \frac{5}{8} = \square\frac{\square}{\square}$

2. Plot 2 on the number line. Then plot the products on the number line.

0 1 2 3 4

3. Compare the products. Circle whether the products are greater than, less than, or equal to 2.

greater than 2 less than 2 equal to 2

Multiplying a number by a fraction less than one results

in a product that is ______________ than the number.

Talk About It

1. Predict whether the product of 3 and $\frac{4}{5}$ is greater than, less than, or equal to 3. Explain.

____

2. Processes & Practices 3 **Draw a Conclusion** Predict whether the product of 2 and $2\frac{1}{5}$ is greater than, less than, or equal to 2. Explain.

____

Name

Practice It

Without multiplying, circle whether each product is greater than, less than, or equal to the whole number.

3. $2 \times \frac{1}{8}$

greater than

less than

equal to

4. $10 \times 1\frac{3}{5}$

greater than

less than

equal to

5. $1\frac{3}{4} \times 4$

greater than

less than

equal to

6. $12 \times \frac{5}{6}$

greater than

less than

equal to

7. $1\frac{1}{3} \times 2$

greater than

less than

equal to

8. $\frac{3}{5} \times 6$

greater than

less than

equal to

Algebra Without multiplying, circle whether the unknown in each equation is greater than, less than, or equal to the whole number.

9. $1\frac{2}{3} \times 4 = b$

greater than

less than

equal to

10. $8 \times \frac{4}{5} = h$

greater than

less than

equal to

11. $\frac{4}{4} \times 5 = k$

greater than

less than

equal to

12. $6 \times \frac{1}{3} = n$

greater than

less than

equal to

Apply It

For Exercises 13–15, analyze each product in the table.

First Factor	Second Factor	Product
$\frac{1}{2}$	$\frac{3}{4}$	$\frac{3}{8}$
1	$\frac{3}{4}$	$\frac{3}{4}$
$\frac{3}{2}$	$\frac{3}{4}$	$\frac{9}{8}$

13. Why is the first product less than $\frac{3}{4}$?

14. Why is the product of 1 and $\frac{3}{4}$ equal to $\frac{3}{4}$?

15. Why is the product of $\frac{3}{2}$ and $\frac{3}{4}$ greater than $\frac{3}{4}$?

16. Processes & Practices 6 **Explain to a Friend** Miranda spent $\frac{1}{5}$ of her time cooking pasta. If she spent 2 hours cooking, did Miranda spend more than, less than, or equal to 2 hours cooking pasta? Explain.

17. Processes & Practices 3 **Which One Doesn't Belong?** Circle the multiplication expression that does not belong based on scaling. Explain.

$\frac{2}{5} \times 3$ | $1\frac{1}{2} \times 3$ | $3 \times \frac{3}{5}$ | $3 \times \frac{1}{2}$

Write About It

18. How can I use scaling to help predict the product of a number and a fraction?

Name

MY Homework

Lesson 8

Hands On: Multiplication as Scaling

Homework Helper

Need help? connectED.mcgraw-hill.com

Multiply the number 3 by three fractions greater than 1 and three fractions less than 1.

1 Multiply 3 by three different fractions greater than 1 and three different fractions less than 1.

$3 \times 1\frac{1}{5} = 3\frac{3}{5}$ $3 \times 1\frac{1}{2} = 4\frac{1}{2}$ $3 \times 1\frac{3}{4} = 5\frac{1}{4}$

$3 \times \frac{1}{4} = \frac{3}{4}$ $3 \times \frac{1}{2} = 1\frac{1}{2}$ $3 \times \frac{5}{8} = 1\frac{7}{8}$

2 Plot 3 on the number line. Then plot the products.

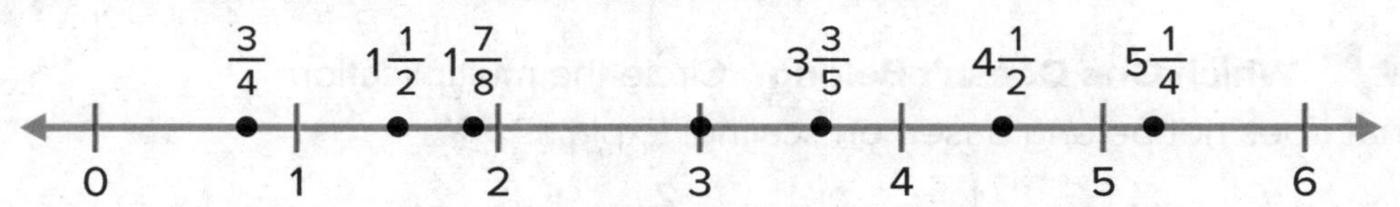

Multiplying a number by a fraction greater than one results in a product greater than the number.

Multiplying a number by a fraction less than one results in a product less than the number.

Practice

Without multiplying, circle whether each product is greater than, less than, or equal to the whole number.

1. $4 \times \frac{1}{7}$

greater than less than equal to

2. $12 \times 2\frac{5}{6}$

greater than less than equal to

Problem Solving

3. Maresol spent $\frac{4}{5}$ of her allowance on a pair of jeans. If she received $15 for allowance, did Maresol spend more than, less than, or equal to $15 on the jeans? Explain.

____

4. Dee used $2\frac{1}{3}$ cups of sugar for a cake recipe. If the amount of sugar the container holds is 3 times the amount she used, does the container hold more than, less than, or equal to the amount Dee used? Explain.

____

5. Derek ran $\frac{3}{8}$ of a race before he had to stop for a break. If the race is 6 miles long, did Derek run more than, less than, or equal to 6 miles before he took a break? Explain.

____

6. **Processes & Practices** 3 **Which One Doesn't Belong?** Circle the multiplication expression that does not belong based on scaling. Explain.

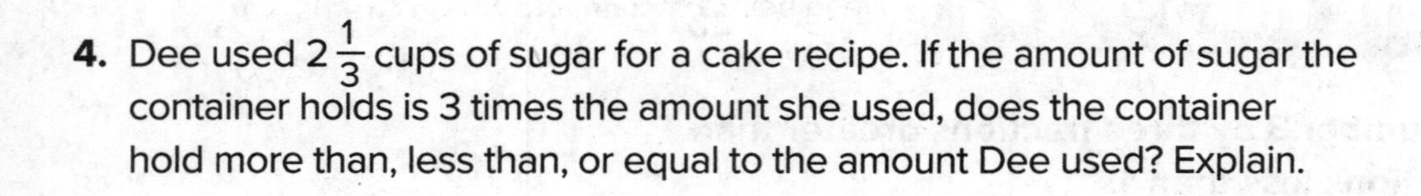

$6 \times 2\frac{1}{2}$ | $6 \times 1\frac{3}{5}$ | $\frac{1}{2} \times 6$ | $2\frac{2}{5} \times 6$

____

____

Vocabulary Check

Vocab

7. Fill in the blank with the correct term or number to complete the sentence.

______________ is the process of resizing a number when you multiply by a fraction that is greater than or less than 1.

Check My Progress

Vocabulary Check

1. Fill in the blank with the correct term or number to complete the sentence.

Scaling is the process of ______________ a number when you multiply by a fraction that is greater than or less than 1.

Concept Check

Shade the models to find each product. Write in simplest form.

2. $\frac{2}{5} \times \frac{1}{2} =$ ________

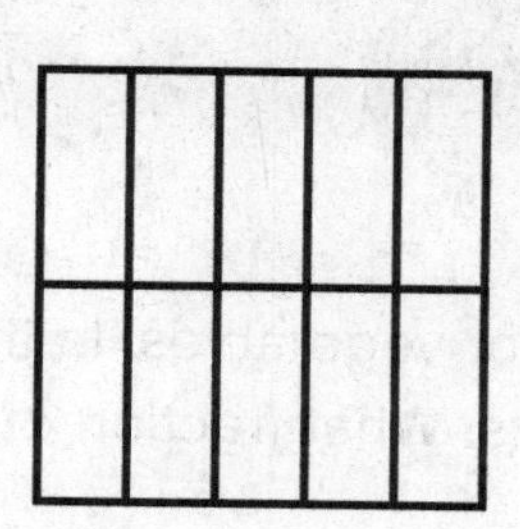

3. $\frac{3}{4} \times \frac{3}{5} =$ ________

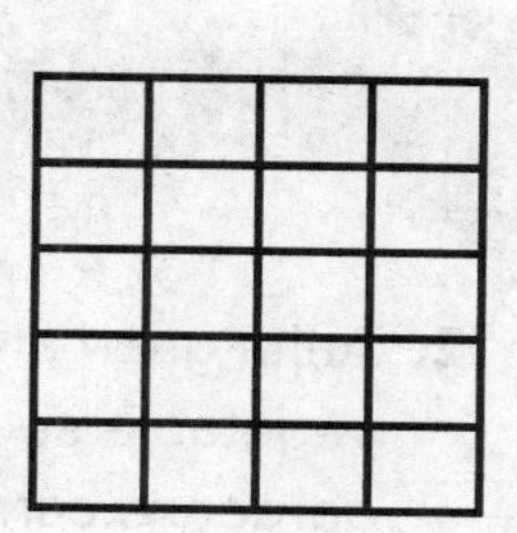

Multiply. Write in simplest form.

4. $\frac{1}{2} \times \frac{4}{9} =$ ________

5. $\frac{4}{5} \times \frac{5}{6} =$ ________

6. $\frac{1}{5} \times 4\frac{2}{3} =$ ________

7. $7\frac{1}{8} \times 2\frac{5}{6} =$ ________

Without multiplying, circle whether each product is greater than, less than, or equal to the whole number.

8. $4\frac{1}{8} \times 7$

greater than

less than

equal to

9. $\frac{3}{5} \times 1$

greater than

less than

equal to

Problem Solving

10. Tom spent $\frac{1}{6}$ of the day working in the flower bed. He spent $\frac{2}{3}$ of that time adding mulch to the flower bed. What fraction of the day did Tom spend adding mulch?

11. Nick spent $1\frac{3}{4}$ hours doing homework. Of that time, $\frac{2}{3}$ was spent on a social studies project. How many hours did Nick spend working on the project?

Brain Builders

12. Anita grew a garden with $\frac{3}{4}$ of the area for vegetables. In the vegetable section, $\frac{1}{6}$ of the area is carrots. What fraction of the total garden is carrots? Draw a model.

13. Test Practice Mr. Taylor collected apples on Friday. On Saturday, he collected $\frac{2}{3}$ of the amount collected on Friday. How many bushels of apples did Mr. Taylor collect on Saturday?

Apples Collected	
Day	**Bushels**
Friday	$3\frac{2}{3}$
Saturday	■

Ⓐ $1\frac{2}{3}$ bushels

Ⓑ $1\frac{4}{9}$ bushels

Ⓒ $2\frac{2}{3}$ bushels

Ⓓ $2\frac{4}{9}$ bushels

Name

Lesson 9
Hands On

Division with Unit Fractions

ESSENTIAL QUESTION
What strategies can be used to multiply and divide fractions?

You know that $6 \div 3$ means to take 6 objects and divide them into 3 equal groups. There would be 2 in each group. You can think of division with unit fractions in the same way. A **unit fraction** is a fraction with a numerator of 1.

Build It

Tools

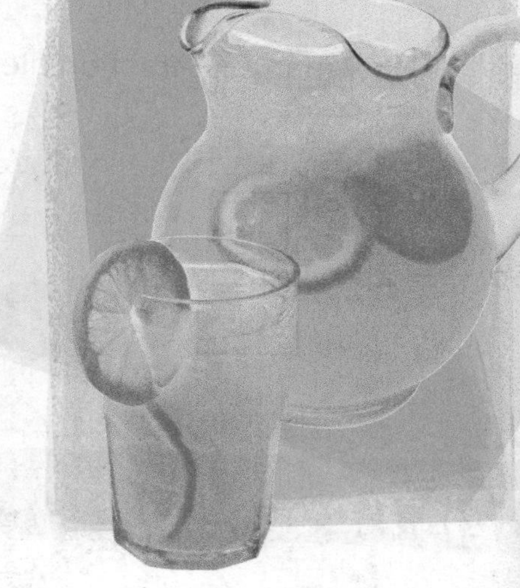

Remy has 2 gallons of lemonade. She needs to divide the lemonade into $\frac{1}{4}$-gallon containers. How many $\frac{1}{4}$-gallon containers does she need?

Find $2 \div \frac{1}{4}$. ← THINK: How many groups of $\frac{1}{4}$ are in 2?

1. Represent the 2 gallons by placing 2 whole fraction tiles.

2. Place enough of the $\frac{1}{4}$-fraction tiles below the 2 whole fraction tiles to represent the same amount. Label the model with $\frac{1}{4}$-fraction tiles.

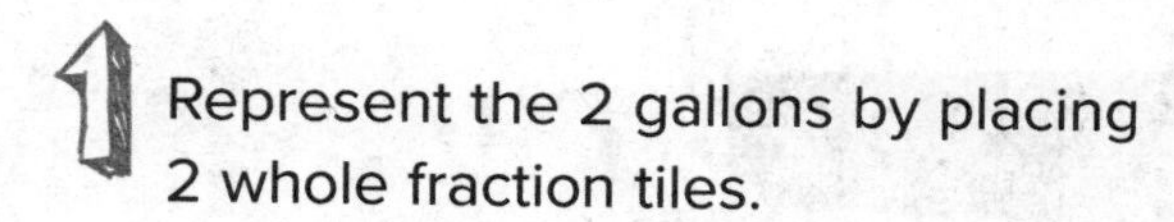

How many $\frac{1}{4}$-fraction tiles did it take to equal the 2 whole fractions tiles? ______

So, $2 \div \frac{1}{4} =$ ______.

Remy will need ______ containers.

Try It

Find $\frac{1}{5} \div 2$.

Place a $\frac{1}{5}$-fraction tile.

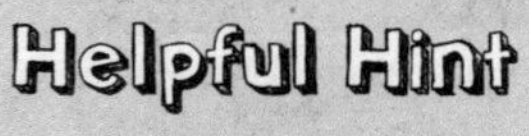

Helpful Hint

$\frac{1}{5}$ represents one out of the five tiles that would make a whole.

2 Since you are dividing by 2, you are dividing $\frac{1}{5}$ into two equal groups.
Find a fraction tile that when placed twice, will be the same size as the $\frac{1}{5}$-fraction tile.
Label the model with the fraction tiles you placed.

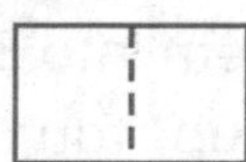

What fraction tile did you place? ________

How many $\frac{1}{10}$-fraction tiles did it take to equal the $\frac{1}{5}$-fraction tile? ________

So, $\frac{1}{5} \div 2 = \frac{\square}{\square}$.

Talk About It

Processes & Practices 2 Reason Determine whether each of the following statements is *true* or *false*. Explain your reasoning.

1. When a whole number greater than one is divided by a unit fraction less than one, the quotient is always greater than the dividend.

____

____

2. When a unit fraction less than one is divided by a whole number greater than one, the quotient is always greater than the dividend.

____

____

Name

Practice It

Processes & Practices 5 **Use Math Tools Use fraction tiles to divide. Draw your models below.**

3. $3 \div \frac{1}{3} =$ ______

4. $2 \div \frac{1}{5} =$ ______

5. $4 \div \frac{1}{2} =$ ______

6. $3 \div \frac{1}{6} =$ ______

7. $\frac{1}{3} \div 4 =$ ______

8. $\frac{1}{2} \div 3 =$ ______

9. $\frac{1}{3} \div 2 =$ ______

10. $\frac{1}{4} \div 2 =$ ______

Apply It

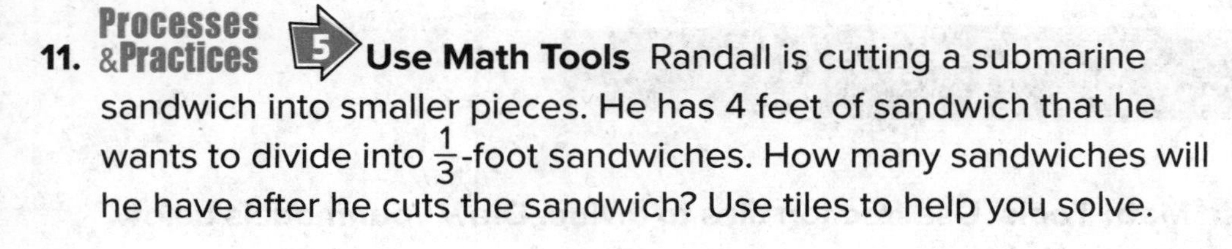

11. Processes & Practices 5 **Use Math Tools** Randall is cutting a submarine sandwich into smaller pieces. He has 4 feet of sandwich that he wants to divide into $\frac{1}{3}$-foot sandwiches. How many sandwiches will he have after he cuts the sandwich? Use tiles to help you solve.

____

12. Brittany is making party favors. She is dividing $\frac{1}{2}$ pound of jelly beans into 6 packages. How many pounds of jelly beans will be in each package? Use tiles to help you solve.

____

13. Processes & Practices 2 **Use Symbols** Circle the missing divisor below from the equation $5 \div ■ = 25$. Explain.

$\frac{1}{2}$ $\frac{2}{5}$ 5 $\frac{1}{5}$

____

____

14. **Building on the Essential Question** How can I use fraction tiles to help divide a whole number by a unit fraction?

____

____

____

Name

MY Homework

Lesson 9

Hands On: Division with Unit Fractions

Homework Helper

Need help? connectED.mcgraw-hill.com

Find $2 \div \frac{1}{8}$. ← THINK: How many groups of $\frac{1}{8}$ are in 2?

1 Represent the 2 by using 2 whole fraction tiles.

2 Place enough of the $\frac{1}{8}$-fraction tiles below the 2 whole fraction tiles to represent the same amount.

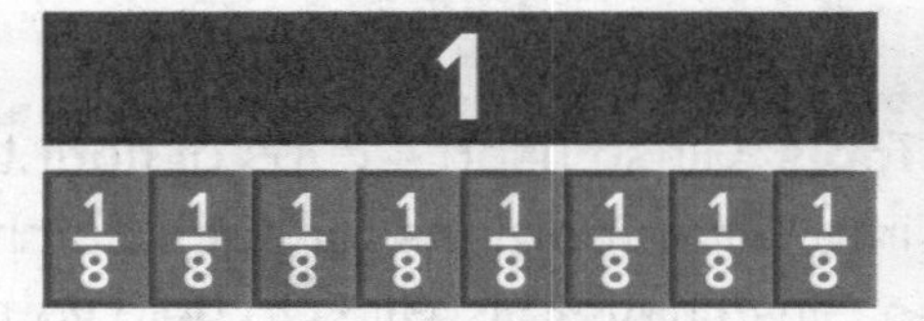

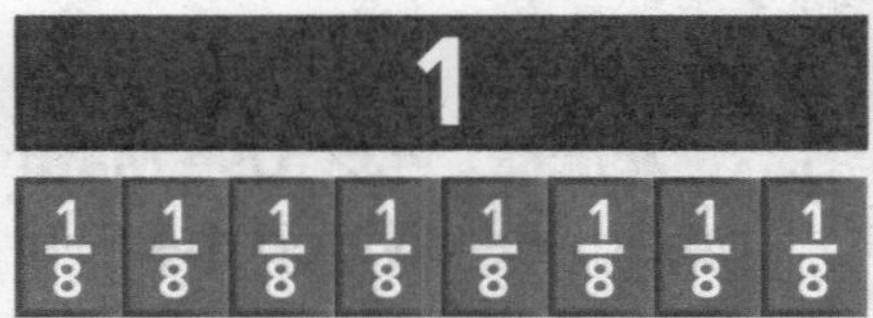

Sixteen $\frac{1}{8}$-fraction tiles were needed to equal the length of 2 whole fraction tiles.

So, $2 \div \frac{1}{8} = 16$.

Practice

Use fraction tiles to divide. Draw your models below.

1. $2 \div \frac{1}{6} =$ ______

2. $\frac{1}{4} \div 3 =$ ______

Problem Solving

3. Harold spent $\frac{3}{4}$ hour milking cows. He milked 3 groups of cows. If he spent an equal amount of time on each group of cows, what fraction of an hour did he spend milking each group? Use tiles to help you solve.

4. Rosanne and her family drove for 8 hours on their vacation trip. Every $\frac{1}{2}$ hour, they played a different game. How many different games did they play? Use tiles to help you solve.

5. Processes & Practices 5 **Use Math Tools** Girish used 4 cups of flour to make bread. He divided the flour into $\frac{1}{3}$-cup portions for each batch. How many batches of bread does Girish have? Use tiles to help you solve.

6. Jason is cutting a roll of sausage into pieces that are $\frac{1}{2}$ inch thick. If the roll is 6 inches long, how many pieces of sausage can he cut? Use tiles to help you solve.

7. Hubert cut a piece of yarn into pieces that are each $\frac{2}{3}$ foot long. If the yarn is 6 feet long, how many pieces of yarn did he cut? Use tiles to help you solve.

Name ....

Lesson 10
Divide Whole Numbers by Unit Fractions

ESSENTIAL QUESTION
What strategies can be used to multiply and divide fractions?

Math in My World

You can divide a whole number by a unit fraction using models. A **unit fraction** is a fraction with a numerator of 1.

Example 1

A sports Web site has score updates every $\frac{1}{4}$ hour. How many times does the Web site have score updates in a 3-hour period?

Find $3 \div \frac{1}{4}$ using models. How many $\frac{1}{4}$s are in 3?

1. The model represents 3.

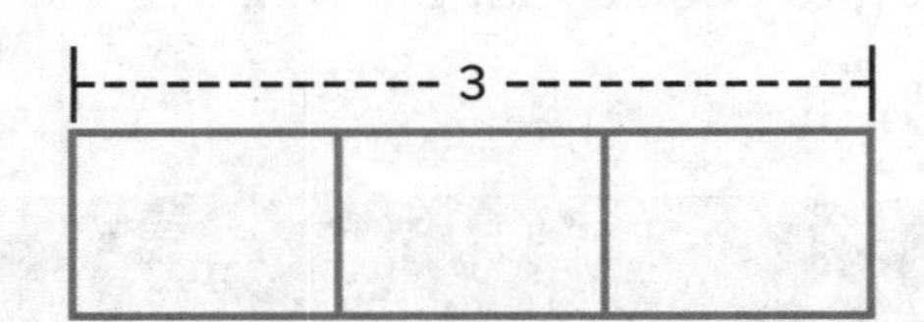

2. Divide each of the three rectangles into fourths.

3. Count the number of fourths.

There are ______ fourths in the model.

$3 \div \frac{1}{4} =$ ______.

So, the Web site has ______ score updates in 3 hours.

Check You can check division problems using multiplication because they are inverse operations.

$12 \times \frac{1}{4} = \frac{\square}{\square}$ or ______

Example 2

Lisa takes four apple pies to her family reunion. Each pie is cut into sixths. How many pieces of pie can Lisa serve? Find the unknown in $4 \div \frac{1}{6} = g$.

 The model represents 4.

4

2 Divide each of the four smaller rectangles into sixths.

 Count the number of sixths. There are ______ sixths in the model.

So, $4 \div \frac{1}{6} =$ ______.

$g =$ ______ ← unknown

Lisa can serve ______ pieces of pie.

Check Use multiplication to check your answer. ______ $\times \frac{1}{6} = 4$

Guided Practice

1. Find the quotient of $2 \div \frac{1}{3}$. Use a model. Check using multiplication.

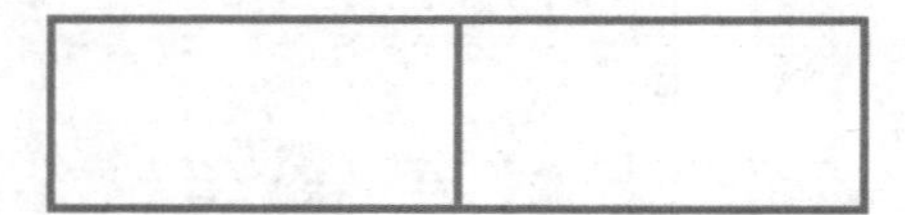

Why can you use multiplication to check your answer to a division problem?

$2 \div \frac{1}{3} =$ ______

Check ______ $\times \frac{1}{3} = \frac{\square}{\square}$ or 2

Name ....

Independent Practice

Find each quotient. Use a model. Check using multiplication.

2. $4 \div \frac{1}{3} =$ ________

Check ________ $\times \frac{1}{3} = \frac{\square}{\square}$ or 4

3. $3 \div \frac{1}{5} =$ ________

Check ________ $\times \frac{1}{5} = \frac{\square}{\square}$ or 3

4. $6 \div \frac{1}{4} =$ ________

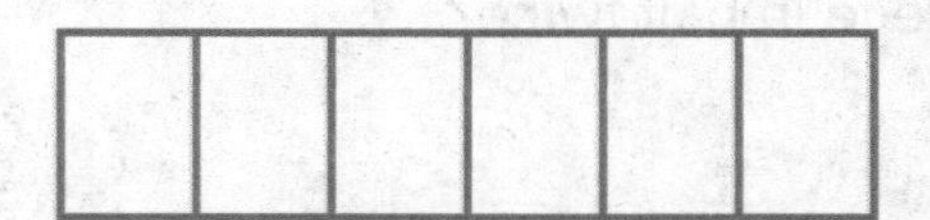

Check ________ $\times \frac{1}{4} = \frac{\square}{\square}$ or 6

5. $5 \div \frac{1}{4} =$ ________

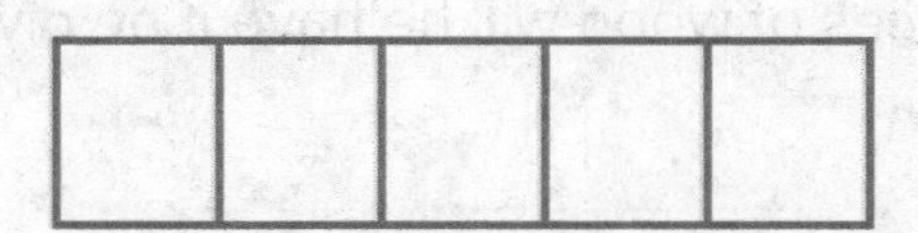

Check ________ $\times \frac{1}{4} = \frac{\square}{\square}$ or 5

6. $2 \div \frac{1}{2} =$ ________

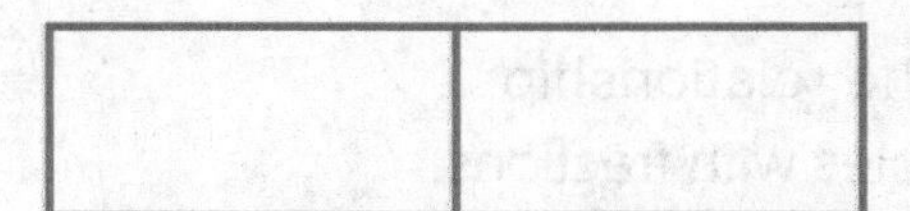

Check ________ $\times \frac{1}{2} = \frac{\square}{\square}$ or 2

7. $3 \div \frac{1}{6} =$ ________

Check ________ $\times \frac{1}{6} = \frac{\square}{\square}$ or 3

Problem Solving

Processes & Practices  **Use Math Tools Draw a model to solve each problem.**

8. Denise has 4 hours to paint crafts. She would like to spend no more than $\frac{1}{4}$ of each hour on each craft. How many crafts can she paint during that time?

____

9. Laura has a 6-foot long piece of ribbon that she wants to cut to make bows. Each piece needs to be $\frac{1}{3}$-foot long to make one bow. How many bows will she be able to make?

____

Brain Builders

10. Ray has 8 pieces of wood to finish building a doghouse. How many pieces of wood will he have if he divides each piece in half twice?

____

11. **Processes & Practices** 6 **Be Precise** Write and solve a real-world problem for $9 \div \frac{1}{8}$. Then explain the meaning of the quotient.

____

____

____

12. **Building on the Essential Question** Explain the relationship between division and multiplication. Include examples with fractions.

____

____

____

Name ____________________

MY Homework

Lesson 10

Divide Whole Numbers by Unit Fractions

Homework Helper

Need help? connectED.mcgraw-hill.com

The recipe Melinda is using to make fruit punch is for one serving. It calls for $\frac{1}{2}$ cup of pineapple juice. Melinda has 5 cups of pineapple juice. How many servings can Melinda make?

The model represents 5 cups of juice. Since each serving uses $\frac{1}{2}$ cup of pineapple juice, find how many $\frac{1}{2}$ cups are in 5 cups.

5

$\frac{1}{2}$	$\frac{1}{2}$	$\frac{1}{2}$	$\frac{1}{2}$	$\frac{1}{2}$	$\frac{1}{2}$	$\frac{1}{2}$	$\frac{1}{2}$	$\frac{1}{2}$	$\frac{1}{2}$

There are ten $\frac{1}{2}$-cup sections in the model, so $5 \div \frac{1}{2} = 10$.

Melinda can make 10 servings of fruit punch.

Check $10 \times \frac{1}{2} = \frac{10}{2}$ or 5

Practice

Find each quotient. Use a model. Check using multiplication.

1. $4 \div \frac{1}{5} =$ ______

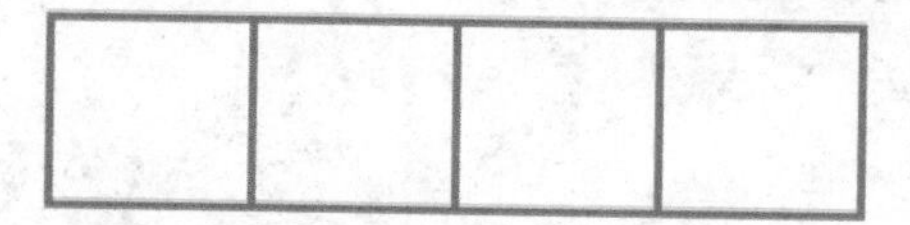

Check ______ $\times \frac{1}{5} = \frac{\square}{\square}$ or 4

2. $6 \div \frac{1}{2} =$ ______

Check ______ $\times \frac{1}{2} = \frac{\square}{\square}$ or 6

Real World Problem Solving

Processes & Practices 5 **Use Math Tools Draw a model to solve each problem.**

3. Chef Erin has 4 pizzas that need to be cut into slices. Each slice represents $\frac{1}{6}$ of a pizza. How many slices of pizza can she serve?

____

4. A baker cuts 3 cakes into pieces. Each piece represents $\frac{1}{8}$ of a cake. What is the total number of pieces?

____

Brain Builders

5. **Processes & Practices** 7 **Identify Structure** Circle the model below that represents $2 \div \frac{1}{3}$. Then write a division problem using a whole number and a fraction for each of the remaining models.

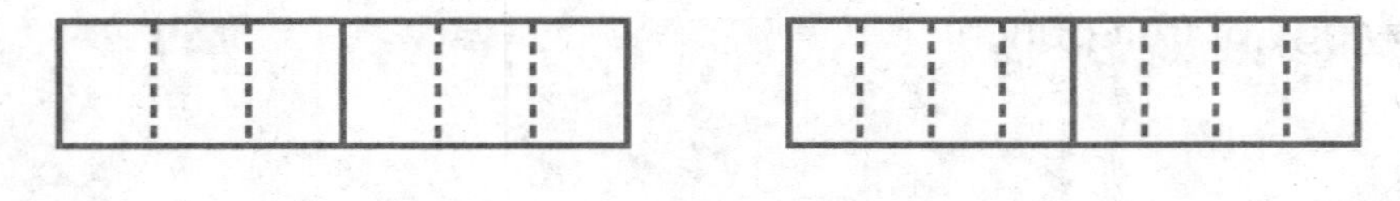

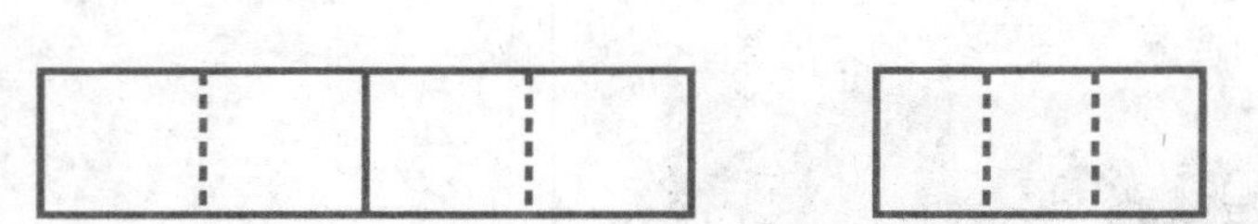

6. **Test Practice** Karen uses five pounds of meat to grill hamburgers. Each pound is divided into thirds to make each hamburger. How many hamburgers can Karen serve?

Ⓐ 10 hamburgers

Ⓑ 12 hamburgers

Ⓒ 15 hamburgers

Ⓓ 18 hamburgers

Name ....

Lesson 11 Divide Unit Fractions by Whole Numbers

ESSENTIAL QUESTION
What strategies can be used to multiply and divide fractions?

Math in My World

Example 1

Elijah is sorting his music into playlists. One-half of his music is rock. He wants to make 3 separate rock playlists. If each playlist is the same size, what fraction of Elijah's music will be on one of the rock playlists?

Find $\frac{1}{2} \div 3$.

1. The model shows the amount of Elijah's music that is rock.

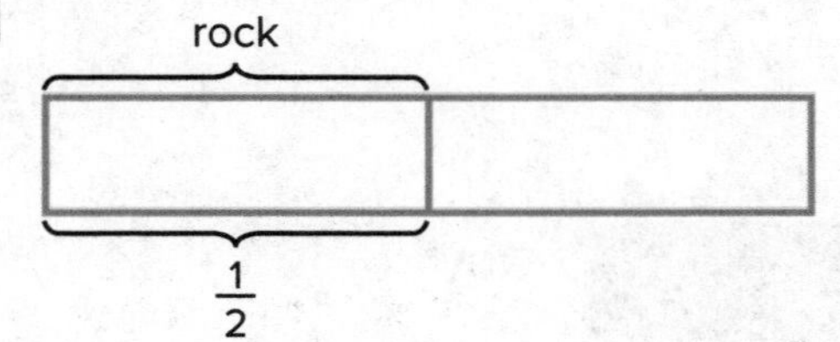

2. Divide each section into 3 equal parts.

How many equal sections are there now? ______

How many of the parts represent the fraction of the music that is on one of the rock playlists? ______

3. Write the fraction.

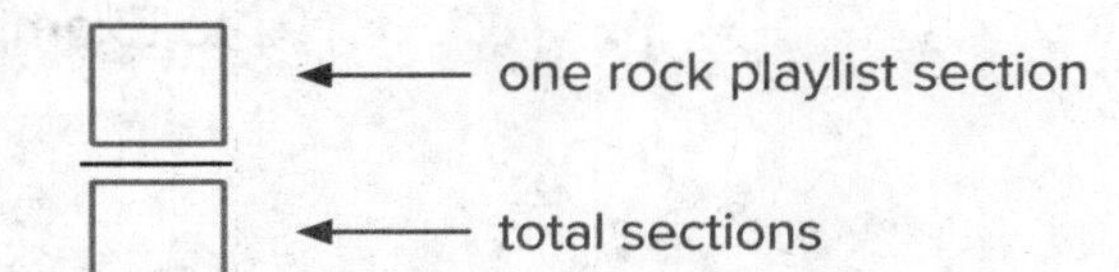

So, $\frac{1}{2} \div 3 = \frac{\square}{\square}$.

Check Use multiplication to check.

$\frac{1}{6} \times 3 = \frac{1}{6} \times \frac{3}{1} = \frac{3}{6}$ or $\frac{1}{2}$

Example 2

In the buffet line, there was only $\frac{1}{4}$ of a pan of spaghetti left. Three friends decide to divide the spaghetti evenly. What fraction of a whole pan of spaghetti did each friend receive? Find the unknown in $\frac{1}{4} \div 3 = s$.

1. The model shows $\frac{1}{4}$.

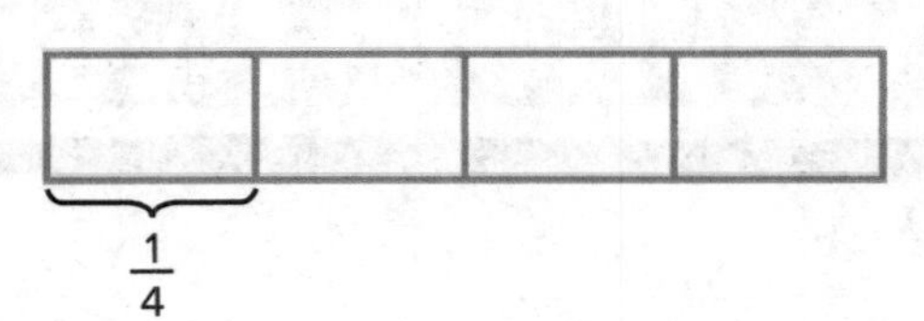

2. Divide each of the four sections into 3 equal parts. There are ______ total sections.

3. Write the fraction.

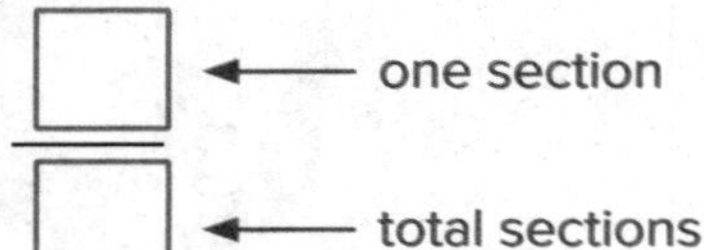

So, $\frac{1}{4} \div 3 = \frac{\square}{\square}$.

$s = \frac{\square}{\square}$ ← unknown

Each friend receives $\frac{\square}{\square}$ of a pan of spaghetti.

Guided Practice

1. Find the quotient of $\frac{1}{3} \div 3$. Use the model. Check using multiplication.

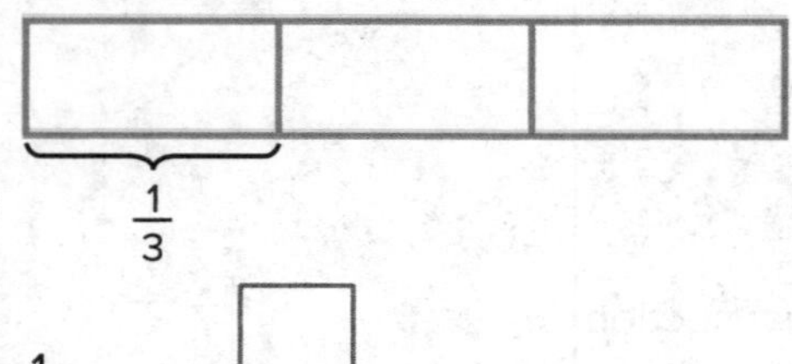

$\frac{1}{3} \div 3 = \frac{\square}{\square}$

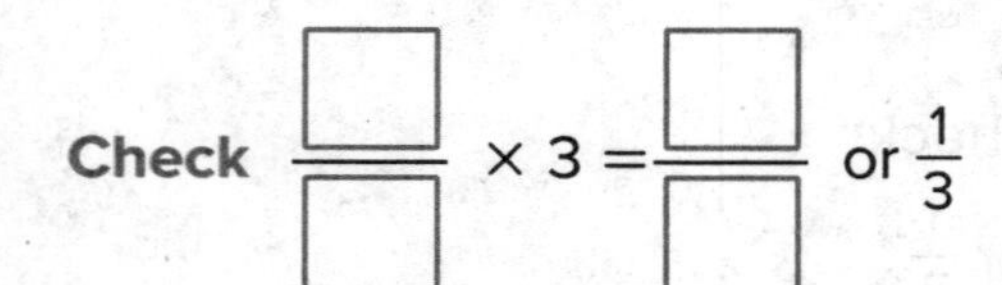

Check $\frac{\square}{\square} \times 3 = \frac{\square}{\square}$ or $\frac{1}{3}$

What multiplication equation can you use to check your answer to Example 2? Explain.

Name ______________________________

Independent Practice

Find each quotient. Use each model. Check using multiplication.

2. $\frac{1}{2} \div 6 =$ ______

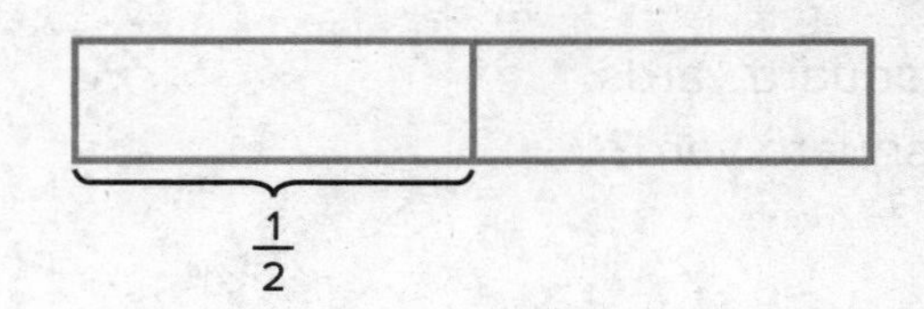

Check ______ $\times 6 =$ ______ or $\frac{1}{2}$

3. $\frac{1}{7} \div 2 =$ ______

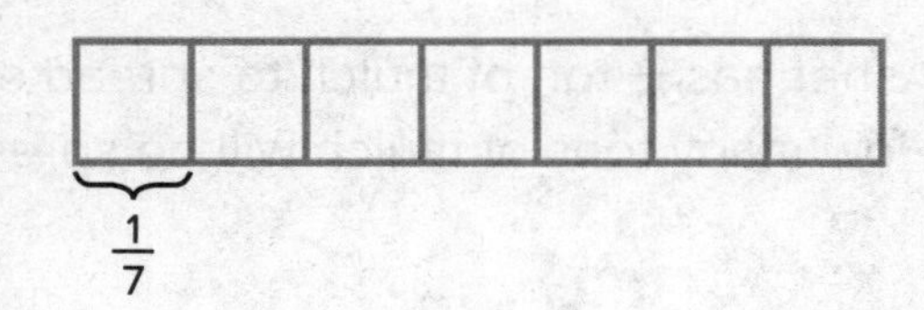

Check ______ $\times 2 =$ ______ or $\frac{1}{7}$

4. $\frac{1}{4} \div 4 =$ ______

Check ______ $\times 4 =$ ______ or $\frac{1}{4}$

5. $\frac{1}{5} \div 2 =$ ______

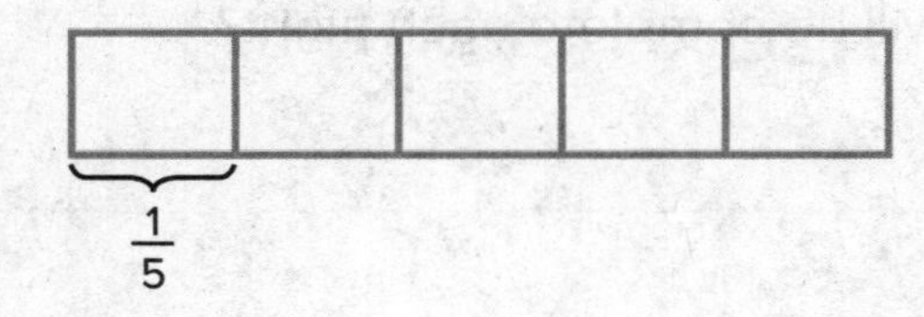

Check ______ $\times 2 =$ ______ or $\frac{1}{5}$

6. $\frac{1}{2} \div 4 =$ ______

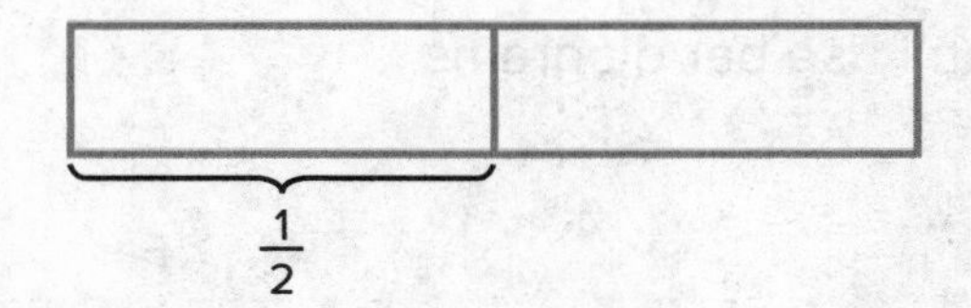

Check ______ $\times 4 =$ ______ or $\frac{1}{2}$

7. $\frac{1}{9} \div 2 =$ ______

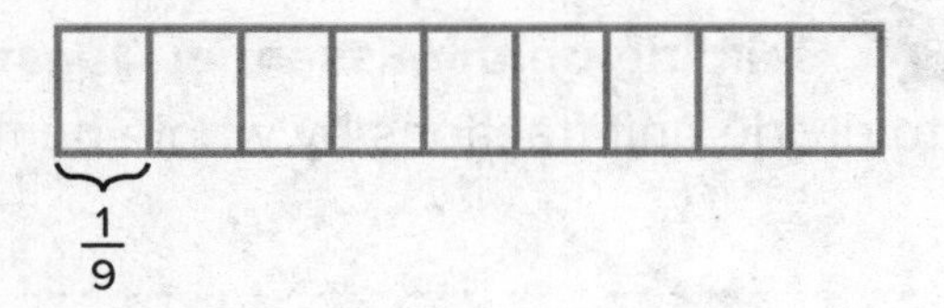

Check ______ $\times 2 =$ ______ or $\frac{1}{9}$

Problem Solving

Processes & Practices  **Use Math Tools Draw a model to solve each problem.**

8. One-fifth of the text messages that Natalie sends are to her family. She sends an equal amount of messages to her mom, dad, and sister. What fraction of all the text messages she sends go to her sister?

____

9. Asher has $\frac{1}{2}$ ton of mulch to spread equally in 8 square yards. How many tons of mulch will be spread in each square yard?

____

Brain Builders

10. Hyun has $\frac{1}{3}$ of his science project to complete still. He wants to spend an equal amount of time on his science project on Monday, Tuesday, and Wednesday night. What fraction of the science project will he complete each night?

____

11. Processes & Practices 6 **Be Precise** Write and solve a real-world problem for $\frac{1}{6} \div 4$. Then explain the meaning of the quotient.

____

____

12. **Building on the Essential Question** How can I use bar diagrams to divide unit fractions by whole numbers?

____

____

____

Name

MY Homework

Lesson 11
Divide Unit Fractions by Whole Numbers

Homework Helper

Need help? connectED.mcgraw-hill.com

Find $\frac{1}{6} \div 2$.

1. The model shows $\frac{1}{6}$.

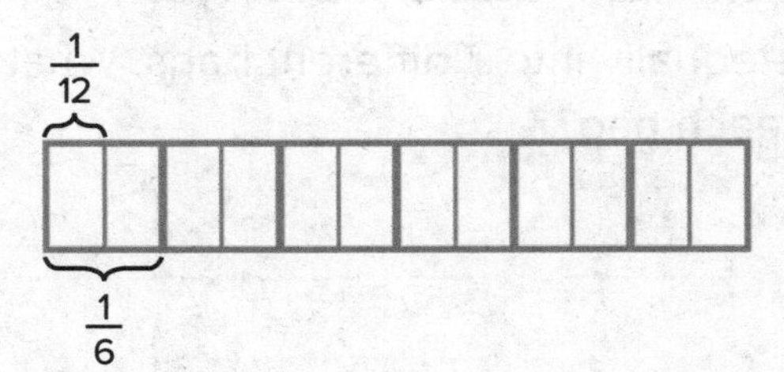

2. Divide each of the six equal sections into 2 equal parts. There are now 12 sections.

3. Write the fraction.

$\frac{1}{12}$ ← one section
← total sections

So $\frac{1}{6} \div 2 = \frac{1}{12}$.

Check $\frac{1}{12} \times 2 = 2/12$ or $\frac{1}{6}$

Practice

Find each quotient. Use each model. Check using multiplication.

1. $\frac{1}{2} \div 3 =$ ______

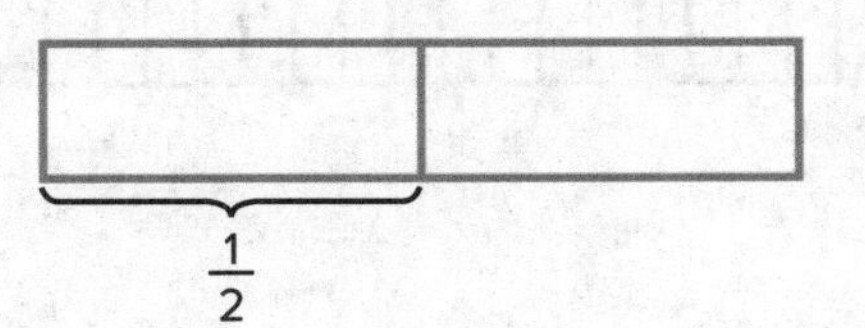

Check ______ $\times 3 =$ ______ or $\frac{1}{2}$

2. $\frac{1}{4} \div 5 =$ ______

Check ______ $\times 5 =$ ______ or $\frac{1}{4}$

Problem Solving

Draw a model to solve Exercises 3 and 4.

3. There is only one granola bar left out of a pan of 10 bars. If Will and Rachel decide to split the last granola bar, what fraction of the entire pan of granola bars will each friend receive?

____

4. Processes &Practices 4 **Model Math** Tyson has $\frac{1}{2}$ pound of raisins to divide equally into 7 different bags. What fraction of a pound will be in each bag?

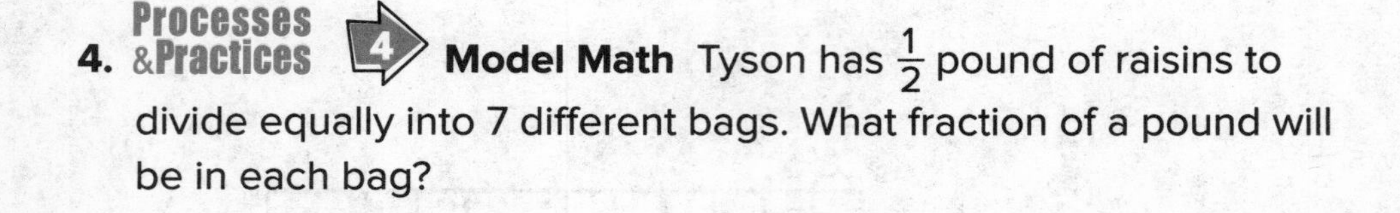

____

Brain Builders

5. Processes &Practices 6 **Explain to a Friend** Write and solve a real-world problem for $\frac{1}{2} \div 3$. Draw a model. Then explain the meaning of the quotient.

____

____

____

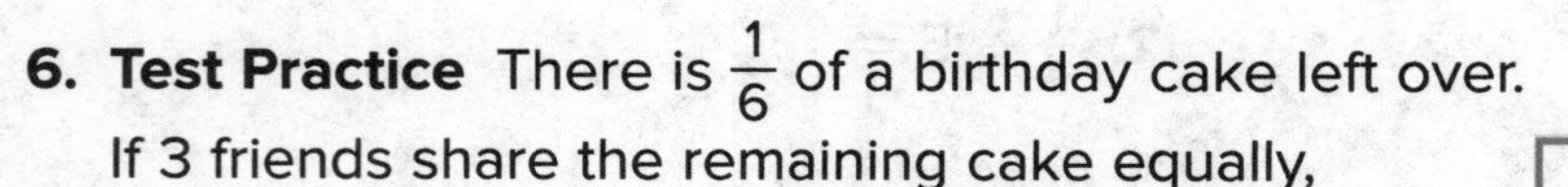

6. **Test Practice** There is $\frac{1}{6}$ of a birthday cake left over. If 3 friends share the remaining cake equally, what fraction of the entire cake will each friend receive? Find the unknown in $\frac{1}{6} \div 3 = p$.

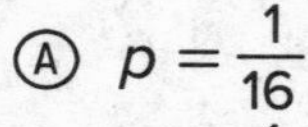

Ⓐ $p = \frac{1}{16}$ Ⓑ $p = \frac{1}{18}$

Ⓒ $p = \frac{1}{21}$ Ⓓ $p = \frac{1}{24}$

Name ______________________________

Lesson 12
Problem-Solving Investigation
STRATEGY: Draw a Diagram

ESSENTIAL QUESTION
What strategies can be used to multiply and divide fractions?

Learn the Strategy

Jaheim visited an aquarium over the weekend and saw 35 species of fish. This was $\frac{1}{5}$ the total number of species of fish. How many total species of fish are at the aquarium?

1 Understand

What facts do you know?

Jaheim saw ________ species of fish and this is ________ of the total amount of species.

What do you need to find?

- the total number of species of ________ in the aquarium

2 Plan

I can solve the problem by drawing a diagram.

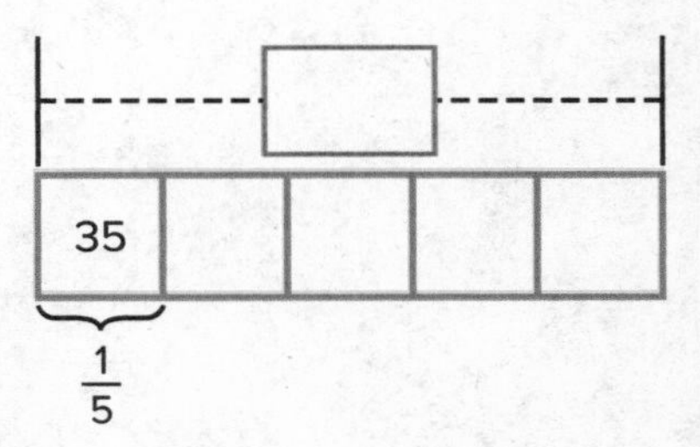

3 Solve

The model is divided into fifths. Since each part represents 35 species, there is a total of 35×5, or ________ species of fish at the aquarium.

4 Check

Is my answer reasonable? ________ $\div 5 = 35$

Practice the Strategy

Victor has \$18 in a piggy bank. He spends $\frac{2}{3}$ of the money on a video game and $\frac{1}{6}$ on candy. How much money will Victor have left?

1 Understand

What facts do you know?

____

____

What do you need to find?

____

2 Plan

____

3 Solve

4 Check

Is my answer reasonable? Explain.

____

Name

Apply the Strategy

Solve each problem by drawing a diagram.

1. Mrs. Vallez purchased sand toys that were originally \$20. She received $\frac{1}{4}$ off of the total price. How much did she save?

2. Sue has four DVDs and Terry has six DVDs. They put all their DVDs together and sold them for \$10 for two DVDs. How much money will they earn if they sell all of their DVDs?

Brain Builders

3. Processes & Practices 4 **Model Math** Jacinda is decorating cookies for a class party. She can decorate $\frac{2}{3}$ of a cookie per minute. At this rate, how many cookies can she decorate between 4:30 P.M. and 5:15 P.M.?

4. At a bird sanctuary, Ricky counted 80 birds. Of the birds he counted, $\frac{1}{4}$ were baby birds. If he counted an equal number of adult males and females, how many adult female birds did Ricky count?

5. Processes & Practices 5 **Use Math Tools** The fourth grade class collected \$56 in donations. This is \$4 more than one-third the amount collected by the fifth grade class. How much money in donations did the fifth grade class collect?

Review the Strategies

Use any strategy to solve each problem.

- Draw a diagram.
- Work backward.
- Guess, check, and revise.
- Act it out.

6. A cook needs 12 pounds of flour. He wants to spend the least amount of money. If a 2-pound bag costs $1.59 and a 5-pound bag costs $2.89, how many bags of each type of flour should he buy? What will be the total cost?

____

7. Processes & Practices 5 **Use Math Tools** A ride at a theme park lasts $1\frac{1}{2}$ minutes. It takes 2 minutes to prepare the ride for each trip. How many times can the ride be completed in 30 minutes?

____

8. On Monday, 21 DVDs were checked out at the library. This is 3 less than half the amount of books checked out that day. How many books were checked out?

____

9. Processes & Practices 4 **Model Math** Leo takes 30 minutes to eat dinner, 15 minutes to change clothes, and 20 minutes to walk to practice. If Leo needs to be at hockey practice at 7:15 P.M., what time does he need to begin?

____

10. Mi-Ling sets up tables for her art class students. Each square table can seat two people on each side. How many people can be seated if 8 square tables are pushed together in a row?

____

Name

MY Homework

Lesson 12
Problem Solving: Draw a Diagram

Homework Helper

Need help? connectED.mcgraw-hill.com

Norah earned \$50 helping her neighbor. She spent $\frac{3}{5}$ of the money on a purse and put $\frac{1}{5}$ in the bank. How much money does Norah have left?

1 Understand

What facts do you know?

Norah earned \$50 and spent $\frac{3}{5}$ on a purse and put $\frac{1}{5}$ in the bank.

What do you need to find?

the amount of money Norah has left

2 Plan

I can solve the problem by drawing a diagram.

3 Solve

The bar diagram represents the total amount earned.

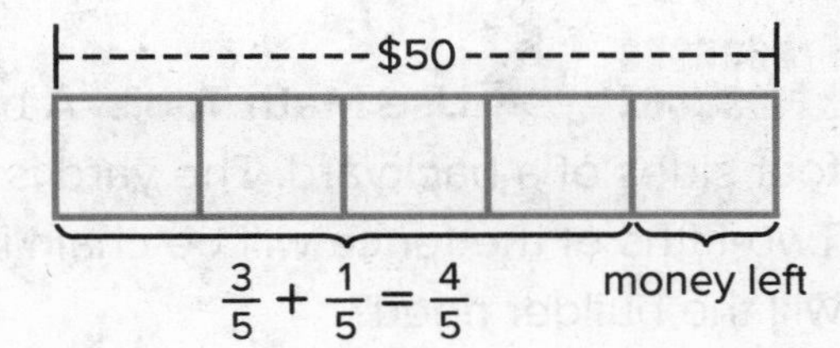

There are 5 equal sections and each section represents \$10.

So, Norah has \$10 left.

4 Check

Is my answer reasonable? Explain.

Use multiplication to check. $\frac{1}{5} \times 50 = 10$

Real World

Problem Solving

Solve each problem by drawing a diagram.

1. Paul is decorating cupcakes for a birthday party. He can decorate $\frac{3}{4}$ of a cupcake per minute. At this rate, how many cupcakes can he decorate in 12 minutes?

____

2. Processes & Practices 4 **Model Math** To make cheesecake, you need about $2\frac{1}{2}$ pounds of cream cheese. How many pounds of cream cheese will you need to make 2 cheesecakes?

____

Brain Builders

3. Scott viewed 72 classic cars at a car show. Of the cars he viewed, $\frac{3}{8}$ were sports cars. How many of the cars Scott viewed were not sports cars?

____

4. Processes & Practices 5 **Use Math Tools** A builder is installing a fence on all four sides of a backyard. The yard is 40 feet long and 50 feet wide. Two-fifths of the fence will be chain link. How many feet of chain link will the builder need?

____

5. Misty purchased two pairs of jeans that were originally $24 each, but were on sale for $\frac{1}{3}$ off of the total price. How much did she save?

____

Review

Vocabulary Check

Read each clue. Fill the corresponding section of the crossword puzzle to answer each clue. Use the words in the word bank.

dividend	**divisor**	**fraction**	**inverse operations**
quotient	**mixed number**	**scaling**	**unit fraction**

Across

1. The result of a division problem.
2. The number that divides the dividend.
3. A number that is being divided.
4. The process of resizing a number when you multiply by a fraction that is greater than or less than 1.
5. Operations that undo each other, such as multiplication and division.

Down

6. A fraction with a numerator of 1.
7. A number that has a whole number part and a fraction part.
8. A number that represents part of a whole or part of a set.

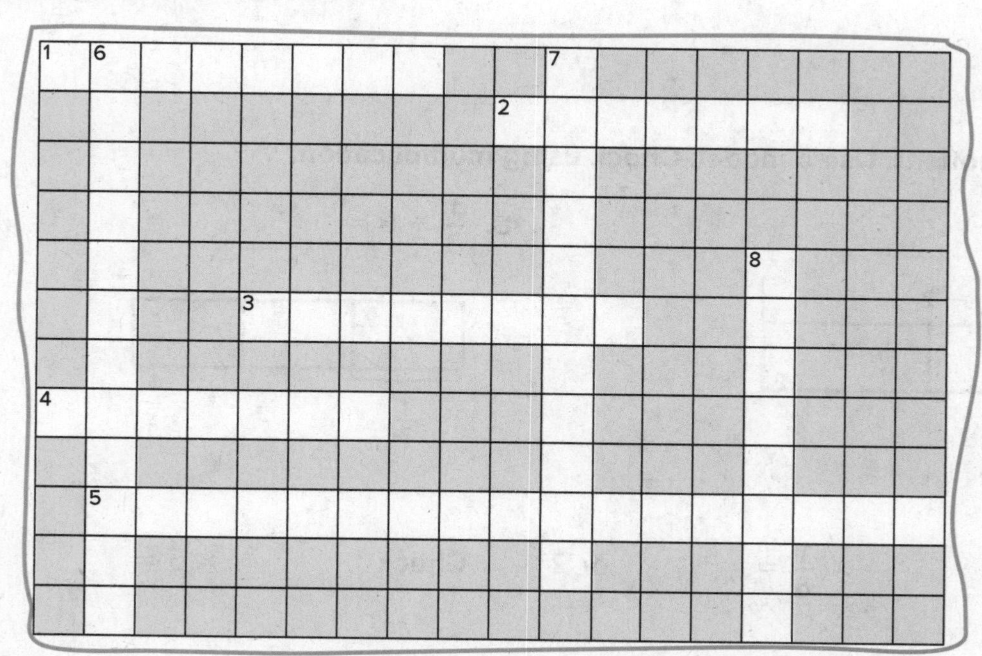

Concept Check

Estimate each product. Draw a bar diagram if necessary.

9. $\frac{3}{4} \times 23$

______

10. $\frac{1}{5} \times 22$

______

Multiply. Write in simplest form.

11. $\frac{1}{3} \times 21 =$ ______

12. $26 \times \frac{1}{2} =$ ______

13. $\frac{1}{5} \times \frac{3}{8} =$ ______

14. $\frac{1}{2} \times \frac{7}{8} =$ ______

15. $3\frac{1}{4} \times \frac{3}{5} =$ ______

16. $1\frac{1}{8} \times 3\frac{2}{3} =$ ______

Find each quotient. Use a model. Check using multiplication.

17. $2 \div \frac{1}{6} =$ ______

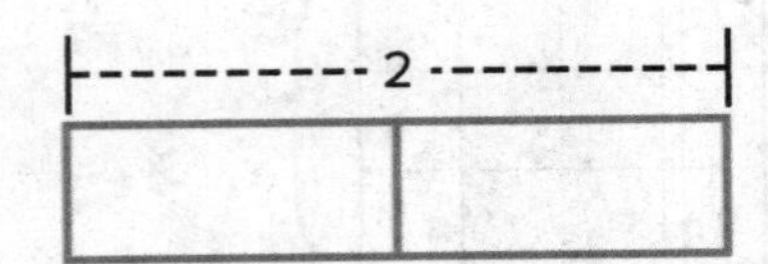

Check ______ $\times \frac{1}{6} =$ ______ or 2

18. $\frac{1}{3} \div 5 =$ ______

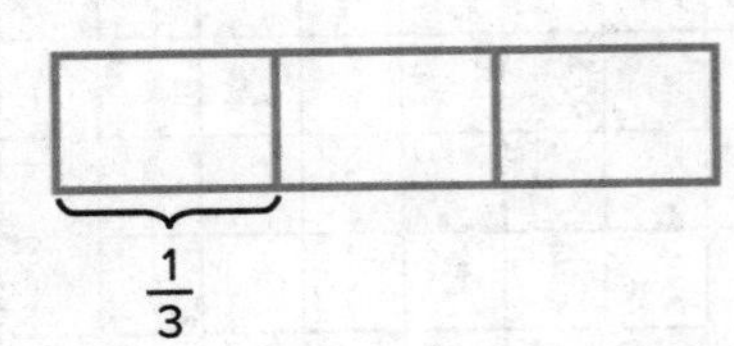

Check ______ $\times 5 =$ ______ or $\frac{1}{3}$

Problem Solving

19. Ken is working on a project for social studies. He has a piece of poster board that needs to be divided equally into $\frac{1}{3}$-foot sections. The poster board is 4 feet wide. How many sections will Ken have on the poster board?

______

20. Samantha measured the dimensions of a rectangular poster frame. It was $\frac{3}{4}$ yard long and $\frac{1}{2}$ yard wide. What is the area of the frame?

______

21. At Middle Avenue School, $\frac{1}{30}$ of the students are on the track team. The track coach offered to buy pizza or subs for everyone on the team. Of the students for whom the coach bought food, $\frac{3}{5}$ ordered pizza. What fraction of the students at Middle Avenue School ate pizza?

______

Brain Builders

22. Sarah is dividing $\frac{1}{4}$ pound of cashews and $\frac{1}{2}$ pound of almonds into 5 plastic bags. What fraction of a pound of nuts will be in each plastic bag? Draw a model to help you solve.

______

23. **Test Practice** The Great Rope Company sells a rope that is 5 feet long. Juan wants to make sections that are $\frac{1}{3}$ foot long. How many sections will Juan be able to make?

Ⓐ 18 sections Ⓑ 15 sections Ⓒ 6 sections Ⓓ 3 sections

Reflect

Chapter 10

Answering the ESSENTIAL QUESTION

Use what you learned about fraction operations to complete the graphic organizer.

Write the Example

Vocabulary

ESSENTIAL QUESTION

What strategies can be used to multiply and divide fractions?

Model

Real-World Example

Now reflect on the ESSENTIAL QUESTION  Write your answer below.

Name

Date

Score

Performance Task

Creating a Floor Plan

A university is adding on to its sports center. The president and athletic director have decided that $\frac{1}{2}$ of the floor plan will be an aqua center for physical therapy. Once that is marked off, they want $\frac{3}{4}$ of the remaining space to be a gym for fitness classes and the rest to be a cycling fitness room.

Show all your work to receive full credit.

Part A

A model of the floor plan is shown below. Divide the floor into areas to find the fraction of the whole that the gym for fitness classes will take up. Shade the area for the gym for fitness classes. Explain your answer and diagram with a number sentence.

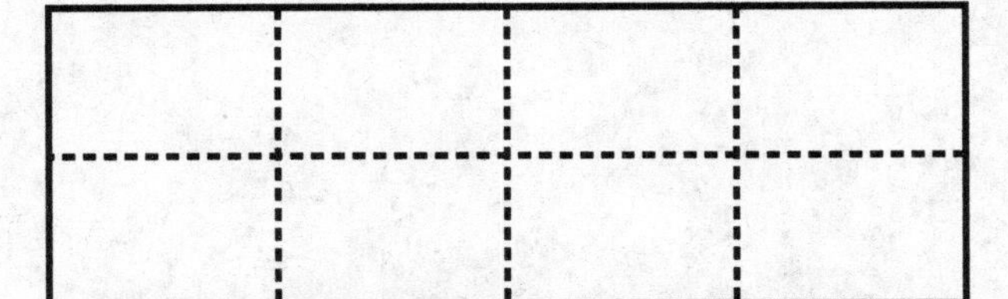

Part B

The new addition is 65 feet by 50 feet. Find the area of the gym for fitness classes in square feet. Explain.

Part C

The cost of building the gym for fitness classes is $98 per square foot. Find the cost of constructing this gym. Explain.

Chapter

11 Measurement

ESSENTIAL QUESTION

How can I use measurement conversions to solve real-world problems?

My Favorite Animals

Watch

Watch a video!

Name ....

Brain Builders

MY Chapter Project

Stepping It Up

1. With your class, visit the rooms in your school that are listed in the table below.
2. Estimate the number of heel-to-toe steps for the length of each room. Write your estimates in the table.
3. Work with your partner or group to measure the number of heel-to-toe steps for the length of each room. Record your measurements in the table.

Room	Estimated Number of Steps	Actual Number of Steps
hallway		
gym		
cafeteria		
classroom		

4. List other units of measure you might use to estimate length.

Name

Am I Ready?

Multiply.

1. $12 \times 3 =$ ______

2. $36 \times 5 =$ ______

3. $1,760 \times 4 =$ ______

4. $6 \times 1,000 =$ ______

5. $15 \times 100 =$ ______

6. $947 \times 100 =$ ______

7. A musical was sold out for three straight shows. If 825 tickets were sold at each performance, how many tickets were sold in all?

______

Divide.

8. $45 \div 3 =$ ______

9. $112 \div 16 =$ ______

10. $39 \div 4 =$ ______

11. $500 \div 100 =$ ______

12. $150 \div 10 =$ ______

13. $7,900 \div 100 =$ ______

14. A box has 144 ounces of grapes. How many 16-ounce packages of grapes can be made?

______

How Did I Do? **Shade the boxes to show the problems you answered correctly.**

1	2	3	4	5	6	7	8	9	10	11	12	13	14

Name ________________________________

MY Math Words

Vocab

Review Vocabulary

capacity   estimate   length   weight

Making Connections

Use the review vocabulary to tell what you would measure for each question. Then provide estimates for each category.

About how much water does a camel's hump hold?

____

What is the approximate distance across a camel's snout?

____

Provide estimates for each measure.

capacity ________________

length ________________

weight ________________

About how heavy is an adult camel?

____

MY Vocabulary Cards

Processes & Practices

Lesson 11-6

capacity

Lesson 11-9

centimeter (cm)

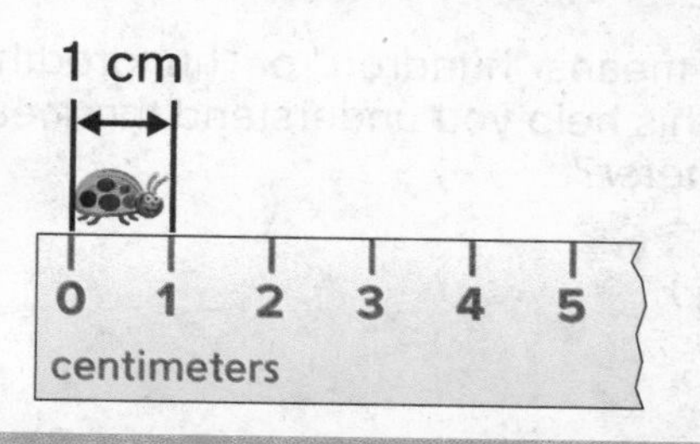

Lesson 11-2

convert

4 quarts = 1 gallon

Lesson 11-6

cup (c)

1 cup

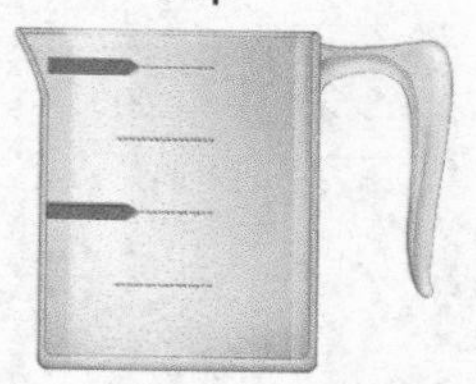

Lesson 11-2

customary system

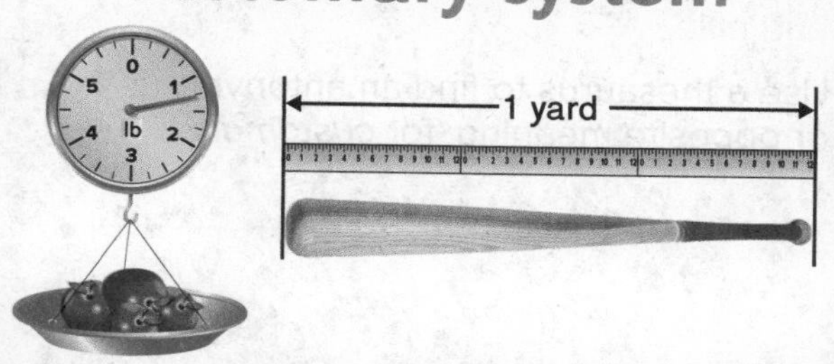

Lesson 11-8

fair share

Lesson 11-7

fluid ounce (fl oz)

8 fluid ounces

Lesson 11-2

foot (ft)

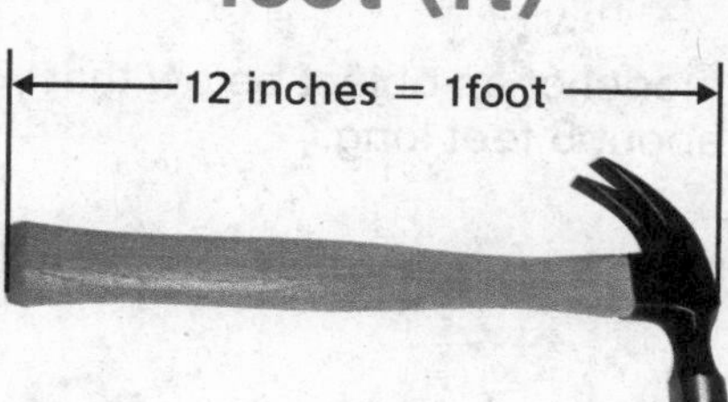

Ideas for Use

- During this school year, create a separate stack of cards for key math verbs, such as *convert.* These verbs will help you in your problem solving.
- Design a crossword puzzle. Use the definition for each word as the clues.

A metric unit for measuring length.
100 centimeters = 1 meter

Centi- means "hundred" or "hundredth." How does this help you understand the meaning of *centimeter?*

The amount a container can hold.

Give an example of a customary unit of capacity.

A customary unit of capacity.
1 cup = 8 fluid ounces

What is a common item you might measure in cups?

To change from one unit of measurement to another.

Give a real-life example of a time when you might need to convert measurements.

An amount divided equally.

Give a real-life example of a time you might need to find a fair share.

The units of measurement most often used in the United States, such as the inch, yard, and mile.

Use a thesaurus to find an antonym, or opposite meaning, for *customary.*

A customary unit for measuring length that is equal to 12 inches.

Draw a model of an object below that you think is about 5 feet long.

A customary unit of capacity.

8 fluid ounces = 1 cup

What clue does this word give you about whether it measures dry or liquid items?

MY Vocabulary Cards

Processes & Practices

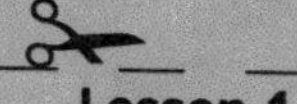

Lesson 11-6

gallon (gal)

Lesson 11-11

gram (g)

5 grams

Lesson 11-1

inch (in)

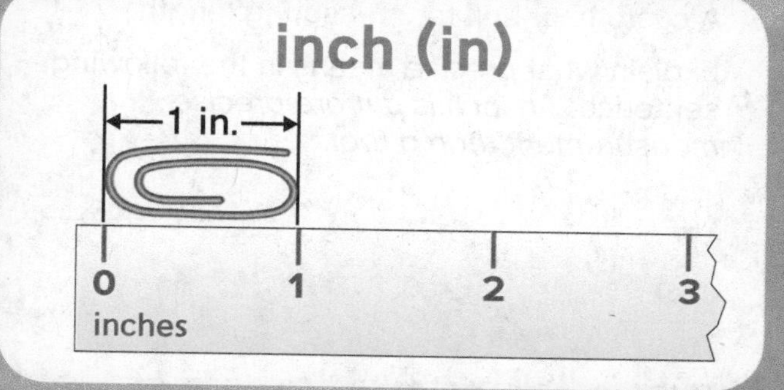

Lesson 11-11

kilogram (kg)

50 kilograms

Lesson 11-10

kilometer (km)

Lesson 11-1

length

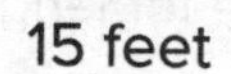

Lesson 11-13

liter (L)

1 liter

5 liters

Lesson 11-11

mass

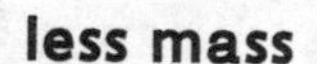

less mass

more mass

Ideas for Use

- Write a tally mark on each card every time you read the word in this chapter or use it in your writing. Challenge yourself to use at least 2 tally marks for each card.
- Work with a partner to name the part of speech of each word. Consult a dictionary to check your answers.

A metric unit for measuring mass.
1 gram = 1,000 milligrams

Decide whether to use grams or kilograms to measure the mass of a dog. Explain.

A customary unit for measuring capacity.
1 gallon = 4 quarts

How do you know if you should measure something in gallons instead of cups?

A metric unit for measuring mass.
1 kilogram = 1,000 grams

The prefix *kilo-* means "thousand." How does that help you understand the meaning of *kilogram?*

A customary unit for measuring length.

Explain what *precise* means in the following sentence: *An inch is a more precise measurement than a foot.*

Distance measured between two points.

Give one metric and one customary example of a unit of measure for length.

A metric unit for measuring longer distances of length.

How many meters are in a kilometer? What part of the word tells you this?

The amount of matter in an object.

Write a sentence using the multiple-meaning word *mass* as an adjective.

A metric unit for measuring volume or capacity. 1 liter = 1,000 milliliters

What are two examples of items that might be measured in liters?

MY Vocabulary Cards

Processes &Practices

Lesson 11-10

meter (m)

1 meter

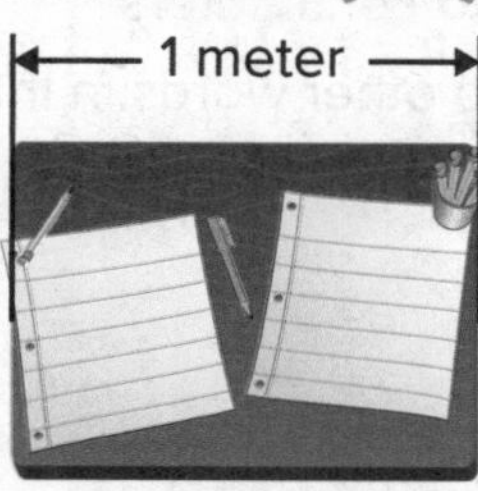

Lesson 11-10

metric system

1 liter

1 gram

Lesson 11-2

mile (mi)

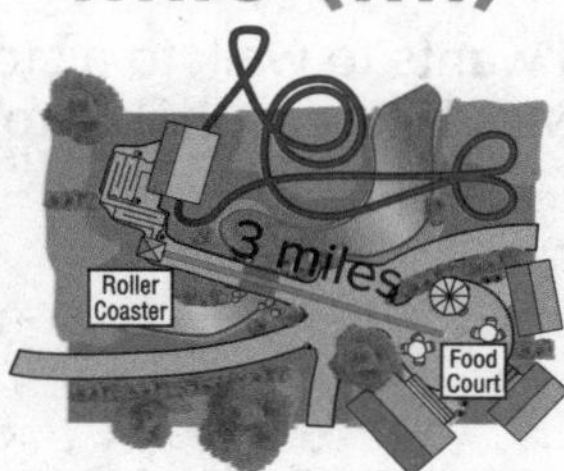

Lesson 11-12

milligram (mg)

1 milligram

Lesson 11-13

milliliter (mL)

250 milliliters

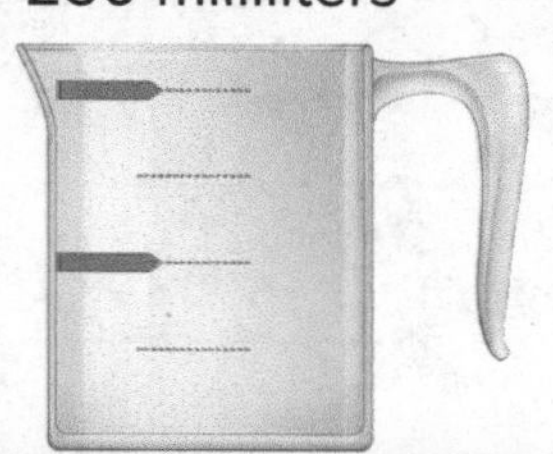

Lesson 11-9

millimeter (mm)

1 millimeter

Lesson 11-4

ounce (oz)

3 ounces

Lesson 11-6

pint (pt)

1 pint

Ideas for Use

- Develop categories for the words. Sort them by category. Ask another student to guess each category.
- Draw or write examples for each card. Be sure your examples are different from what is shown on each card.

A decimal system of measurement. Includes units such as meter, gram, and liter.

Use the dictionary to define *system* as it is used in metric *system.*

A metric unit used to measure length.
1 meter = 100 centimeters

What are two other words in this chapter that include the root word *meter?*

A metric unit for measuring mass.
1,000 milligrams = 1 gram

The prefix *milli-* means "thousand." How can this help you remember the meaning of *milligram?*

A customary unit for measuring length equal to 5,280 feet.

Your friend wants to walk to a store that is 15 miles away. Is it reasonable to walk this distance?

A metric unit used for measuring length.
1,000 millimeters = 1 meter

What is another word that begins with the prefix *milli-?* What does it mean?

A metric unit used for measuring capacity.
1,000 milliliters = 1 liter

How are the words *milliliter* and *millimeter* related?

A customary unit for measuring capacity.
1 pint = 2 cups

Explain whether 1 pint or 2 cups is a more familiar unit of measure.

A customary unit for measuring weight.
16 ounces = 1 pound

Write a sentence using *ounce* to estimate the weight of something.

MY Vocabulary Cards

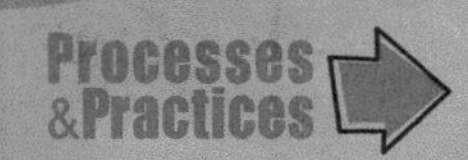

Lesson 11-4

pound (lb)

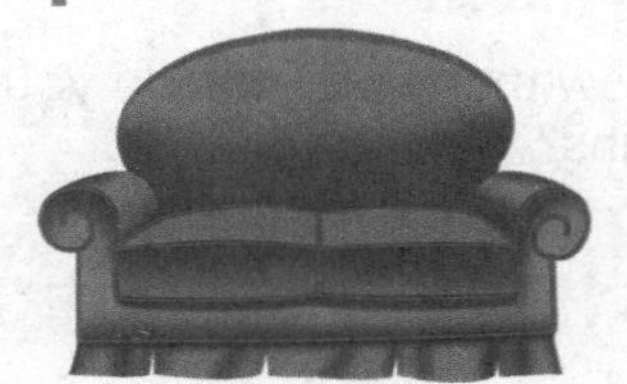

150 pounds

Lesson 11-7

quart (qt)

Lesson 11-5

ton (T)

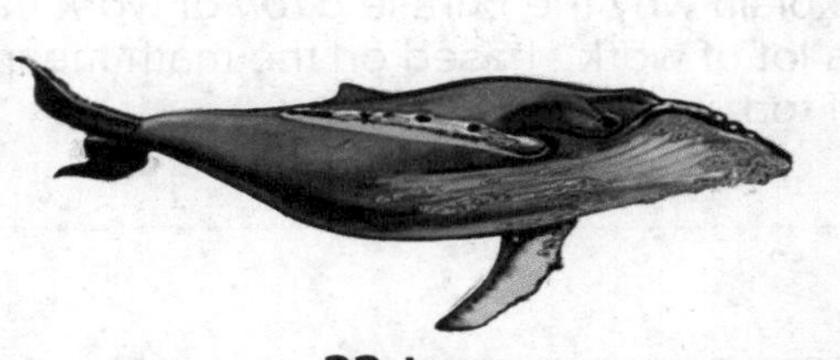

33 tons

Lesson 11-4

weight

Lesson 11-2

yard (yd)

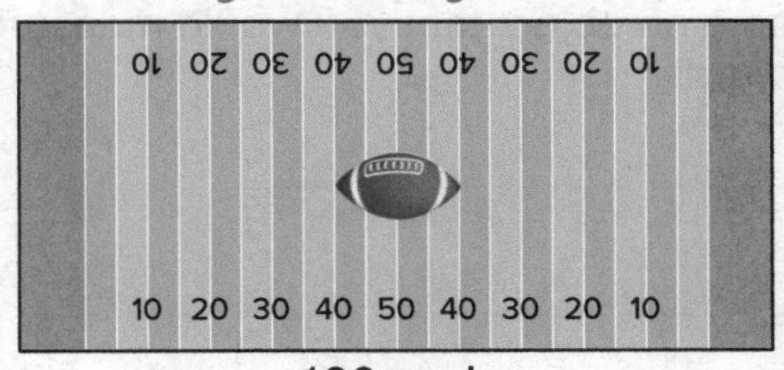

100 yards

Ideas for Use

- Write the name of a lesson you would like to review on the front of a blank card. Write a few study tips to help you on the back of the card.
- Use a blank card to write this chapter's essential question. Use the back of the card to write or draw examples that help you answer the question.

A customary unit for measuring capacity.
1 quart = 2 pints

How can the word *quarter* help you remember the meaning of *quart?*

___

___

___

A customary unit for measuring weight.
1 pound = 16 ounces

About how many pounds do you think a cat weighs?

___

A measurement that tells how heavy an object is.

An *idiom* is a group of words that has a special meaning. What does the idiom "to pull one's weight" mean?

___

___

A customary unit for measuring weight.
1 ton = 2,000 pounds

Explain why the phrase *a ton of work* means "a lot of work", based on the math meaning of ton.

___

___

A customary unit for measuring length that is equal to 3 feet.

Name an object that you would measure in yards.

___

___

MY Foldable

FOLDABLES® Follow the steps on the back to make your Foldable.

1 gallon

4 quarts

1

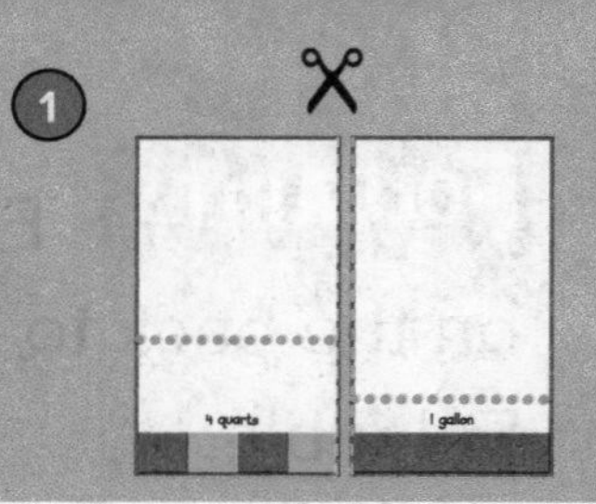

2

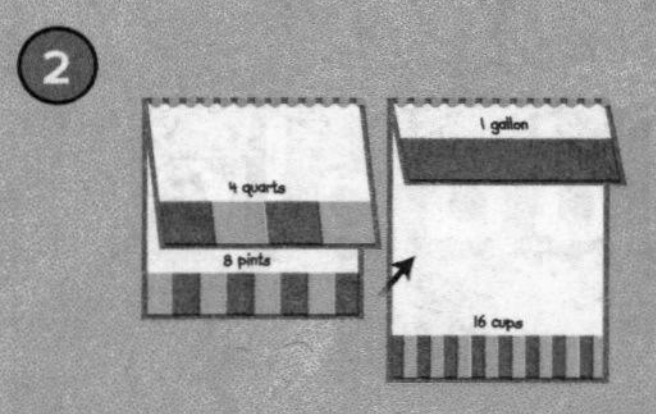

3

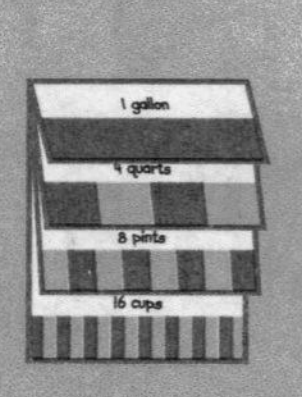

16 cups

8 pints

Name ______________________

Lesson 1 Hands On
Measure with a Ruler

ESSENTIAL QUESTION
How can I use measurement conversions to solve real-world problems?

Length is the measurement of distance between two points. You can use a ruler to measure the length of objects to the nearest half inch or quarter inch.

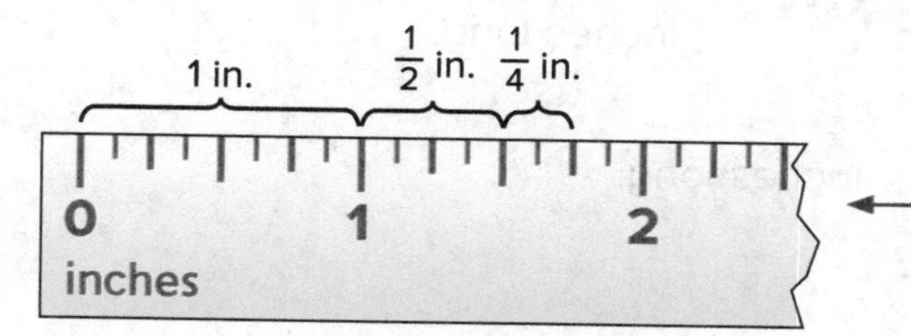

An **inch** is a unit of measurement of length.

If you measure an object to a smaller unit of measure, you get a more precise or accurate measurement.

Cut and use this inch ruler.

Measure It

The width of a button is the length of the widest part of the button. Find the width of the button to the nearest half inch and quarter inch.

1. Place the ruler against one edge of the object. Line up the zero on the ruler with the end of the object.

2. Find the half-inch mark that is closest to the other end. Repeat for the quarter-inch mark.

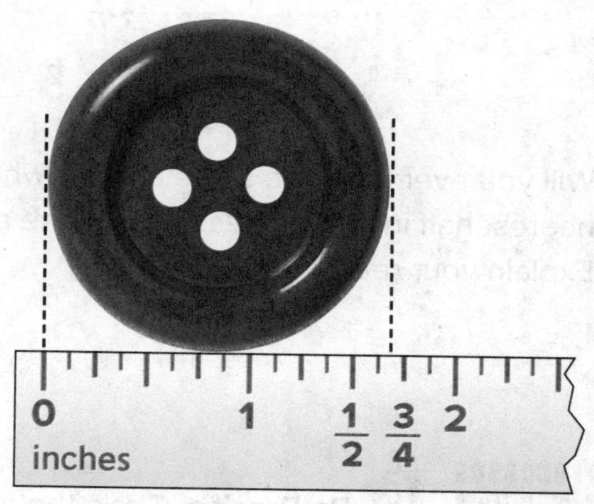

To the nearest half inch, the button is ______ inches wide.

To the nearest quarter inch, it is ______ inches wide.

Try It

Find the length of a paper clip to the nearest half inch and quarter inch.

1. Place the ruler against one edge of the object. Line up the zero on the ruler with the end of the object.

2. Find the half-inch mark that is closest to the other end. Repeat for the quarter-inch mark.

Helpful Hint

All measurements are approximations. However, if you use smaller units, you will get a more precise measure, or one that is closer to the exact measure.

To the nearest half inch, the paper clip is ________ inches long.

To the nearest quarter inch, it is ________ inches long.

Talk About It

1. Explain how you can tell the difference between the half-inch and quarter-inch marks when measuring an object with a ruler.

____

____

2. Will you ever have the same answer when measuring to the nearest half inch and measuring to the nearest quarter inch? Explain your reasoning.

____

3. Processes & Practices 6 **Be Precise** Circle the more precise measurement of the button in the first activity.

$1\frac{1}{2}$ inches $1\frac{3}{4}$ inches

Name

Practice It

Measure the length of each to the nearest half inch and quarter inch.

4.

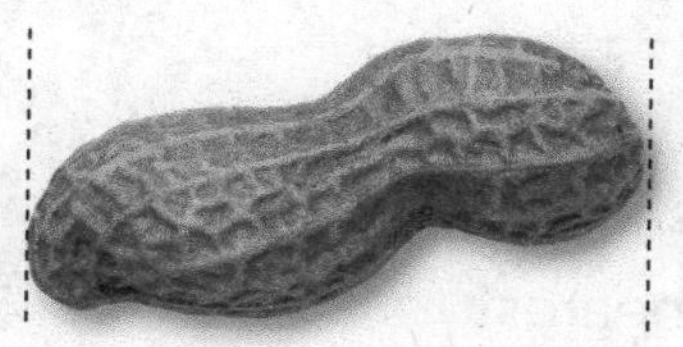

Half Inch: ____________

Quarter Inch: ____________

5.

Half Inch: ____________

Quarter Inch: ____________

Find the length of each object to the nearest half inch and quarter inch.

6. width of a book

____

7. length of a pencil

____

8. width of a calculator

____

9. length of a tape dispenser

____

Draw a line segment with each of the following lengths.

10. $1\frac{1}{4}$ inches

11. $2\frac{1}{2}$ inches

12. $3\frac{3}{4}$ inches

Apply It

13. The length of a remote controlled car is $8\frac{1}{2}$ inches to the nearest half inch and $8\frac{1}{4}$ inches to the nearest quarter inch. Which measurement is more precise?

14. Processes & Practices 6 **Be Precise** Carrie and Sam each measured the length of their cat's tail. Carrie measured the length to the nearest half inch. Her measurement was $6\frac{1}{2}$ inches. Sam measured the length to the nearest quarter inch. His measurement was $6\frac{3}{4}$ inches. Circle the correct description of the actual measurement of the cat's tail.

less than $6\frac{1}{2}$ in. between $6\frac{1}{2}$ in. and $6\frac{3}{4}$ in. greater than $6\frac{3}{4}$ in.

15. Processes & Practices 3 **Find the Error** James used a ruler to measure a piece of string. James said the string is $2\frac{1}{2}$ inches long. Find his mistake and correct it.

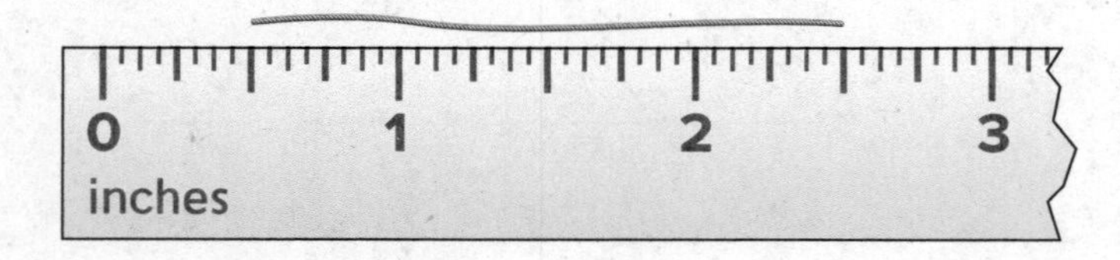

Write About It

16. When might you want to measure the length of an object to the nearest quarter inch as opposed to the nearest half inch?

Name ....

MY Homework

Lesson 1

Hands On: Measure with a Ruler

Homework Helper

Need help? connectED.mcgraw-hill.com

Find the length of the push pin to the nearest half inch and quarter inch.

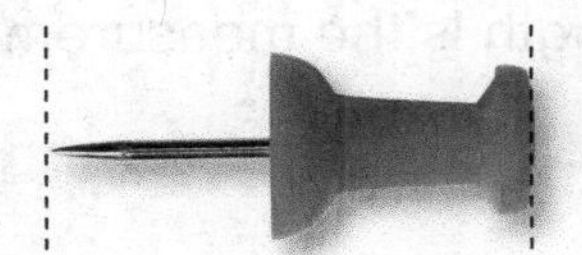

1. Place the ruler against one edge of the object. Line up the zero on the ruler with the end of the object.

2. Find the half-inch mark that is closest to the other end. Repeat for the quarter-inch mark.

To the nearest half inch, the push pin is 1 inch long. To the nearest quarter inch, it is $1\frac{1}{4}$ inches long.

Practice

Measure the length of each to the nearest half inch and quarter inch.

1.

2.

______________________ ______________________

Cut out and use this inch ruler.

in. 0 1 2 3

Find the length of each object to the nearest half inch and quarter inch.

3. width of an MP3 player

____

4. length of a stapler

____

Draw a line segment with each of the following lengths.

5. $2\frac{1}{4}$ inches

6. $\frac{3}{4}$ inch

Vocabulary Check

7. Choose the correct word(s) to complete the sentence below.

Length is the measurement of ______________ between two points.

Problem Solving

8. The height of Josie's dog is $11\frac{1}{4}$ inches to the nearest quarter inch and 11 inches to the nearest half inch. Which measurement is more precise?

____

9. Allison has a ruler that is marked in fourths of an inch and a tape measure that is marked in halves of an inch. Which measuring tool will give Allison a more precise measure?

____

10. Processes & Practices 6 **Be Precise** Gretchen measured the length of her toaster to the nearest half inch to be 9 inches. Steve measured the toaster to the nearest quarter inch and found that it measured $9\frac{1}{4}$ inches. Who used a more precise measurement?

____

Name ....

Lesson 2
Convert Customary Units of Length

ESSENTIAL QUESTION
How can I use measurement conversions to solve real-world problems?

Math in My World

AAHHH!

Example 1

For many of the thrill rides at amusement parks, riders must be at least 48 inches tall. Cheng is 4 feet tall. Is he tall enough to ride a thrill ride at an amusement park?

Convert 4 feet to inches.

Since 1 foot = 12 inches, multiply 4 by 12.

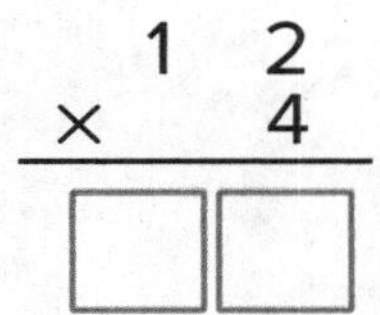

So, 4 feet = ________ inches.

Is Cheng tall enough to ride the thrill rides? ________

Check Use division to check your answer. ________ ÷ 4 = 12

The units of length most often used in the United States are the inch, foot, yard, and mile. These units are part of the **customary system.**

Key Concept Customary Units of Length

1 **foot** (ft) = 12 **inches** (in.)
1 **yard** (yd) = 3 ft or 36 in.
1 **mile** (mi) = 5,280 ft or 1,760 yd

When you **convert** measurements, you change from one unit to another. To convert a larger unit to a smaller unit, multiply. To convert a smaller unit to a larger unit, divide.

Example 2 

Convert 42 inches to feet.

Since you are changing a smaller unit to a larger unit, divide.

$$\begin{array}{r} \square \text{ R } \square \\ 12\overline{)\,4\quad 2} \\ -\,\square\,\square \\ \hline \square \end{array}$$

Interpret the remainder.

One Way **Write the remainder in inches.**

The remainder ________ means there are ________ inches left over.

42 inches = ________ feet ________ inches

Another Way **Write the remainder as a fraction or decimal.**

The fraction of a foot is ________ or ________.

42 inches = ________ feet

So, 42 inches is equal to ________ feet ________ inches, or ________ feet, or 3.5 feet.

Guided Practice

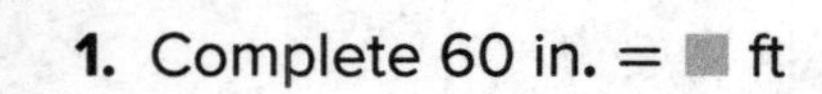

1. Complete 60 in. = ■ ft

   60 ÷ 12 = ________

   So, 60 inches equals ________ feet.

2. Complete 16 yd = ■ ft

   16 × 3 = ________

   So, 16 yards equals ________ feet.

Explain how to convert units from feet to inches.

Name ______

Independent Practice

Complete.

3. 72 in. = ______ ft

4. 19 yd = ______ in.

5. 40 in. = ______ ft

6. 7,920 ft = ______ mi

7. 6 yd = ______ in.

8. 26,400 ft = ______ mi

9. 9 ft = ______ in.

10. 22 ft = ____ yd ____ ft

11. 51 in. = ____ ft ____ in.

Compare. Use <, >, or = to make a true statement.

12. 24 in. ◯ 2 ft 3 in.

13. 7 yd ◯ 20 ft

14. 2 mi ◯ 3,500 yd

15. $11\frac{1}{2}$ ft ◯ 4 yd

16. 72 in. ◯ 2 yd

17. 5.5 yd ◯ 192 in.

Problem Solving

18. Processes & Practices 3 **Draw a Conclusion** Dana ran $\frac{1}{4}$ mile. Trish ran 445 yards. Who ran the greater distance? Explain.

19. Carlos is 63 inches tall. What is his height in feet? Use a fraction.

What is his height in feet and inches?

20. The U.S.S. Harry Truman is an aircraft carrier that is 1,092 feet long. Find the length in yards.

Brain Builders

21. Order the following distances from least to greatest: 2 miles, 4,800 feet, 4,400 yards. Explain.

22. Processes & Practices 2 **Use Number Sense** There are 320 rods in a mile. Find the length of a rod in feet. Explain.

23. **Building on the Essential Question** Explain, using an example, why you need to multiply when converting from a larger unit to a smaller unit.

Name ________________________________

Lesson 2

Convert Customary Units of Length

Homework Helper

Need help? connectED.mcgraw-hill.com

The average height of a female giraffe is 15 feet. What is the average height in yards?

Convert 15 feet to yards.

Since 3 feet = 1 yard, divide 15 by 3.

$$\begin{array}{r} 5 \\ 3\overline{)\,15} \\ -\,15 \\ \hline 0 \end{array}$$

So, 15 feet = 5 yards.

The average height of the female giraffe is 5 yards.

Check Use multiplication to check your answer.
$5 \times 3 = 15$

Practice

Complete.

1. 5 mi = ________ yd

2. 7 yd = ________ in.

3. 150 in. = ________ yd

4. 110 in. = ______ ft ______ in.

5. 8 yd = ________ in.

6. 13,200 ft = ________ mi

Vocabulary Check 

Fill in the correct circle that corresponds to the best answer.

7. Which of the following is not a common unit of measurement for the customary system?

Ⓐ feet Ⓒ meter

Ⓑ inches Ⓓ miles

8. Which operation is necessary to convert a smaller unit to a larger unit?

Ⓕ addition Ⓗ multiplication

Ⓖ subtraction Ⓘ division

Brain Builders

9. Processes & Practices 6 **Be Precise** Ty has the two pieces of wood shown in the table. What is the combined length of the pieces, in inches?

Pieces	Length
1	1 yd 9 in.
2	44 in.

____

10. Lily built an orange shelf and a brown shelf to hang in her room. The orange shelf is 5 feet 8 inches long. The brown shelf is twice as long. What is the length of the brown shelf?

____

11. **Test Practice** The table shows the name of each volunteer and the height of the baby elephant each volunteer measured. Who measured the tallest elephant?

Elephant Heights	
Volunteer	**Height of Elephant**
George	2 yd
Kelsey	68 in.
Mariah	5 ft
Bryan	5 ft 10 in.

Ⓐ George Ⓒ Mariah

Ⓑ Kelsey Ⓓ Bryan

Name ....

Lesson 3
Problem-Solving Investigation
STRATEGY: Use Logical Reasoning

ESSENTIAL QUESTION
How can I use measurement conversions to solve real-world problems?

Learn the Strategy

Three friends each measured their height. Their heights are 4 feet 10 inches, 4 feet 9 inches, and 4 feet 7 inches. Use the clues to determine the height, in inches, of each person.

- **Elliot is taller than Jorge.**
- **Nicole is 3 inches taller than the shortest person.**
- **Elliot is 57 inches tall.**

6 ft 6 ft 6 ft
5 ft 5 ft 5 ft

1 Understand

What facts do you know?

the clues that are listed above

What do you need to find?

the ________ of each person

2 Plan

I can use logical reasoning to find the height of each person.

3 Solve

Convert the measurements to inches to compare.

$4 \times 12 + 10 = 48 + 10 =$ ________ inches   Nicole is ________ inches tall.

$4 \times 12 + 9 = 48 + 9 =$ ________ inches   Elliot is ________ inches tall.

$4 \times 12 + 7 = 48 + 7 =$ ________ inches   Jorge is ________ inches tall.

4 Check

Is my answer reasonable?

Since all of the answers match the clues, the solution is reasonable.

Practice the Strategy

Three dogs are sitting in a line. Rocky is not last. Coco is in front of the tallest dog. Marley is sitting directly behind Rocky. List the dogs in order from first to last.

1 Understand

What facts do you know?

What do you need to find?

2 Plan

3 Solve

4 Check

Is my answer reasonable?

Name

Apply the Strategy

Solve each problem by using logical reasoning.

1. An after-school club is building a clubhouse that has a rectangular floor that is 8 feet by 6 feet. What is the area, in square inches, of the floor of the clubhouse?

2. There is a red, a green, and a yellow bulletin board hanging in the hallway. All of the bulletin boards are rectangular with a height of 4 feet. Their lengths are 6 feet, 5 feet, and 3 feet. The red bulletin board has the largest area and the yellow one has the smallest area. What is the area of the green bulletin board?

3. Processes &Practices 8 **Look for a Pattern** If the pattern below continues, how many pennies will be in the fourth figure?

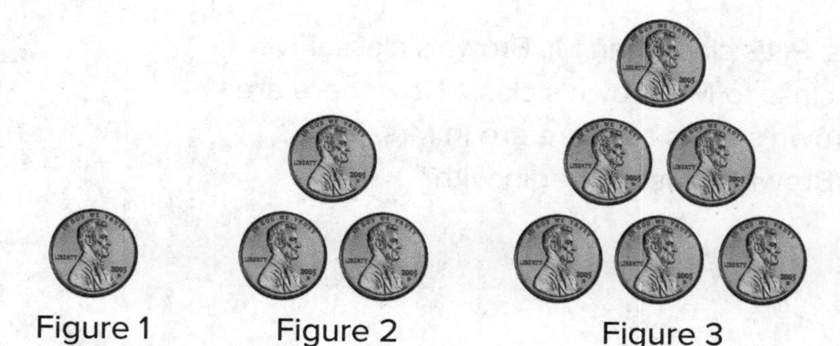

Figure 1 Figure 2 Figure 3

Brain Builders

4. Hiking trail A is 200 yards longer than trail B. Trail C is 900 feet shorter than trail B. If trail A is 1 mile long, how many yards long is trail C?

5. Ethan has $1.25 in dimes, nickels, and pennies. He has twice as many dimes as pennies, and the number of nickels is one less than the number of pennies. How many dimes, nickels, and pennies does he have?

Review the Strategies

Use any strategy to solve each problem.

- Use logical reasoning.
- Draw a diagram.
- Look for a pattern.
- Solve a simpler problem.

6. Madeline has 2 times the number of games as Paulo. Paulo has 4 more games than Tyler. If Tyler has 9 games, how many games are there among the 3 friends?

7. When Cheryl goes mountain climbing, she rests 5 minutes for every 15 minutes that she climbs. If Cheryl's combined time is 2 hours, how many minutes does she rest?

8. There are 8 girls for every 7 boys on a field trip. If there are 56 girls on the trip, how many students are on the trip?

9. There are 4 more girls in Mrs. Pitt's class than Mr. Brown's class. Five girls moved from Mrs. Pitt's class to Mr. Brown's class. Now there are twice as many girls in Mr. Brown's class as there are in Mrs. Pitt's. How many girls were in Mr. Brown's class to begin with?

10. A storage room measures 48 inches by 60 inches. What is the total area, in square feet, of the closet?

11. **Processes & Practices** 1 **Make Sense of Problems** Five friends go to a batting cage. Andrea bats after Daniel and before Jessica. Juwan bats after Andrea and before Jessica and Filipe. Jessica always bats immediately after Juwan. Who bats last?

Name

MY Homework

Lesson 3

Problem Solving: Use Logical Reasoning

Homework Helper

Need help? connectED.mcgraw-hill.com

A family has three guinea pigs. Tiger is 8 years old, which is 2 years older than Max. Max is 3 years older than Patches. List the guinea pigs in order from oldest to youngest.

1 Understand

What facts do you know?

Tiger is 8 years old.

Tiger is 2 years older than Max, and Max is 3 years older than Patches.

What do you need to find?

the order of the guinea pigs from oldest to youngest

2 Plan

Use logical reasoning to find the age of each guinea pig.

Make a table to help organize the information.

3 Solve

Place an "X" in each box that cannot be true.

You know that Tiger is 8 years old.

Subtract 2 from Tiger's age to find Max's age. Max is 6 years old.

Subtract 3 from Max's age to find Patches' age. Patches is 3 years old.

	oldest	2nd oldest	youngest
Tiger	yes	X	X
Max	X	yes	X
Patches	X	X	yes

So, Tiger is the oldest, Max is the next oldest, and the youngest is Patches.

4 Check

Is my answer reasonable?

Since all of the answers match the clues, the solution is reasonable.

Problem Solving

Solve each problem by using logical reasoning.

1. **Processes & Practices** 5 **Use Math Tools** Mr. Toshi's fifth grade class sold containers of popcorn and peanuts. If each day they sold 25 fewer containers of peanuts than popcorn, how many containers of popcorn and peanuts did they sell in all? Complete the table and solve.

	Day 1	Day 2	Day 3	Day 4
Popcorn	225	200	150	300
Peanuts				

2. A shelter house has a floor area of 400 square feet. If three identical shelter houses are built, what is the combined floor area of the shelter houses?

Brain Builders

3. Meghan earns money babysitting. She charges \$5 an hour before 8 P.M. and \$8 an hour after 8 P.M. She earned \$26 for her last babysitting job. Between what hours did Meghan babysit?

4. Alana is 4 years older than her brother Ernie. Ernie is 2 years older than their sister Amelia. Amelia is 10 years younger than their brother Mazo. If Mazo is 17 years old, how old is Alana?

5. A rectangular garden measures 15 feet by 30 feet. What is the total area, in square yards, of the garden? Explain.

Name

Lesson 4 Hands On
Estimate and Measure Weight

ESSENTIAL QUESTION
How can I use measurement conversions to solve real-world problems?

The **weight** of an object is a measure of how heavy it is. **Ounces** and **pounds** are examples of units of measurement for weight.

Measure It

Object	Estimate	Actual
Eraser		
Glue bottle		
Math book		
Pencil		

1. Estimate the weight of each object in ounces or pounds. Record your results in the table.

2. Measure the weight of each object. Place the eraser on one side of a balance. Set ounce or pound weights on the other side until the sides are level. Record the actual weight. Repeat this step for the other objects.

Try It

1. Place a 1-pound weight on one side of a balance.

2. Place 1-ounce weights on the opposite side of the balance. Continue placing 1-ounce weights until the balance is equal. Use your results to complete the table.

Pounds (lb)	Ounces (oz)
1	
2	
3	
4	

So, a 1-pound weight is equal to ______ ounces.

Talk About It

1. Order the four objects you weighed in the activity from greatest to least weight.

____

2. Processes & Practices 6 **Explain to a Friend** Is the total weight of the four objects you measured in the first activity greater than 2 pounds? Explain.

____

____

____

3. Write a sentence that describes the relationship that is usually found between an object's size and weight.

____

____

____

Name ....

Practice It

4. Identify four objects in your classroom that you can use a balance to find their weights. Estimate each object's weight. Then weigh each object and record the actual weight in the table below.

Object	Estimate	Actual

Compare. Use >, <, or = to make a true statement.

5. 46 ounces ◯ 3 pounds

6. 96 ounces ◯ 6 pounds

7. 130 ounces ◯ 8 pounds

8. 113 ounces ◯ 7 pounds

9. 5 pounds ◯ 78 ounces

10. 9 pounds ◯ 145 ounces

11. 10 pounds ◯ 160 ounces

12. 12 pounds ◯ 196 ounces

Apply It

13. Tim and Cameron measured the weight of the same calculator. Tim measured the calculator as 1 pound. Cameron measured the calculator as 13 ounces. Circle the greater measurement.

1 pound　　13 ounces

14. Cara measured a dictionary twice. Her first measurement was 1 pound. The second measurement was 28 ounces. Circle the greater measurement.

1 pound　　28 ounces

15. If you are measuring a bag of sugar, would ounces or pounds give you a more precise measurement? Explain.

____

____

____

16. Processes & Practices 6 **Be Precise** Jeffrey measured the weight of his kitten. His first measurement was 34 ounces. His second measurement was 2 pounds. Use >, <, or = to make a true statement.

34 ounces ◯ 2 pounds

17. Processes & Practices 3 **Draw a Conclusion** Compare and contrast pounds and ounces.

____

____

Write About It

18. Why is estimation of weight important in everyday life?

____

____

____

Name

MY Homework

Lesson 4

Hands On: Estimate and Measure Weight

Homework Helper

Need help? connectED.mcgraw-hill.com

Sixteen 1-ounce weights are equal to 1 pound. Use the table below to determine a pattern. Then find the weight, in ounces, that is equal to 4 pounds.

Pounds (lb)	Ounces (oz)
1	16
2	32
3	48
4	64

+16
+16
+16

So, a 4-pound weight is equal to 48 + 16, or 64 ounces.

Practice

Compare. Use >, <, or = to make a true statement.

1. 30 ounces ◯ 2 pounds

2. 98 ounces ◯ 6 pounds

3. 11 pounds ◯ 176 ounces

4. 7 pounds ◯ 109 ounces

Vocabulary Check

5. Fill in the blank with the correct word to complete the sentence.

Weight is a measure of how ________ an object is.

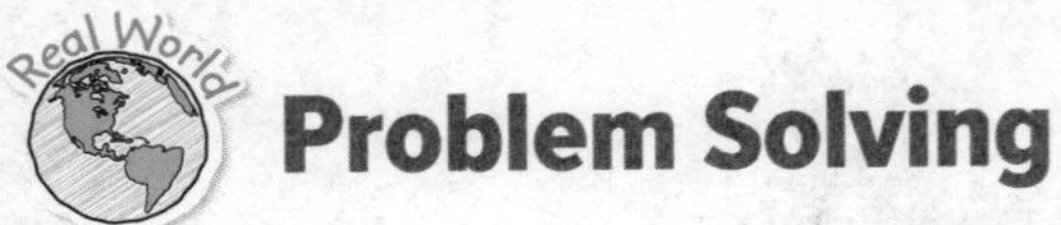

Problem Solving

6. Kurt measured the weight of his ball python snake. His first measurement was 48 ounces. His second measurement was 3 pounds. Compare the two measurements. Use >, <, or = to make a true statement.

____

7. Lori and Mindy measured the weight of their book bags. Lori measured her bag using pounds. Mindy measured her bag using ounces. Circle the measure that is more appropriate to measure the weight of the book bag.

pounds　　　　ounces

8. Tracy measured the weight of her hamster. Her first measurement was 13 ounces. Her second measurement was 1 pound. Which measurement is more precise?

____

9. **Processes & Practices 6** **Be Precise** Quinton measured the weight of his laptop. His first measurement was 94 ounces. His second measurement was 6 pounds. Compare the two measurements. Use >, <, or = to make a true statement.

____

10. If you are selling bags of candy, would you want to weigh the candy in ounces or pounds? Explain.

____

____

____

Name ....

Lesson 5
Convert Customary Units of Weight

ESSENTIAL QUESTION
How can I use measurement conversions to solve real-world problems?

Weight is a measure of how heavy an object is. Customary units of weight are ounce, pound, and ton.

Math in My World

Example 1

A newborn lion cub weighs about 5 pounds. How many ounces does the cub weigh?

Convert 5 pounds to ounces.

$$\begin{array}{r} 16 \\ \times \ 5 \\ \hline \square\square \end{array}$$

Since 1 pound = 16 ounces, multiply 5 by 16.

So, 5 pounds = ______ ounces.

A newborn lion cub weighs about ______ ounces.

Check Use division to check your answer.

______ ÷ 5 = 16

Key Concept Customary Units of Weight

1 **pound** (lb) = 16 **ounces** (oz)   1 **ton** (T) = 2,000 pounds

1 ounce

1 pound

1 ton

As with units of length, units of weight in the customary system can also be expressed using different units or as fractions and decimals.

Example 2

Convert 56 ounces to pounds.

Since you are changing a smaller unit to a larger unit, divide.

$$\begin{array}{r} \square \text{ R } \square \\ 16 \overline{)\,5 \quad 6} \\ -\,\square\,\square \\ \hline \square \end{array}$$

The remainder _______ means there are _______ ounces left over.

The fraction of a pound is $\frac{8}{16}$ or $\frac{1}{2}$.

So, 56 ounces = _______ pounds _______ ounces, $3\frac{1}{2}$ pounds, or 3.5 pounds.

Guided Practice

Complete.

1. 3 lb = ■ oz

3 × 16 = _______

So, 3 pounds equals _______ ounces.

Explain how to compare 22 ounces to 2 pounds.

2. $2\frac{1}{2}$ lb = ■ oz

$2\frac{1}{2} \times 16 =$ _______

So, $2\frac{1}{2}$ pounds equals _______ ounces.

3. 32 oz = ■ lb

32 ÷ 16 = _______

So, 32 ounces equals _______ pounds.

Name

Independent Practice

Complete.

4. 96 oz = ________ lb

5. 7 T = ________ lb

6. 1.5 T = ________ lb

7. 11,000 lb = ________ T

8. 50 oz = ________ lb ________ oz

9. $1\frac{1}{4}$ lb = ________ oz

10. 7,000 lb = ________ T

11. $\frac{3}{4}$ T = ________ lb

Compare. Use <, >, or = to make a true statement.

12. 7,500 lb ◯ 4 T

13. $\frac{1}{2}$ T ◯ 1,000 lb

14. 1,200 oz ◯ 72 lb

15. 6 lb 11 oz ◯ 117 oz

Problem Solving

16. Mia combines the items in the table to make potting soil. Order the items according to weight, from least to greatest.

Item	Amount
Topsoil	3 lb
Fertilizer	2 lb 9 oz
Bone meal	43 oz

17. How many 4-ounce bags of peanuts can be filled from a 5-pound bag of peanuts?

18. Alfonso mails a package that weighs 9 pounds. How many ounces is this package?

Brain Builders

19. Processes & Practices 2 **Use Number Sense** One puppy weighs 12 ounces. Another puppy weighs $\frac{5}{8}$ pound. Which puppy weighs more? Explain.

20. Processes & Practices 2 **Use Number Sense** A baby weighs 8 pounds 10 ounces. If her weight doubles in 6 months, will she weigh more or less than 17 pounds? Explain.

21. **Building on the Essential Question** How can I apply converting customary units of weight in everyday life? Provide an example and show the unit conversion.

Name ______________________________

Lesson 5

Convert Customary Units of Weight

Homework Helper

Need help? connectED.mcgraw-hill.com

A white rhinoceros weighs about $2\frac{1}{2}$ tons. How many pounds does the white rhinoceros weigh?

Convert $2\frac{1}{2}$ tons to pounds.

You know that 2 tons = 4,000 pounds and $\frac{1}{2}$ ton = 1,000 pounds.

So, $2\frac{1}{2}$ tons = 4,000 + 1,000 or 5,000 pounds.

A white rhinoceros weighs about 5,000 pounds.

Practice

Complete.

1. 64 oz = __________ lb

2. 5 T = __________ lb

3. 13 lb = __________ oz

4. 9,000 lb = __________ T

5. 45 oz = __________ lb __________ oz

6. 3.25 T = __________ lb

Problem Solving

7. A hippopotamus weighs 1.5 tons. What is its weight in ounces?

8. A restaurant serves a 20-ounce steak. How much does the steak weigh in pounds and ounces?

Brain Builders

9. Processes & Practices 2 **Use Number Sense** What fractional part of a pound is an ounce?

What fractional part of a ton is a pound?

Vocabulary Check

Circle the correct term to complete each sentence.

10. (Ounces, Pounds) would be the most appropriate customary unit to measure the weight of a paper clip.

11. (Weight, Length) is a measure of how heavy an object is.

12. (Tons, Pounds) would be the most appropriate customary unit to measure the weight of a bag of flour.

13. **Test Practice** One kitten weighs 2 pounds 4 ounces. Another kitten weighs 2 ounces less. What is the combined weight of the two kittens in ounces?

1 pound = 16 ounces

Ⓐ 70 oz
Ⓑ 36 oz
Ⓒ 34 oz
Ⓓ 16 oz

Check My Progress

Vocabulary Check

Choose the correct word that completes each sentence.

foot	inches	length	mile
pound	ounce	ton	yard

1. The __________ is an appropriate unit to measure the distance run over a one-week period.

2. The __________ is an appropriate unit to measure the weight of a bag of sugar.

3. The __________ is an appropriate unit to measure the distance run on a football field.

Concept Check

Draw a line segment with each of the following lengths.

4. $2\frac{3}{4}$ inches

5. $\frac{1}{2}$ inch

Complete.

6. 72 in. = __________ ft

7. 11,880 ft = __________ mi

8. 3.75 yd = __________ in.

9. 17 ft = ________ yd ________ ft

Complete.

10. 124 oz = ______________ lb

11. 8 T = ______________ lb

12. 7 T 500 lb = ______________ lb

13. $1\frac{1}{2}$ lb = ______________ oz

Compare. Use >, <, or = to make a true statement.

14. 74 ounces ◯ 4 pounds

15. 5 pounds ◯ 81 ounces

Problem Solving

16. An African lion weighs 450 pounds. What is its weight in ounces?

____

17. Lin and Pazi each think of a number. Lin's number is 7 more than Pazi's number. The sum of the two numbers is 49. What is Pazi's number?

____

Brain Builders

18. Janna jumped 156 inches in the long jump competition at the high school track meet. Her personal record is 13 feet 1 inch. Did Janna beat her personal record at the track meet? Explain.

____

____

19. **Test Practice** How many 8-ounce bags of trail mix can be filled from an 8-pound bag of trail mix?

Ⓐ 1 bag

Ⓑ 2 bags

Ⓒ 8 bags

Ⓓ 16 bags

Name

Lesson 6 Hands On
Estimate and Measure Capacity

ESSENTIAL QUESTION
How can I use measurement conversions to solve real-world problems?

Capacity is the measure of how much a container can hold. **Cups**, **pints**, and **gallons** are examples of units of measurement for capacity.

Measure It

Container	Estimate	Actual
Pint	______ cups	______ cups
Quart	______ pints	______ pints

1. Estimate the capacity of a pint container. Record your estimate in the table.

2. Fill one cup with water and pour its contents into the pint container. Repeat until the pint container is full.

   How many cups does it take to fill the pint container? ______

   So, there are ______ cups in a pint.

   Record the measure in the table.

3. Estimate the capacity of a quart container. Record your estimate in the table.

4. Fill one pint with water and pour its contents into the quart container. Repeat until the quart container is full.

   How many pints does it take to fill the quart container? ______

   So, there are ______ pints in a quart.

   Record the measure in the table.

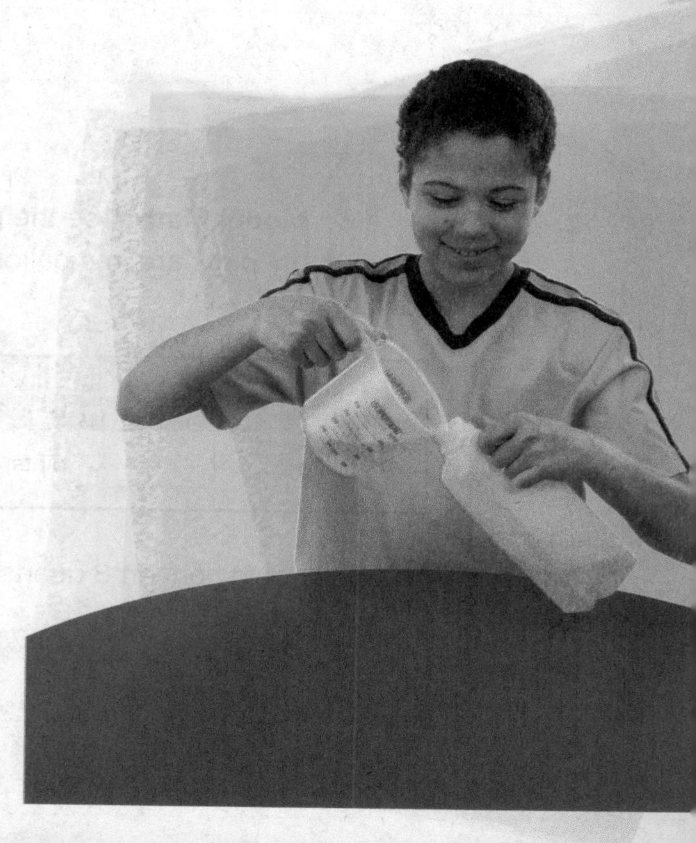

Try It

1 Estimate the capacity of a gallon container. Record your estimate in the table.

Container	Number of Quarts	
	Estimate	Actual
Gallon	_____ quarts	_____ quarts

2 Fill one quart container with water and pour its contents into the gallon container. Repeat until the gallon container is full.

How many quarts does it take to fill the gallon container?

_____

So, there are _____ quarts in a gallon.
Record the measure in the table.

Talk About It

1. Is the capacity of the gallon container less than or greater than the capacity of the pint container? Explain.

_____

_____

_____

2. **Processes & Practices** 4 **Model Math** Use the pint container to estimate and measure how many pints are in a gallon. Record your results in the table below.

Container	Number of Pints	
	Estimate	Actual
Gallon	_____ pints	_____ pints

3. Explain how you could convert 8 quarts to gallons without measuring.

_____

_____

_____

Name ....

Practice It

Complete using your findings from the activities.

4. 3 pints = ________ cups

5. 4 gallons = ________ pints

6. 12 cups = ________ pints

7. 3 gallons = ________ quarts

8. 24 quarts = ________ gallons

9. 20 cups = ________ quarts

10. 3 quarts = ________ cups

11. 2 gallons = ________ pints

Compare. Use >, <, or = to make a true statement.

12. 8 pints ◯ 15 cups

13. 12 cups ◯ 8 pints

14. 4 gallons ◯ 30 pints

15. 2 quarts ◯ 8 cups

Apply It

16. Seth made 4 quarts of fruit punch for a birthday party. How many cups of fruit punch did he make?

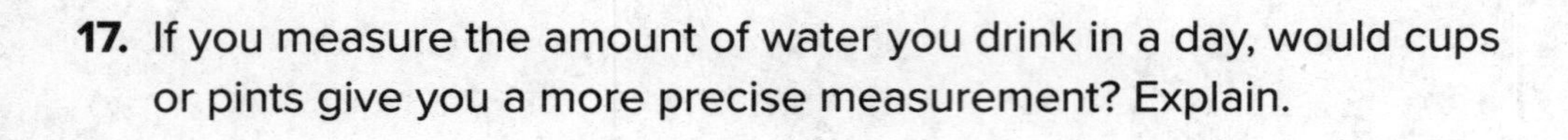

17. If you measure the amount of water you drink in a day, would cups or pints give you a more precise measurement? Explain.

18. **Processes & Practices 6** **Be Precise** Peggy measured the capacity of her aquarium. Her first measurement was 9 gallons. Her second measurement was 35 quarts. Compare the two measurements. Use >, <, or = to make a true statement.

19. **Processes & Practices 3** **Justify Conclusions** Is it faster to water two large flower pots using a one-cup pitcher or a one-quart pitcher? Explain.

Write About It

20. How can I use measuring tools to find capacity?

Name ....

MY Homework

Lesson 6

Hands On: Estimate and Measure Capacity

Homework Helper

Need help? connectED.mcgraw-hill.com

The table shows the measures you found in the activities.

Customary Units of Capacity
1 pint = 2 cups
1 quart = 2 pints
1 gallon = 4 quarts

How many quarts equal 10 pints?

Since 2 pints = 1 quart, divide by 10 by 2.

$10 \div 2 = 5$

So, 5 quarts is equal to 10 pints.

Practice

Complete.

1. 4 cups = ________ pints

2. 16 quarts = ________ gallons

3. 8 quarts = ________ pints

4. 7 pints = ________ cups

5. 5 gallons = ________ pints

6. 28 cups = ________ quarts

Vocabulary Check

7. Fill in the blank with the correct word to complete the sentence below.

________ is the measure of how much a container can hold.

Problem Solving

8. Processes & Practices 2 **Reason** Sophie drank 8 cups of water one day. How many pints of water did she drink?

9. A container holds 26 pints of water. Circle whether 26 pints is *greater than, less than,* or *equal to* 12 quarts.

greater than   less than   equal to

10. Coach Cole measured the capacity of his water cooler. His first measurement was 6 gallons. His second measurement was 25 quarts. Compare the two measurements. Use >, <, or = to make a true statement.

11. If you are measuring the amount of water evaporated from a swimming pool, would gallons or quarts give you a more precise measurement? Explain.

12. Jason is a zookeeper and needs to fill a large tub of water for a zebra. Would it be faster to use a one-quart container or a one-gallon container? Explain.

Name

Lesson 7
Convert Customary Units of Capacity

ESSENTIAL QUESTION
How can I use measurement conversions to solve real-world problems?

Capacity is the measure of how much a container can hold. Just like length and weight, the smaller the unit you use to measure capacity, the more precise the measurement.

Math in My World

Example 1

Sondra tries to drink 9 cups of water each day. How many fluid ounces of water is 9 cups?

Convert 9 cups to fluid ounces.

Since 1 cup = 8 fluid ounces, multiply 9 by 8.

$$\begin{array}{r} 9 \\ \times\ 8 \\ \hline \square\square \end{array}$$

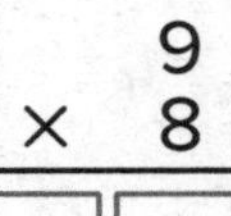

So, 9 cups = ______ fluid ounces.

Key Concept Customary Units of Capacity

Helpful Hint
A fluid ounce is different from the ounce used to measure weight.

1 **cup** (c) = 8 **fluid ounces** (fl oz)
1 **pint** (pt) = 2 c = 16 fl oz
1 **quart** (qt) = 2 pt = 32 fl oz
1 **gallon** (gal) = 4 qt = 128 fl oz

1 fl oz

1 cup

1 pint

1 quart

1 gallon

As with units of length and units of weight, measurements of capacity in the customary system can also be expressed as fractions.

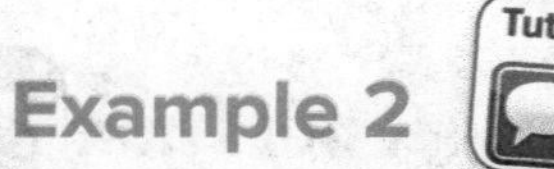

Example 2

How many quarts can be made from 7 pints?

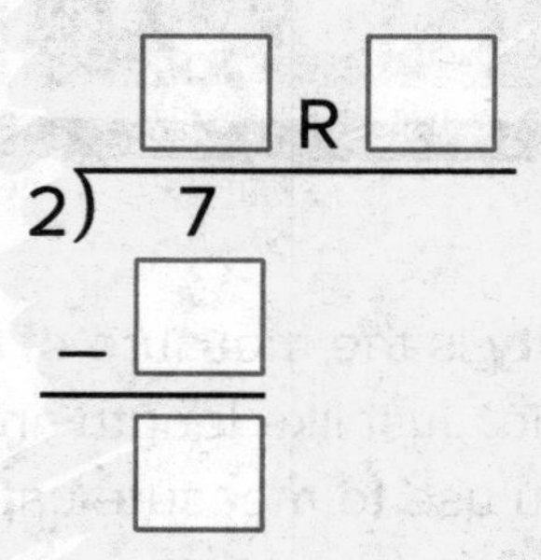

Since 2 pints = ________ quart, divide 7 by ________.

The remainder ________ means there is ________ pint left over.

The fraction of a quart is ________.

So, 7 pints = ________ quarts ________ pint, $3\frac{1}{2}$ quarts, or 3.5 quarts.

Guided Practice

Complete.

1. 3 c = ■ fl oz

3 × 8 = ______

So, 3 cups equals ______ fluid ounces.

2. 4 qt = ■ c

4 × 2 = ______ pt

8 × 2 = ______ c

So, 4 quarts equals ______ cups.

3. 18 pt = ■ qt

18 ÷ 2 = ______

So, 18 pints equals ______ quarts.

Explain how to compare 18 fluid ounces to 2 pints.

Name ______________________________

Independent Practice

Complete.

4. 5 c = __________ fl oz

5. 19 qt = __________ gal __________ qt

6. 50 c = __________ pt

7. 22 fl oz = __________ c __________ fl oz

8. 2 gal = __________ fl oz

9. 7 c = __________ pt

10. 19 c = __________ fl oz

11. 25 gal = __________ qt

12. 68 pt = __________ gal

13. 56 qt = __________ gal

Compare. Use <, >, or = to make a true statement.

14. 4 c ◯ 40 fl oz

15. 1.75 gal ◯ 10 pt

16. $2\frac{3}{4}$ qt ◯ $5\frac{1}{2}$ pt

17. 3 gal ◯ 13 qt

18. 5 qt ◯ 10 pt

19. 300 fl oz ◯ 2 gal

Problem Solving

20. Processes & Practices 6 **Explain to a Friend** A bucket contains $1\frac{1}{2}$ gallons of water. Is the amount of water in the bucket *greater than, less than,* or *equal to* 6 quarts? Explain.

____

21. Zach had 1 quart of milk. He used 1 pint to make pancakes and 1 cup to make scrambled eggs. How many cups of milk were left?

____

Brain Builders

22. A cow can drink up to 35 gallons of water in a day. How many 8-ounce glasses could a cow drink in a day?

____

23. Processes & Practices 6 **Be Precise** Trevor needs to use the most precise measurement when measuring orange juice to add to a mix. Which unit of measure should he use? Explain.

____

____

24. **Building on the Essential Question** How is converting from smaller to larger units of capacity different from converting larger to smaller units of capacity?

____

____

____

____

Name

Lesson 7

Convert Customary Units of Capacity

Homework Helper

Need help? connectED.mcgraw-hill.com

Paco has 3 pints of juice plus 1 cup of juice. How many cups of juice does he have in all?

Convert 3 pints to cups.

Since 1 pint = 2 cups, multiply 3 by 2.

$$\begin{array}{r} 3 \\ \times\ 2 \\ \hline 6 \end{array}$$

So, 3 pints = 6 cups.

Then add the remaining 1 cup. $6 + 1 = 7$

So, Paco has 7 cups of juice.

Practice

Complete.

1. 16 c = ________ pt

2. 64 fl oz = ________ c

3. 5 qt = ________ gal ________ qt

4. 18 qt = ________ gal

5. 7 c = ________ pt

6. 16 pt 1 c = ________ c

Problem Solving

7. The table shows the amount of paint left in each jar. Which jar contains the greatest amount of paint? The least?

Jar	Amount
blue	2 pt 4 fl oz
purple	5 cups
green	39 fl oz

Brain Builders

8. Processes & Practices  **Model Math** Write about a real-world situation that can be solved by converting customary units of capacity. Then solve.

Vocabulary Check

Circle the correct term to complete each sentence.

9. (Cups, Gallons) would be the most appropriate customary unit to measure the capacity of a hot chocolate mug.

10. (Gallons, Pints) would be the most appropriate customary unit to measure the capacity of a swimming pool.

11. (Weight, Capacity) is a measure of how much a container can hold.

12. **Test Practice** The average person drinks 16 ounces of milk a day. At this rate, how many gallons will a person drink in a leap year (366 days)?

Ⓐ 45 gallons

Ⓑ $45\frac{1}{2}$ gallons

Ⓒ $45\frac{3}{4}$ gallons

Ⓓ 46 gallons

Name ________________________________

Lesson 8
Display Measurement Data on a Line Plot

ESSENTIAL QUESTION
How can I use measurement conversions to solve real-world problems?

Math in My World

Example 1

Six friends shared several foot-long submarine sandwiches. The table shows the amount each friend ate. Make a line plot of the lengths in the table.

Sandwich Lengths (ft)		
$\frac{1}{3}$	$\frac{1}{3}$	$\frac{1}{3}$
$\frac{1}{2}$	$\frac{1}{4}$	$\frac{1}{4}$

1. Count the number of times each fraction appears in the table.

$\frac{1}{4}$ appears ________ times.

$\frac{1}{3}$ appears ________ times.

$\frac{1}{2}$ appears ________ time.

2. Place the correct number of Xs above each fraction on the number line.

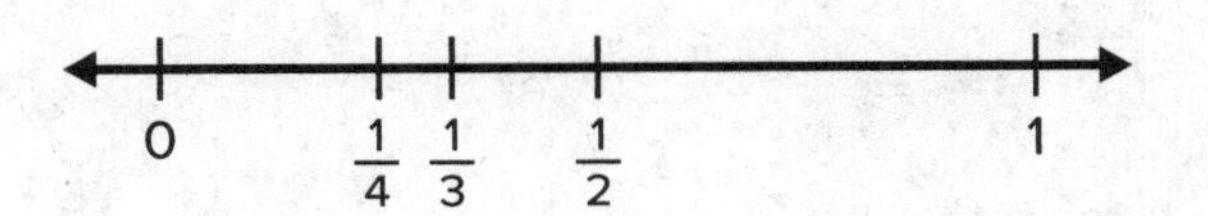

3. Add a title to the line plot.

You can find the **fair share**, or the amount each friend would receive if the sandwiches were divided equally. First add the measurements to find the whole. Then divide the whole by the number of measurements.

Example 2

Find the fair share using the line plot from Example 1.

Add the fractions to find the total amount of sandwiches eaten. Add the fractions with like denominators first.

2 Xs above $\frac{1}{4}$: $\frac{1}{4} + \frac{1}{4} = \frac{2}{4}$ or $\frac{1}{2}$

3 Xs above $\frac{1}{3}$: $\frac{1}{3} + \frac{1}{3} + \frac{1}{3} = \frac{3}{3}$ or 1

1 X above $\frac{1}{2}$: $\frac{1}{2}$

So, $\frac{1}{2} + 1 + \frac{1}{2}$ or ______ whole sandwiches were eaten.

Divide the total amount by the number of Xs on the line plot. To find 2 ÷ 6, you can draw a model.

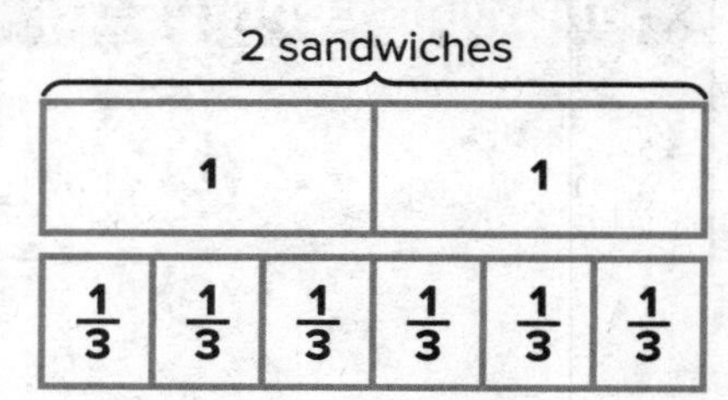

Draw 2 rectangles to show 2 whole sandwiches.

Divide the entire amount into 6 equal pieces.

Each piece represents ______ of a sandwich. So, if the sandwiches were divided equally, each person would have eaten ______ of a sandwich.

Guided Practice

1. Make a line plot of the measurements in the table. Then find the fair share.

Amount of Juice (gal)							
$\frac{1}{4}$	$\frac{1}{4}$	$\frac{1}{4}$	$\frac{1}{2}$	$\frac{1}{2}$	$\frac{1}{2}$	$\frac{1}{4}$	$\frac{1}{2}$

fair share: ______

Talk MATH

Describe a situation in everyday life in which you would want to find a fair share.

Name ....

Independent Practice

Make a line plot of the measurements in each table. Then find the fair share.

2.

Yarn Lengths (ft)					
$\frac{1}{4}$	$\frac{1}{2}$	$\frac{1}{3}$	$\frac{1}{3}$	$\frac{1}{4}$	$\frac{1}{3}$
$\frac{1}{4}$	$\frac{3}{4}$	$\frac{1}{2}$	$\frac{1}{2}$	$\frac{3}{4}$	$\frac{1}{4}$

fair share: ____________

3.

Iced Tea (qt)								
$\frac{1}{2}$	$\frac{1}{4}$	$\frac{1}{8}$	$\frac{1}{4}$	$\frac{1}{2}$	$\frac{1}{4}$	$\frac{1}{2}$	$\frac{1}{8}$	$\frac{1}{2}$

fair share: ____________

4.

Amount of Sliced Turkey (lb)							
$\frac{1}{8}$	$\frac{1}{4}$	$\frac{1}{8}$	$\frac{1}{4}$	$\frac{1}{2}$	$\frac{1}{2}$	$\frac{1}{8}$	$\frac{1}{8}$
$\frac{7}{8}$	$\frac{1}{4}$	$\frac{3}{4}$	$\frac{3}{8}$	$\frac{1}{4}$	$\frac{3}{8}$	$\frac{1}{2}$	$\frac{5}{8}$

fair share: ____________

5.

Distance Swam (mi)				
$\frac{1}{2}$	$\frac{1}{2}$	$\frac{1}{3}$	$\frac{1}{3}$	$\frac{1}{4}$
$\frac{1}{3}$	$\frac{1}{4}$	$\frac{1}{2}$	$\frac{1}{2}$	$\frac{1}{2}$

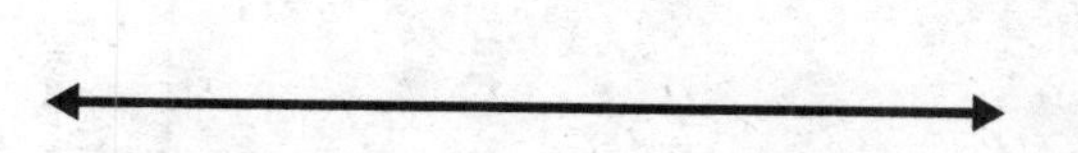

fair share: ____________

Problem Solving

For Exercises 6–7, use the line plot that shows the amount of rainfall over a twelve-month period in the city of Middleton.

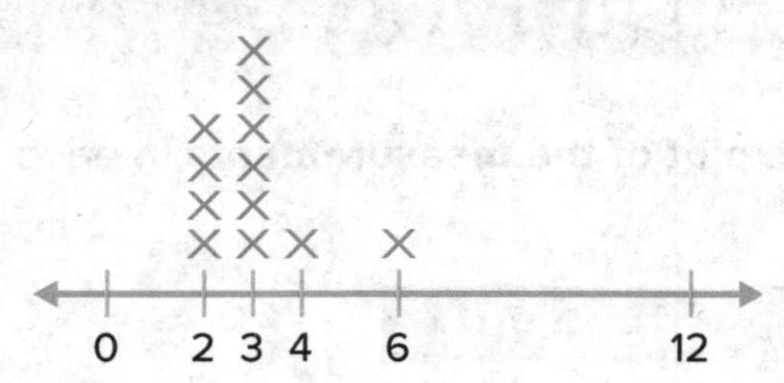

6.  **Model Math** Convert the rainfall amounts to feet and make a new line plot.

7. What is the fair share, in feet, if the same amount of rain fell each month?

Brain Builders

8. Processes & Practices 2 **Use Number Sense** Five friends each ate a fraction of a cup of fruit salad. Explain how you can determine whether the size of a fair share of fruit salad would be less than or greater than 1 cup.

9. **Building on the Essential Question** How can I find the fair share of a set of measurements? Give an example.

Name

MY Homework

Lesson 8

Display Measurement Data on a Line Plot

Homework Helper

Need help? connectED.mcgraw-hill.com

The zoo lists the weights of several animals in the table. Make a line plot of the weights in the table.

Animal Weights (T)				
$\frac{1}{8}$	$\frac{1}{2}$	$\frac{1}{8}$	$\frac{1}{8}$	$\frac{1}{2}$
$\frac{1}{8}$	$\frac{1}{2}$	$\frac{1}{2}$	$\frac{1}{4}$	$\frac{1}{4}$

1 Count the number of times each fraction appears in the table.

$\frac{1}{8}$ appears 4 times.

$\frac{1}{4}$ appears 2 times.

$\frac{1}{2}$ appears 4 times.

2 Place the correct number of Xs above each fraction on the number line.

3 Then, use the title from the table to add a title to the line plot.

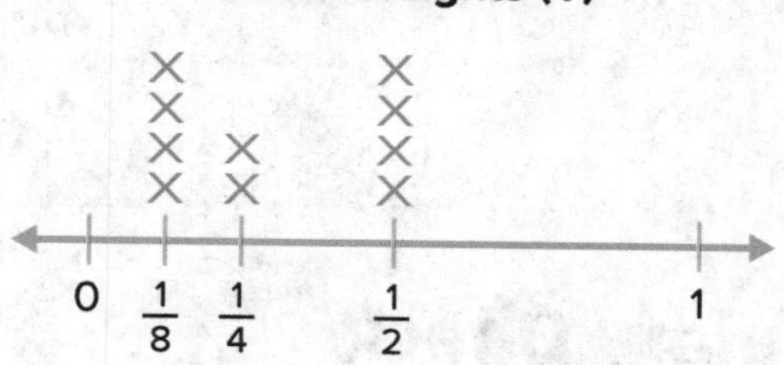

Practice

Refer to the Homework Helper to answer Exercises 1 and 2.

1. Which weight(s) occurred the most often?

____

2. Find the fair share.

____

Problem Solving

3. Make a line plot of the measurements in the table.

Amount of Cashews (lb)					
$\frac{1}{2}$	$\frac{1}{4}$	$\frac{1}{4}$	$\frac{3}{4}$	$\frac{1}{2}$	$\frac{3}{4}$
$\frac{1}{2}$	$\frac{1}{2}$	$\frac{3}{4}$	$\frac{1}{4}$	$\frac{1}{4}$	$\frac{3}{4}$

4. Refer to the table in Exercise 3. What is the fair share, in pounds, of the cashews?

____

Brain Builders

5. Processes & Practices  **Use Math Tools** How would the fair share in Exercise 3 change if the table included two more $\frac{1}{2}$ lb measures of cashews? Explain.

____

____

Vocabulary Check

6. Fill in the correct circle that corresponds to the best answer. Which of the following could be used to find the amount each person would receive if it was divided equally?

Ⓐ fair share
Ⓑ customary system
Ⓒ length
Ⓓ weight

7. **Test Practice** Which is the correct fair share for the measurements shown in the line plot?

Ⓐ $\frac{1}{6}$ mile
Ⓑ $\frac{1}{3}$ mile
Ⓒ $\frac{1}{2}$ mile
Ⓓ $\frac{2}{3}$ mile

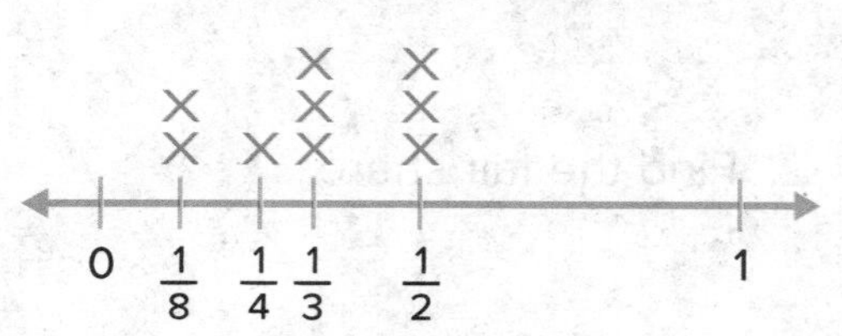

Name

Lesson 9
Hands On

Metric Rulers

ESSENTIAL QUESTION
How can I use measurement conversions to solve real-world problems?

Use a ruler like the one shown to measure objects to the nearest centimeter or to the nearest millimeter.

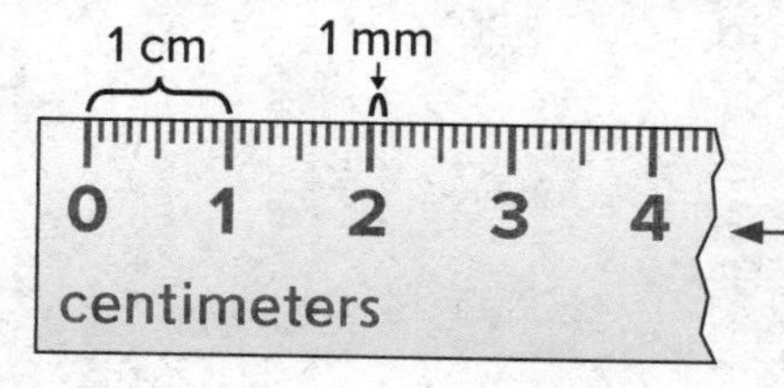

Centimeters and **millimeters** are units of length.

1 centimeter = 10 millimeters

Measure It

Find the length of the piece of chalk to the nearest centimeter.

Place the ruler against the piece of chalk. Line up the zero on the ruler with the end of the piece of chalk.

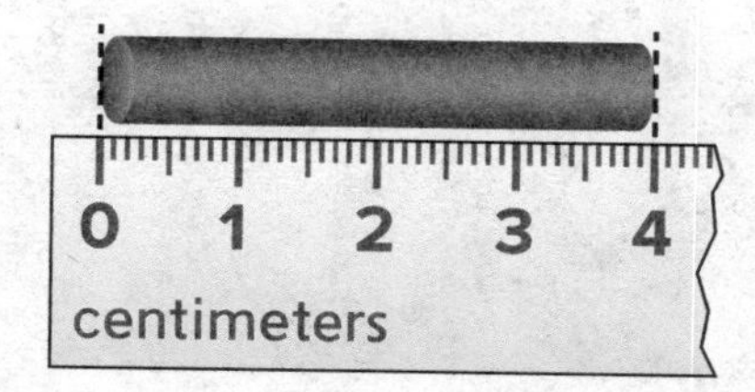

2 Find the centimeter mark that is closest to the other end.

To the nearest centimeter, the length of the piece of chalk is ______ centimeters long.

Cut out and use this centimeter ruler.

cm 0 1 2 3 4 5

Try It

Find the length of the toy car to the nearest millimeter.

Place the ruler against one edge of the car. Line up the zero on the ruler with the end of the car.

2 Find the millimeter mark that is closest to the other end.

To the nearest millimeter, the toy car is ______ millimeters long.

Talk About It

1. Explain how you can tell the difference between the centimeter and millimeter marks when measuring an object with a metric ruler.

______

______

2. Is it easier to measure objects to the nearest centimeter or to the nearest millimeter? Explain.

______

______

3. Processes & Practices 3 **Justify Conclusions** Should you measure the length across a penny to the nearest centimeter or millimeter? Explain your reasoning.

______

______

Name

Practice It

Measure the length of each object to the nearest centimeter and millimeter.

4. 

5. 

Find the length of each object to the nearest centimeter and millimeter.

6. width of a book

7. length of a pencil

8. width of a calculator

9. length of a tape dispenser

Draw a line segment with each of the following lengths.

10. 6 centimeters

11. 27 millimeters

12. 5 centimeters

Apply It

13. Compare the units of length you would use to measure the following: the length of a bicycle and the width of a dime. Explain your reasoning.

14. The length of a cell phone is 8 centimeters to the nearest centimeter and 81 millimeters to the nearest millimeter. Which measurement is more precise?

15. Processes & Practices 3 **Find the Error** Jessie used a ruler to measure a colored pencil. Jessie said the pencil is 14.3 millimeters long. Find her mistake and correct it.

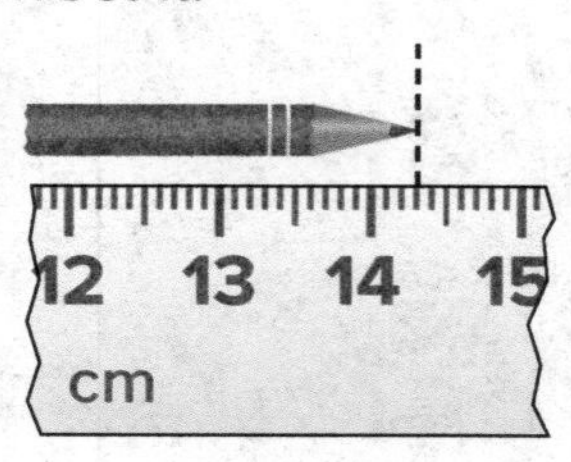

Write About It

16. Will I get a more precise measurement if I measure an object to the nearest centimeter or to the nearest millimeter? Explain your reasoning.

Name

MY Homework

Lesson 9

Hands On: Metric Rulers

Homework Helper

Need help? connectED.mcgraw-hill.com

Find the length of the leaf to the nearest centimeter and millimeter.

1 Place the ruler against one edge of the object. Line up the zero on the ruler with the end of the object.

2 Find the centimeter and millimeter mark that is closest to the other end.

To the nearest centimeter, the leaf is 5 centimeters long. To the nearest millimeter, it is 48 millimeters long.

Practice

Measure the length of each object to the nearest centimeter and millimeter.

1.

2.

Cut out and use this centimeter ruler.

cm 0 1 2 3 4 5

Find the length of each object to the nearest centimeter and millimeter.

3. length of a pen

4. length of a paper clip

Draw a line segment with each of the following lengths.

5. 7 centimeters

6. 105 millimeters

Problem Solving

7. Processes & Practices 6 **Be Precise** The length of Manny's hamster is 114 millimeters to the nearest millimeter and 11 centimeters to the nearest centimeter. Which measurement is more precise?

8. Lauren has a ruler that is marked in millimeters and a tape measure that is marked in centimeters. Which measuring tool will give Lauren a more precise measurement?

9. Hanley measured the height of his glass to be 13 centimeters. Sally measured the same glass and found that it measured 132 millimeters. Who had a more precise measurement?

Name

Lesson 10
Convert Metric Units of Length

ESSENTIAL QUESTION
How can I use measurement conversions to solve real-world problems?

The **metric system** is a decimal system of measurement. To convert metric units, multiply or divide by powers of 10.

Math in My World

Example 1

One of the largest recorded pythons measured 7.3 meters long. What is the length of the python in centimeters?

Convert 7.3 meters to centimeters.

Since 1 meter = 100 centimeters, multiply 7.3 by 100.

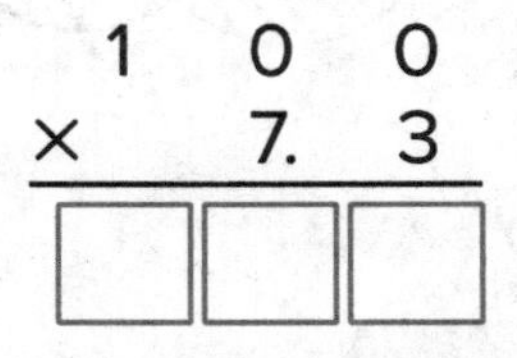

To multiply by 10, 100, or 1,000, use basic facts and count the number of zeros in the factors.

So, 7.3 meters = ______ centimeters.

The python is ______ centimeters long.

Key Concept Metric Units of Length

1 **centimeter** (cm) = 10 **millimeters** (mm)
1 **meter** (m) = 100 cm or 1,000 mm
1 **kilometer** (km) = 1,000 m

1 millimeter thickness of a dime	**1 centimeter** width of pinky finger	**1 meter** height of a doorknob	**1 kilometer** 6 city blocks

Example 2 

Roshonda has 50 dominoes. Each domino is 4 centimeters long. She lines them up end to end as shown. How many meters long is the line of dominoes?

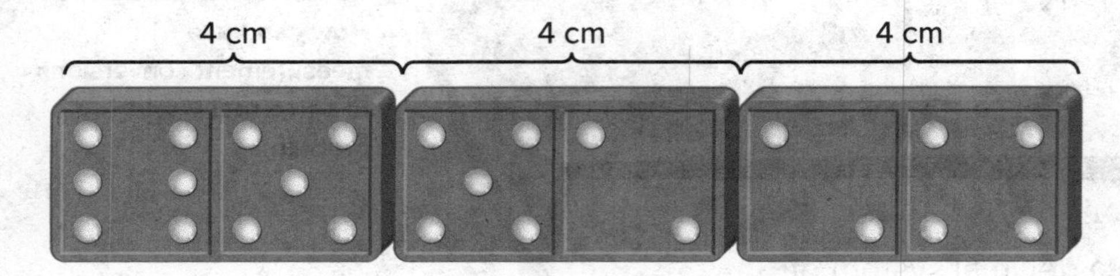

1 Find the length in centimeters.

50 × 4 centimeters = ________ centimeters

2 Convert ________ centimeters to meters.

Since 1 meter = ________ centimeters,

divide ________ by ________.

________ ÷ ________ = ________

So, ________ centimeters = ________ meters.

The line of 50 dominoes is ________ meters long.

Helpful Hint

To divide by 10, 100, or 1,000, cross out the same number of zeros in both the dividend and divisor.

Guided Practice

Complete.

1. 5 m = ■ cm

5 × 100 = ________

So, 5 meters equals ________ centimeters.

2. 9,000 m = ■ km

9,000 ÷ 1,000 = ________

So, 9,000 meters equals

________ kilometers.

How can you use mental math to convert 7.38 kilometers to meters?

Name ....

Independent Practice

Complete.

3. 700 cm = ________ m

4. 8,500 mm = ________ m

5. 15 km = ________ m

6. 73,000 m = ________ km

7. 2.71 m = ________ mm

8. 9.2 m = ________ cm

9. 17.5 mm = ________ cm

10. 0.509 km = ________ m

Complete. Use <, >, or = to make a true statement.

11. 30 cm ◯ 300 mm

12. 4.8 km ◯ 4,800 m

13. 25 mm ◯ 3 cm

14. 9 km ◯ 8,500 m

15. 1.5 m ◯ 145 cm

16. 17 m ◯ 116 cm

Problem Solving

17. Measure the distance across the sunflower to the nearest centimeter. How many centimeters shorter than 1 meter is the width of the sunflower?

18. Processes & Practices 1 **Check for Reasonableness** Which is the most reasonable estimate for the depth of a lake: 6 millimeters, 6 centimeters, or 6 meters? Explain.

Brain Builders

19. How many 5-millimeter-long spiders would it take to make a line of spiders 1 meter long?

20. Processes & Practices 3 **Which One Doesn't Belong?** Circle the measure that does not belong with the other three. Explain your reasoning.

35 km	3.5 m	350 cm	3,500 mm

21. **Building on the Essential Question** Compare and contrast converting customary units of length and converting metric units of length.

Name ____________________

Lesson 10

Convert Metric Units of Length

Homework Helper

Need help? connectED.mcgraw-hill.com

The average length of a great white shark is about 4 meters. What is the average length in centimeters?

Convert 4 meters to centimeters.

Since 1 meter = 100 centimeters, multiply 4 by 100.

$4 \times 100 = 400$

So, 4 meters = 400 centimeters.

The average length of the great white shark is about 400 centimeters.

Practice

Complete.

1. 300 cm = ________ m

2. 500 mm = ________ cm

3. 1.7 km = ________ cm

4. 2 km = ________ m

5. 6 cm = ________ mm

6. 238 cm = ________ m

7. 2,400 mm = ________ m

8. 175 mm = ________ m

Problem Solving

9. When completed, the tunnel will be 1,500 meters long. What is this length in kilometers?

____

Brain Builders

10. Processes & Practices  **Use Number Sense** The depth of a swimming pool is 8.5 meters. What is half of the depth in millimeters? Explain.

____

____

Vocabulary Check

Choose the correct word(s) that completes each sentence.

millimeter **centimeter** **meter**
kilometer **metric system**

11. The ______________ is an appropriate unit to measure the length of a ladybug.

12. The ______________ is an appropriate unit to measure the distance between two cities.

13. The ______________ is a decimal system of measurement.

14. **Test Practice** Kaelyn is reading a book. The book's thickness is 25 millimeters. How many of these books would fit on a shelf that is 1 meter long?

Ⓐ 4 books

Ⓑ 10 books

Ⓒ 40 books

Ⓓ 100 books

Check My Progress

Vocabulary Check

Choose the correct word(s) that completes each sentence.

capacity	centimeter	cup	fluid ounces	gallon
kilometer	pint	quart	meter	millimeter

1. The __________ is an appropriate unit to measure the height of an oak tree.

2. The __________ is an appropriate unit to measure the capacity of a gasoline tank.

Concept Check

Compare. Use >, <, or = to make a true statement.

3. 7 c ◯ $3\frac{1}{4}$ pt

4. 5 gal ◯ 18 qt

5. 45 cm ◯ 450 mm

6. 4.5 km ◯ 5,000 m

Complete.

7. 7 c = ________ fl oz

8. 17 gal = ________ qt

9. 22 qt = ________ gal ________ qt

10. 835 cm = ________ m

11. 88,000 m = ________ km

12. 49.3 mm = ________ cm

13. Make a line plot of the measurements in the table. Then find the fair share.

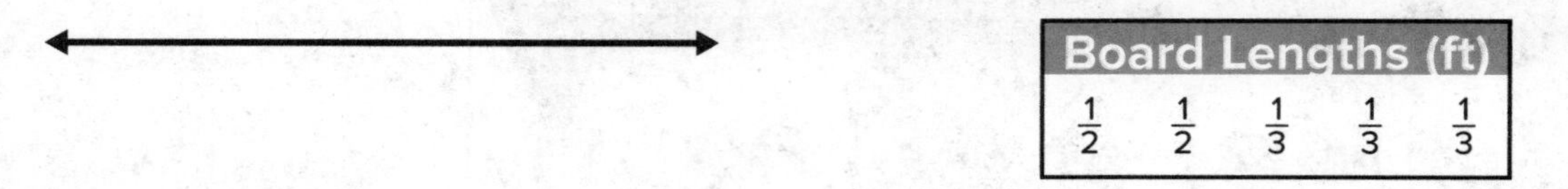

Board Lengths (ft)				
$\frac{1}{2}$	$\frac{1}{2}$	$\frac{1}{3}$	$\frac{1}{3}$	$\frac{1}{3}$

fair share: ___________

Problem Solving

14. Which is the most reasonable estimate for the height of a two-story house: 15 centimeters, 15 meters, or 15 kilometers? Explain.

____

15. Torri measured the capacity of the punch bowl. Her first measurement was 2 gallons. Her second measurement was 9 quarts. Compare the two measurements. Use >, <, or = to make a true statement.

____

Brain Builders

16. Phil has 7 quarts of hot chocolate to give to his classmates. How many of Phil's classmates can have one cup of hot chocolate? Explain.

____

17. **Test Practice** The depth of a lake is 100 centimeters less than 1,401 meters. What is the depth in kilometers?

Ⓐ 0.14 kilometers

Ⓑ 1.4 kilometers

Ⓒ 14 kilometers

Ⓓ 140 kilometers

Name

Lesson 11 Hands On 

Estimate and Measure Metric Mass

ESSENTIAL QUESTION
How can I use measurement conversions to solve real-world problems?

The **mass** of an object is the amount of matter it has.
A **gram** is a metric unit of measurement of mass.

Measure It

1 Estimate the mass of each object in grams.
Record your results in the table.

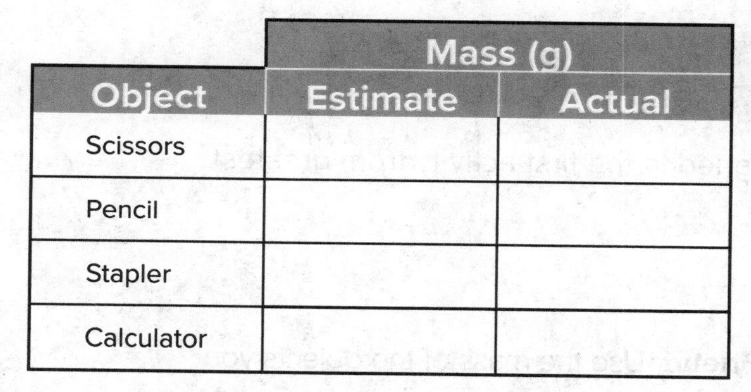

Object	Mass (g) Estimate	Mass (g) Actual
Scissors		
Pencil		
Stapler		
Calculator		

2 Measure the mass of each object.

Place the scissors on one side of a balance. Set gram weights on the other side until the sides are level. Record the actual mass. Repeat this step for the other objects.

Try It

A **kilogram** is also a metric unit of measurement of mass. One kilogram is equal to 1,000 grams. Use this to complete the table below.

Kilograms	Grams
1	1,000
2	
3	
4	
5	

Look for a pattern in the table.

How many grams are in 6 kilograms? __________

How many grams are in 9 kilograms? __________

Talk About It

1. Order the four objects you weighed in the first activity from greatest to least mass.

2. Processes & Practices 6 **Explain to a Friend** Use the mass of the objects you found to estimate the mass of two other objects in your classroom. Then find the mass of the objects. Were your estimates close?

3. Could a larger object have less mass than a smaller object? Explain.

4. Explain how you can use mental math in order to convert kilograms to grams.

Name

Practice It

5. Identify three objects in your classroom that you can use the balance to find their masses. Estimate each object's mass. Then find the mass of each object and record the exact mass in the table.

	Mass (g)	
Object	Estimate	Actual

Compare. Use >, <, or = to make a true statement.

6. 1,500 grams ◯ 1 kilogram

7. 3,000 grams ◯ 3 kilograms

8. 4,000 grams ◯ 3 kilograms

9. 3,700 grams ◯ 4 kilograms

10. 5 kilograms ◯ 6,000 grams

11. 3.5 kilograms ◯ 3,000 grams

12. 2.5 kilograms ◯ 2,500 grams

13. 3.25 kilograms ◯ 3,300 grams

Apply It

14. Randy and Salma measured the mass of the same chinchilla. Randy measured the chinchilla as 1 kilogram. Salma measured the chinchilla as 945 grams. Circle the more precise measurement.

945 grams   1 kilogram

15. Edita measured the mass of her books. She measured the mass as 2 kilograms. Her second measurement was 2,050 grams. Use >, <, or = to make a true statement.

2 kilograms ◯ 2,050 grams

16. Processes & Practices 6 **Be Precise** If you are measuring the mass of a container of salt, would grams or kilograms give you a more precise measurement? Explain.

____

____

____

17. Processes & Practices 3 **Draw a Conclusion** Compare and contrast grams and kilograms.

____

____

____

Write About It

18. How can I convert grams to kilograms without measuring?

____

____

____

Name ....

Lesson 11

Hands On: Estimate and Measure Metric Mass

Homework Helper

Need help? connectED.mcgraw-hill.com

One kilogram is equal to 1,000 grams. Use this information to complete the table. How many grams are in 6 kilograms?

For every increase of one kilogram, increase the number of grams by 1,000.

Kilograms	Grams
1	1,000
2	2,000
3	3,000
4	4,000
5	5,000
6	6,000

+ 1,000
+ 1,000
+ 1,000
+ 1,000
+ 1,000

So, 6 kilograms are equal to 6,000 grams.

Practice

Compare. Use >, <, or = to make a true statement.

1. 2,300 grams ◯ 2 kilograms

2. 4,840 grams ◯ 5 kilograms

3. 4 kilograms ◯ 4,150 grams

4. 1.75 kilograms ◯ 1,750 grams

Vocabulary Check

5. Fill in the blank with the correct word to complete the sentence below.

The ________ of an object is the amount of matter it has.

Real World!

Problem Solving

6. Jason and Scott measured the masses of their cell phones. Jason measured his cell phone using kilograms. Scott measured his cell phone using grams. Which measure would be more appropriate to measure a cell phone?

____

7. Processes & Practices 6 **Be Precise** Gavin has a ten-year old cat, Shadow. Is Shadow's mass more likely to be 6 kilograms or 6 grams? Explain.

____

____

8. Norton measured the mass of his luggage. His luggage mass was 21,530 grams. The airline will only allow luggage that has a mass under 23 kilograms. Will Norton be allowed to fly with his luggage? Explain.

____

9. Chasity measured the mass of her new puppy. Her first measurement was 2,350 grams. Her second measurement was 2.3 kilograms. Circle the measurement that is more precise.

2,350 grams   2.3 kilograms

10. Billy measured the mass of his iguana. His first measurement was 4,100 grams. His second measurement was 4 kilograms. Compare the two measurements. Use >, <, or = to make a true statement.

____

Name ....

Lesson 12
Convert Metric Units of Mass

ESSENTIAL QUESTION
How can I use measurement conversions to solve real-world problems?

Mass is a measure of the amount of matter an object has.

Math in My World

Example 1

A white-tailed deer has a mass of 136 kilograms. What is the mass of the deer in grams?

Convert 136 kilograms to grams.

Since 1 kilogram = 1,000 grams, multiply 136 by 1,000.

$$\begin{array}{r} 1{,}000 \\ \times \quad 136 \\ \hline 136{,}000 \end{array}$$

So, 136 kilograms = ________ grams.

The mass of the white-tailed deer is ________ grams.

Check Use division to check your answer.

________ ÷ 1,000 = 136

Key Concept Metric Units of Mass

1 **gram** (g) = 1,000 **milligrams** (mg)   1 **kilogram** (kg) = 1,000 g

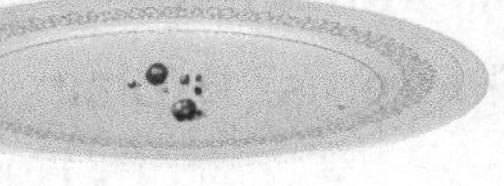

1 milligram
a bread crumb

1 gram
a paper clip

1 kilogram
a loaf of bread

Example 2

Tutor

Convert 1,500 grams to kilograms.

Since you are converting a smaller unit to a larger unit, divide.

```
            ☐  R ☐☐☐
1,000 ) 1 , 5  0  0
      − ☐   ☐  ☐  ☐
        ─────────────
            ☐  ☐  ☐
```

The remainder ________ means there are ________ grams left over.

The decimal part of a kilogram is ________.

So, 1,500 grams = ________ kilogram ________ grams or ________ kilograms.

Guided Practice

Complete.

1. 5,000 mg = ■ g

5,000 ÷ 1,000 = ________

So, 5,000 milligrams equals

________ grams.

2. 5 kg = ■ g

5 × 1,000 = ________

So, 5 kilograms equals

________ grams.

3. 4,000 g = ■ kg

4,000 ÷ 1,000 = ________

So, 4,000 grams equals

________ kilograms.

4. 9 g = ■ mg

9 × 1,000 = ________

So, 9 grams equals

________ milligrams.

Which is a more reasonable estimate for the mass of a baseball: 140 milligrams, 140 grams, or 140 kilograms? Explain.

Name ....

Independent Practice

Complete.

5. 2,000 mg = ________ g

6. 80 g = ________ mg

7. 0.75 kg = ________ mg

8. 6 kg = ________ g

9. 3,100 g = ________ kg

10. 0.05 kg = ________ mg

11. 4.07 g = ________ mg

12. 9 kg = ________ g

Compare. Use <, >, or = to make a true statement.

13. 2,300 mg ◯ 2 g

14. 3 kg ◯ 3,000 g

15. 4.5 kg ◯ 4,050 g

16. 4,120 mg ◯ 4.12 g

17. 75 g ◯ 800 mg

18. 814 g ◯ 8.14 kg

Problem Solving

Use the table shown for Exercises 19 and 20.

Macaws	
Species	**Mass (grams)**
Blue and Gold	800
Green-winged	900
Red-footed	525
Yellow-collared	250

19. How many yellow-collared macaws would have a combined mass of 1 kilogram?

20. Processes & Practices 6 **Explain to a Friend** Is the combined mass of two red-footed macaws and three blue and gold macaws closer to 3 kilograms or 4 kilograms? Explain.

Brain Builders

21. Three different rock samples have masses of 583 grams, 256 milligrams, and 102 kilograms. List the masses of the rock samples from least to greatest.

22. Processes & Practices 2 **Use Number Sense** One pound is approximately equal to 0.5 kilograms. About how many kilograms is 3 pounds? Explain.

23. **Building on the Essential Question** How is converting metric units of mass different from converting customary units of weight?

Name

Lesson 12

Convert Metric Units of Mass

Homework Helper

Need help? connectED.mcgraw-hill.com

Mr. Benavides bakes muffins that have a mass of about 50,000 milligrams. What is the mass in grams?

Convert 50,000 milligrams to grams.

Since 1,000 milligrams = 1 gram, divide 50,000 by 1,000.

So, 50,000 milligrams = 50 grams.

The muffins have a mass of about 50 grams.

Practice

Complete.

1. 7,000 mg = ________ g

2. 4.7 kg = ________ g

3. 18,500 g = ________ kg

4. 8.3 kg = ________ g

5. 22 g = ________ mg

6. 135,000 mg = ________ kg

Problem Solving

7. One highlighter has a mass of 11 grams. Another highlighter has a mass of 10,800 milligrams. Which highlighter has the greater mass?

Brain Builders

8. Processes & Practices 6 **Be Precise** One computer has a mass of 0.8 kilogram and another has a mass of 800 grams. Compare the masses of the computers. Use >, <, or = to make a true statement. Explain.

Vocabulary Check

Fill in the correct circle that corresponds to the best answer.

9. Which of the following is not a common unit of measurement for the metric system?

Ⓐ milligram
Ⓑ kilogram
Ⓒ gram
Ⓓ ounce

10. Which operation is necessary to convert a larger unit to a smaller unit?

Ⓕ addition
Ⓖ subtraction
Ⓗ multiplication
Ⓘ division

11. **Test Practice** For a science experiment, Noel is using two pieces of metal. One has a mass of 3,000 grams and the other has a mass of 500 grams. What is the combined mass of the metals in kilograms?

Ⓐ 0.35 kilogram
Ⓑ 3.5 kilograms
Ⓒ 35 kilograms
Ⓓ 350 kilograms

Name ________________________________

Lesson 13
Convert Metric Units of Capacity

ESSENTIAL QUESTION
How can I use measurement conversions to solve real-world problems?

In the metric system, the common units of capacity are liter and milliliter.

Math in My World

Example 1

A dripping faucet wastes about 90 liters of water every week. How many milliliters of water is this?

Convert 90 liters to milliliters.

Since 1 liter = 1,000 milliliters, multiply 90 by 1,000.

$$\begin{array}{r} 1{,}000 \\ \times \quad 90 \\ \hline 90{,}000 \end{array}$$

So, 90 liters = __________ milliliters.

The dripping faucet wastes __________ milliliters of water.

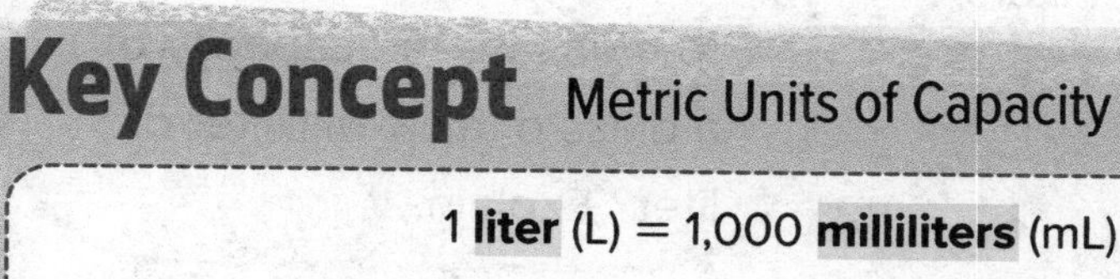

Key Concept Metric Units of Capacity

1 **liter** (L) = 1,000 **milliliters** (mL)

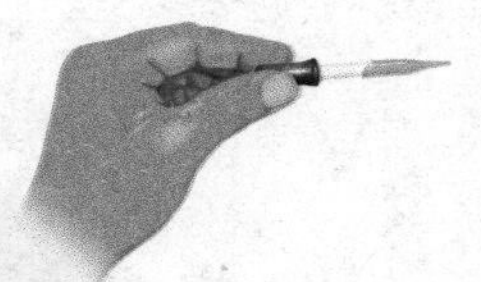

1 milliliter
amount of liquid in an eyedropper

1 liter
a medium-sized sports drink

Example 2 

A container of orange juice holds 580 milliliters. How many liters is 580 milliliters?

Since 1 liter = ________ milliliters, divide 580 by ________.

580 ÷ ________ = ________ ← Move the decimal point 3 places to the left.

So, 580 milliliters = ________ liter.

The container holds ________ liter of orange juice.

Guided Practice

Complete.

1. 6 L = ■ mL

6 × 1,000 = ________

So, 6 liters equals ________ milliliters.

2. 4 L = ■ mL

4 × 1,000 = ________

So, 4 liters equals ________ milliliters.

3. 7,000 mL = ■ L

7,000 ÷ 1,000 = ________

So, 7,000 milliliters equals ________ liters.

4. 42 mL = ■ L

42 ÷ 1,000 = ________

So, 42 milliliters equals ________ liter.

Talk MATH

Which unit would you use to measure the capacity of a glass of milk: milliliter or liter? Explain.

Name ....

Independent Practice

Complete.

5. 70 L = ________ mL

6. 10 mL = ________ L

7. 1.2 L = ________ mL

8. 3,500 mL = ________ L

9. 4 L = ________ mL

10. 230 mL = ________ L

11. 6.21 L = ________ mL

12. 5,000 mL = ________ L

Compare. Use <, >, or = to make a true statement.

13. 2 L ◯ 1,000 mL

14. 390 mL ◯ 0.39 L

15. 82 L ◯ 825 mL

16. 834 mL ◯ 8.34 L

17. 0.34 L ◯ 430 mL

18. 87 mL ◯ 0.087 L

Problem Solving

19. The Nail Shop purchases nail polish in 13-milliliter bottles. Find the total capacity, in liters, of 1,000 bottles.

____

20. Paulene measures the amount of water in a container to be 2,732 milliliters. Justin measures the water in the same container to be 3 liters. Circle the greater measurement.

2,732 milliliters   3 liters

Brain Builders

21. Processes & Practices 1 **Make Sense of Problems** Emanuel is challenging himself to fill his canteen with water using an eye dropper that drops 1 milliliter of water at a time. How many drops will it take him to fill his 1.5-liter canteen?

____

____

____

22. Processes & Practices 2 **Reason** Name three items that have a capacity greater than 10 liters. Estimate the capacity of each in milliliters.

____

____

23. **Building on the Essential Question** Why is it important to be able to convert metric units of capacity?

____

____

Name ______________________

Lesson 13

Convert Metric Units of Capacity

Homework Helper

Need help? connectED.mcgraw-hill.com

A cough syrup bottle contains 120 milliliters of cough syrup. How many liters is 120 milliliters?

Since 1 liter = 1,000 milliliters, divide 120 by 1,000.

$120 \div 1{,}000 = 0.12$ ← Move the decimal point 3 places to the left.

So, 120 milliliters = 0.12 liter.

The bottle holds 0.12 liter of cough syrup.

Practice

Complete.

1. 6 L = ________ mL

2. 13 L = ________ mL

3. 54,000 mL = ________ L

4. 23,500 mL = ________ L

5. 11,000 mL = ________ L

6. 0.201 L = ________ mL

Problem Solving

7. Yesterday, Audrey drank the liquids shown in the table. How many liters of liquids did she drink in all?

Liquid	Amount
Juice	210 mL
Milk	480 mL
Water	1.2 L

____

8. One serving of punch is 250 milliliters. Will ten servings fit in a 2-liter bowl? Explain.

____

____

Brain Builders

9. Processes & Practices  **Make Sense of Problems** Nara received a measles immunization at Dr. Arroyo's office. The vaccine was measured in cubic centimeters. A cubic centimeter has the same capacity as a milliliter. If each immunization is 3.5 cubic centimeters, how many individual immunizations could Dr. Arroyo provide with 0.35 liters of vaccine?

____

Vocabulary Check

Fill in the blank with the correct word(s) that completes each sentence.

10. The ____________ is an appropriate unit to measure the capacity of a hand sanitizer bottle.

11. The ____________ is an appropriate unit to measure the capacity of the water in a fountain.

12. **Test Practice** A soup bowl can hold about 400 milliliters of soup. A restaurant has 8 liters of vegetable soup. How many bowls of soup can they serve?

Ⓐ 500 bowls
Ⓑ 200 bowls
Ⓒ 50 bowls
Ⓓ 20 bowls

Review

Chapter 11

Measurement

Vocabulary Check

Fill in the circle next to the best answer.

1. The **capacity** of a container is which of the following?

Ⓐ the elapsed time

Ⓑ the customary unit

Ⓒ the metric unit

Ⓓ the amount that it can hold

2. Customary units of **length** are measured in which of the following?

Ⓕ meters and centimeters only

Ⓖ inches, feet, yards, and miles

Ⓗ minutes and hours

Ⓘ days and weeks

3. The **metric system** is based on which of the following?

Ⓐ fractions

Ⓑ decimals

Ⓒ inches

Ⓓ gallons

4. When you **convert** from feet to inches, you are doing which of the following?

Ⓕ changing the measurement unit

Ⓖ determining capacity

Ⓗ determining mass

Ⓘ determining volume

5. When finding the **mass** of an object, you determine which of the following?

Ⓐ the quantity of matter in the object

Ⓑ its weight

Ⓒ its height

Ⓓ its length

Concept Check

Complete.

6. 84 in. = ____________ ft

7. 9 yd = ____________ in.

8. 7,920 yd = ____________ mi

9. 64,000 lb = ____________ T

10. $7\frac{1}{2}$ lb = ____________ oz

11. 62 oz = ________ lb ________ oz

12. 7 pt = ____________ c

13. 12 c = ____________ qt

14. 72 pt = ____________ gal

15. 120 mm = ____________ cm

16. Make a line plot of the measurements in the table. Then find the fair share.

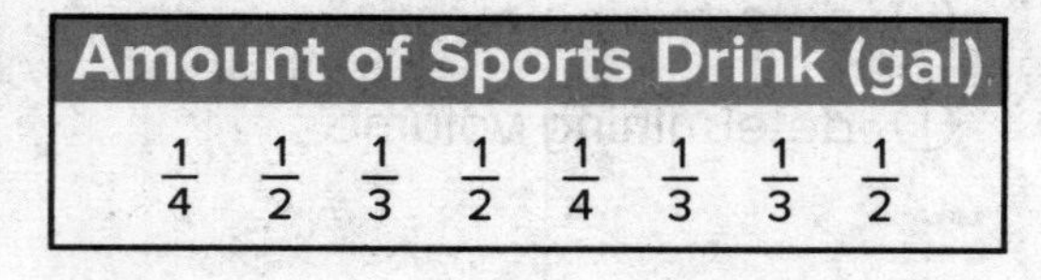

Amount of Sports Drink (gal)							
$\frac{1}{4}$	$\frac{1}{2}$	$\frac{1}{3}$	$\frac{1}{2}$	$\frac{1}{4}$	$\frac{1}{3}$	$\frac{1}{3}$	$\frac{1}{2}$

fair share: ____________

Problem Solving

17. Breanna has quarters, dimes, and nickels in her purse. She has 3 fewer nickels than dimes, but she has 2 more nickels than quarters. If Breanna has 2 quarters, how much money does she have?

____

18. A detergent bottle holds 700 milliliters. Find the capacity in liters.

____

Brain Builders

19. Tyler rides his bike from his house to the library $1\frac{1}{2}$ miles away. Then he rides 1,320 feet to the park. If he returns home the same way, what is the total distance Tyler will travel?

____

20. Deka measured the mass of 100 sheets of paper as 1,500 grams. How many kilograms would 1 sheet of paper weigh?

____

21. **Test Practice** Maisha is using special paint for her artwork. The art supply store charges $1.50 per cup of paint. Maisha needs 2 pints of blue paint, 3 cups of green paint, $1\frac{1}{2}$ quarts of orange paint, and $\frac{1}{2}$ cup of yellow paint. How much will she pay?

Ⓐ $10.50 Ⓒ $14.25

Ⓑ $11.25 Ⓓ $20.25

Reflect

Chapter 11

Answering the ESSENTIAL QUESTION 

Use what you learned about measurement to complete the graphic organizer below.

ESSENTIAL QUESTION

How can I use measurement conversions to solve real-world problems?

0 1 2 3 inches **Customary System**	0 1 2 3 4 5 mm **Metric System**
Vocabulary	**Vocabulary**
Conversions	**Conversions**

Now reflect on the ESSENTIAL QUESTION  Write your answer below.

Name Date

Score

Performance Task

Brain Builders

Researching Rivers

A group of kayakers want to compare information on three different rivers in the United States. Research ahead of time the length in miles of the Yukon River, Yellowstone River, and the Mississippi River.

Show all your work to receive full credit.

Part A

From your research, fill in the lengths in miles of the three rivers. Then convert the measurements to yards.

Rivers	Length in Miles	Length in Yards
Yukon River		
Yellowstone River		
Mississippi River		

Part B

From your research, fill in the lengths in kilometers of the three rivers. Then convert the measurements to meters.

Rivers	Length in Kilometers	Length in Meters
Yukon River		
Yellowstone River		
Mississippi River		

Part C

The weight of a one-person, whitewater kayak is 792 ounces. Convert the measurement to pounds and ounces. Show your calculations.

Part D

It is recommended that a kayaker drink 710 milliliters of water an hour before traveling down a river. How many liters is this? Explain.

Chapter

12 Geometry

ESSENTIAL QUESTION

How does geometry help me solve problems in everyday life?

Let's Travel!

Watch a video!

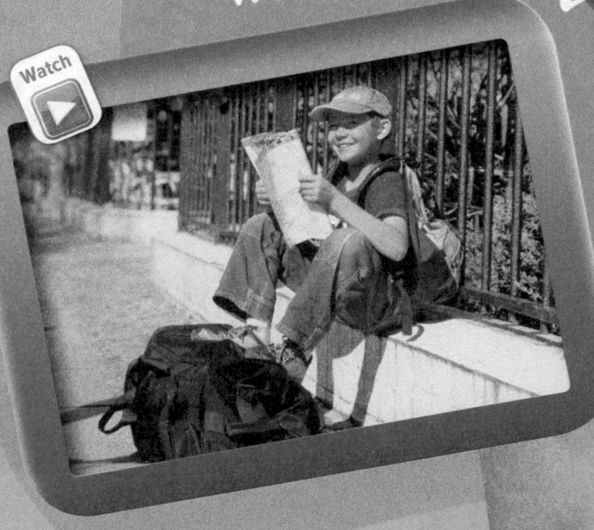

Name

MY Chapter Project

Geo-Ville

1. Your group will collect three-dimensional objects from home or from your classroom. The shapes will be decorated and used to create a model of a city.

2. At the end of the chapter, think about the concepts you learned in the chapter. Use these concepts to create and answer five questions about five different buildings in your city. List the questions and answers below:

Question:

Answer:

Question:

Answer:

Question:

Answer:

Question:

Answer:

Question:

Answer:

3. Work as a group to write each question on an index card. Write the question on one side and the answer on the other side.

Name ....

Am I Ready?

Name the number of sides and the number of angles in each figure.

1. 

_______ sides and _______ angles

2. 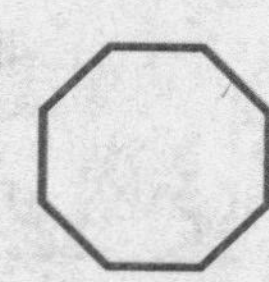

_______ sides and _______ angles

3. 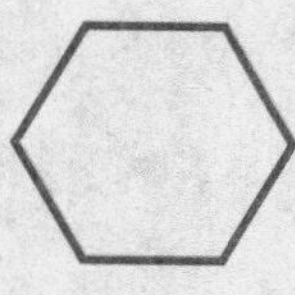

_______ sides and _______ angles

4. 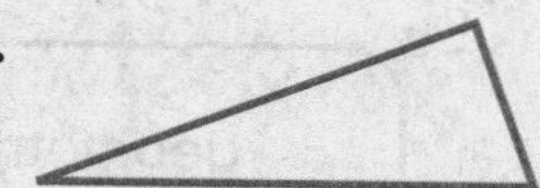

_______ sides and _______ angles

Use the figure below for Exercises 5 and 6.

5. Which side appears to have the same length as side *AD*? _______

6. At which point do sides *AB* and *BC* meet? _______

7. Kai is drawing a triangle with all three sides that are equal. Draw a sketch of this triangle.

How Did I Do?

Shade the boxes to show the problems you answered correctly.

1	2	3	4	5	6	7

Name

MY Math Words

Vocab

Review Vocabulary

acute angle angles lines obtuse angle

parallel perpendicular right angle

Making Connections

Use the review words to classify geometric shapes.

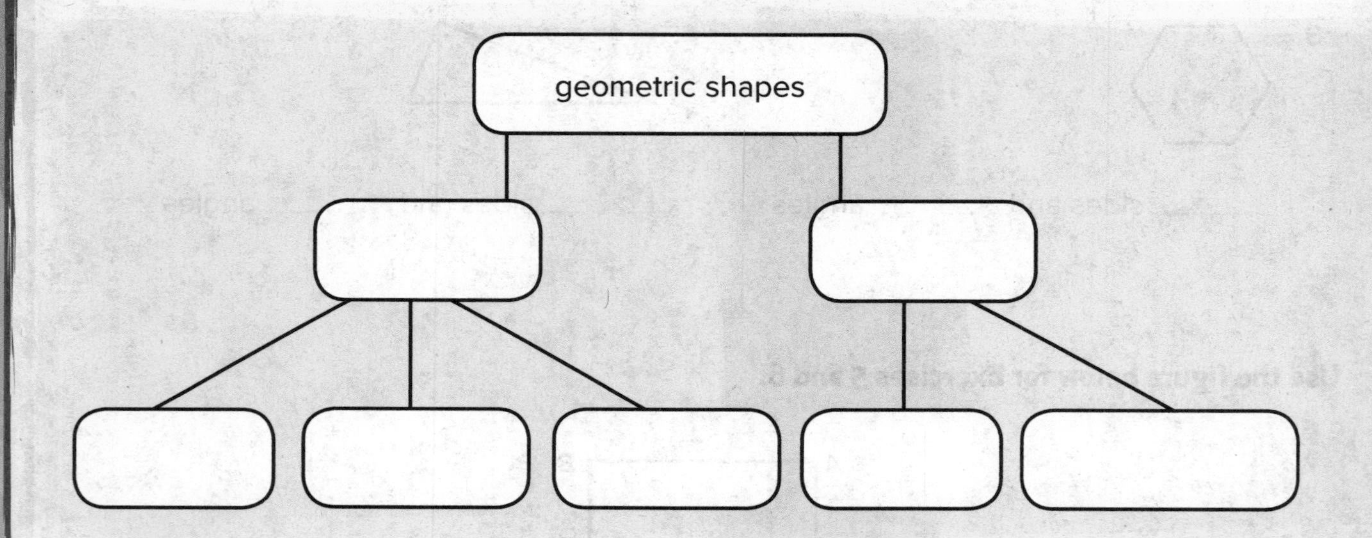

Draw an example of two of the words used above.

MY Vocabulary Cards

Processes & Practices

Lesson 12-3

acute triangle

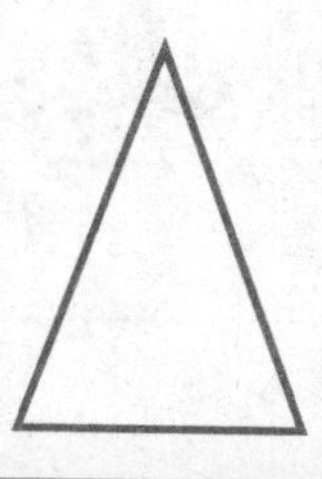

Lesson 12-3

attribute

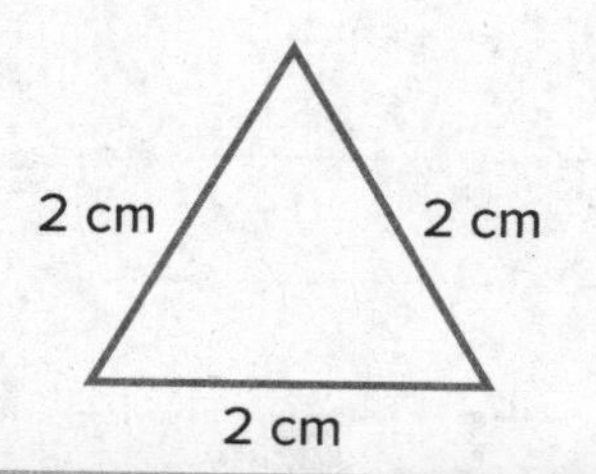

Lesson 12-7

bases

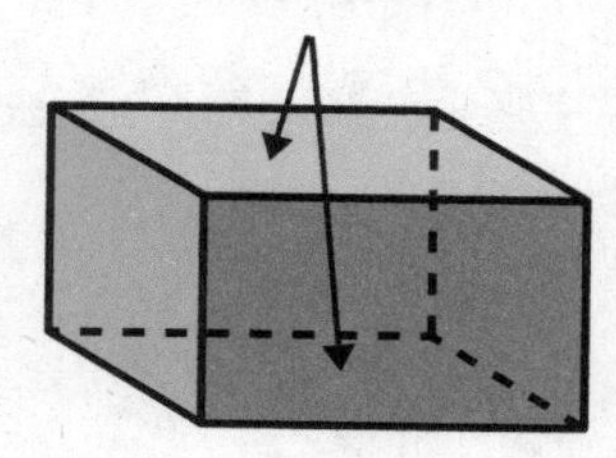

Lesson 12-10

composite figures

Lesson 12-1

congruent angles

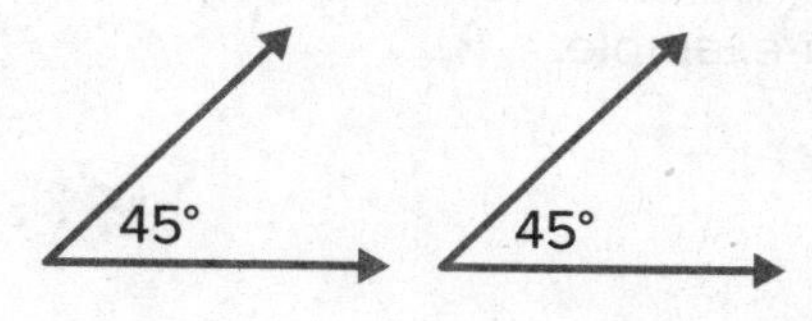

Lesson 12-6

congruent figures

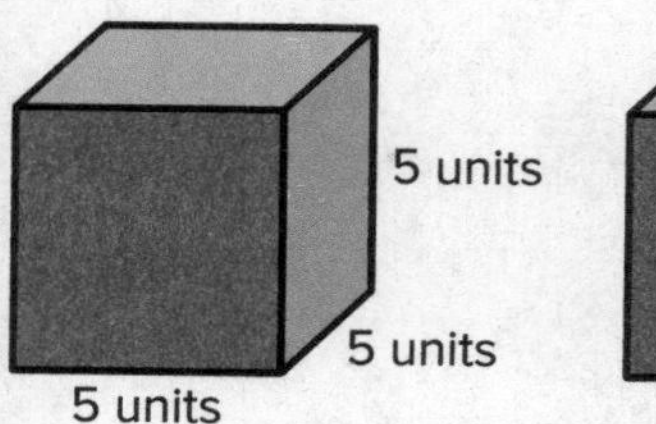

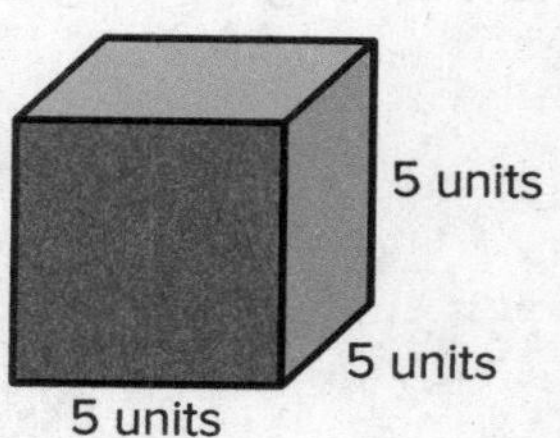

Lesson 12-1

congruent sides

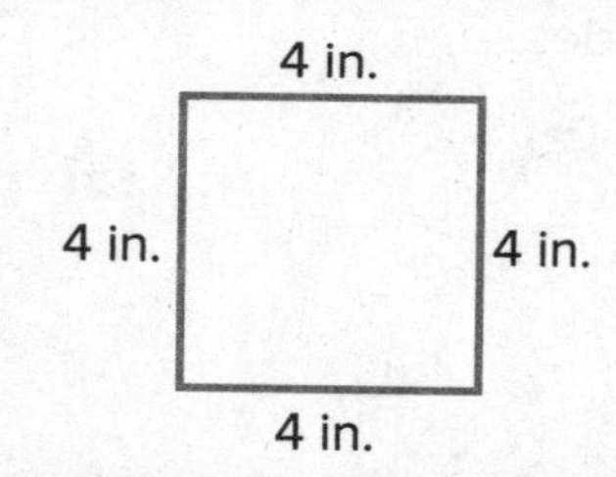

Lesson 12-6

cube

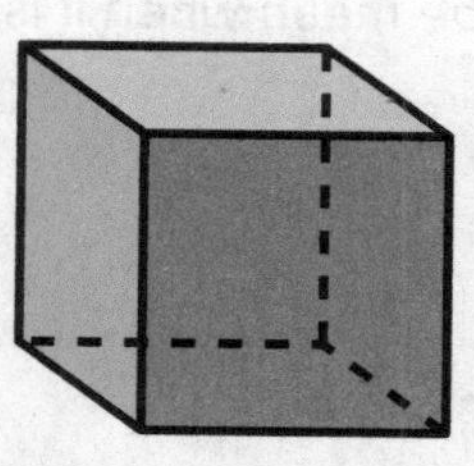

Ideas for Use

- Design a crossword puzzle. Use the definition for each word as the clues.
- Group 2 or 3 common words. Add a word that is unrelated to the group. Then work with a friend to name the unrelated word.

A characteristic of a figure.

Use *attribute* to describe the sides or angles of a rectangle.

A triangle with 3 acute angles.

Explain how to determine if a triangle is an acute triangle.

A figure that is made of two or more three-dimensional figures.

Composite comes from *compose,* "to put together." How does this help you understand a composite figure?

Two parallel congruent faces in a prism.

Describe the *base* of a rectangular prism.

Two figures that have the same size and shape.

Draw 2 congruent figures in the space below.

Angles of a figure that are equal in measure.

Draw an example of a figure with congruent angles. Then draw a non-example. Cross out the non-example.

A three-dimensional figure with six faces that are congruent squares.

What does *cube* mean when it is used as a verb?

Sides of a figure that are equal in length.

Draw two figures that each have at least two congruent sides.

MY Vocabulary Cards

Processes & Practices

Lesson 12-8

cubic unit

Lesson 12-7

edge

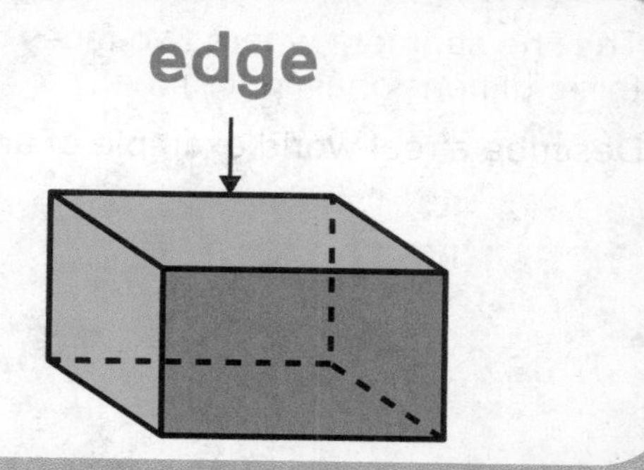

Lesson 12-3

equilateral triangle

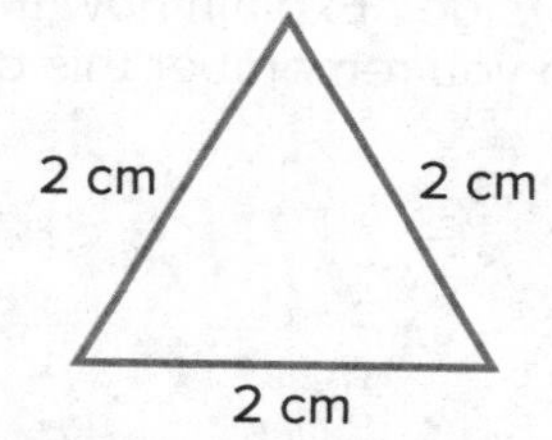

Lesson 12-6

face

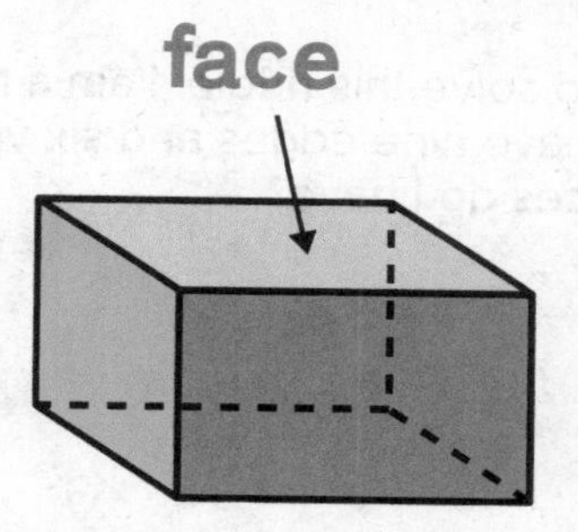

Lesson 12-1

hexagon

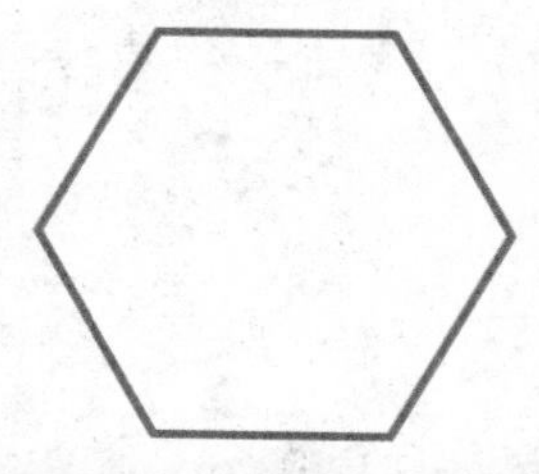

Lesson 12-3

isosceles triangle

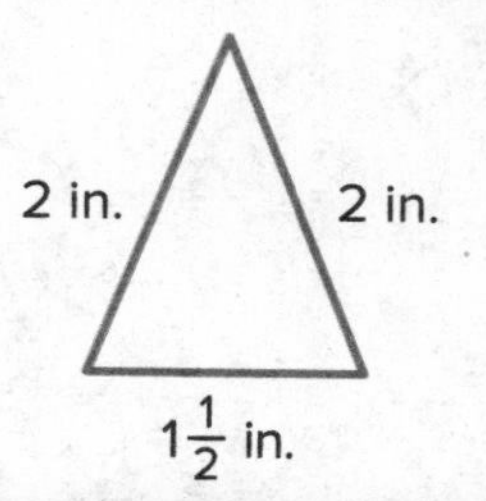

Lesson 12-6

net

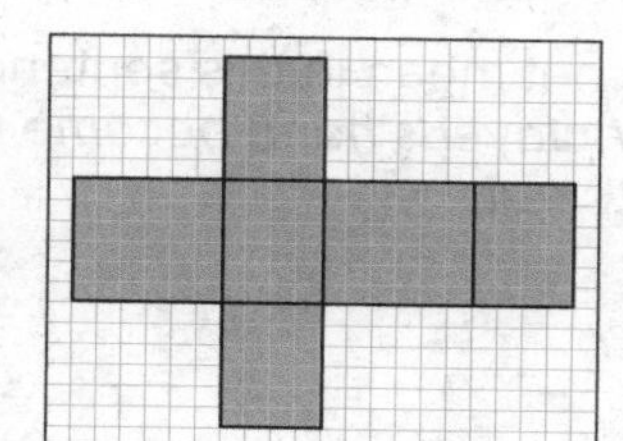

Lesson 12-3

obtuse triangle

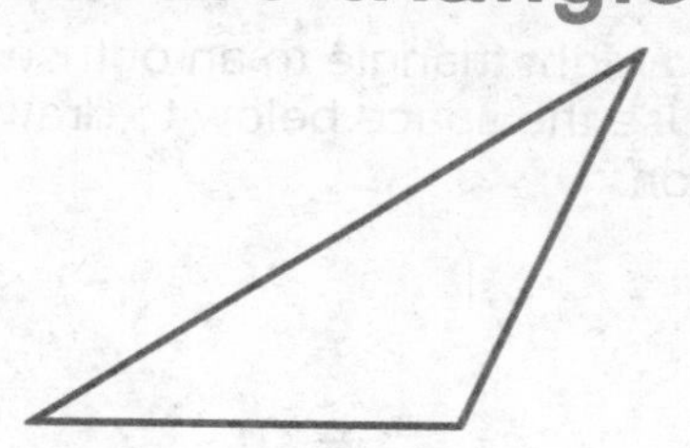

Ideas for Use

- Write a tally mark on each card every time you read the word in this chapter or use it in your writing. Challenge yourself to use at least 10 tally marks for each word card.
- Draw or write additional examples for each card. Be sure your examples are different from what is on the front of the card.

The line segment where two faces of a three-dimensional figure meet.

Describe a real-world example of an edge.

The unit of measure for volume.

Name three other units of measurement in math and what they measure.

A flat surface.

Read and solve this riddle: I am a triangular prism. I have nine edges and six vertices. How many faces do I have?

A triangle with three congruent sides.

The prefix *equi-* means "equal." *Lat* is a Latin root meaning "side." Explain how these word parts can help you remember this definition.

A triangle with at least two congruent sides.

Draw an example of an isosceles triangle.

A polygon with six sides and six angles.

How are hexagons a subcatergory of polygons?

A triangle with 1 obtuse angle and 2 acute angles.

Compare a right triangle to an obtuse triangle. Use the space below to draw your comparison.

A two-dimensional pattern of a three-dimensional figure.

What does net mean in this sentence? *The soccer player's hand became tangled in the goal's net.*

MY Vocabulary Cards

Lesson 12-1

octagon

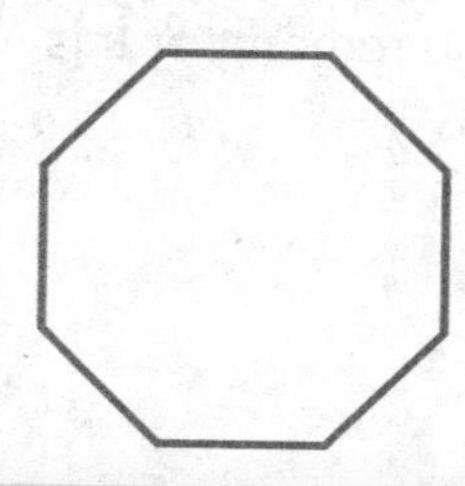

Lesson 12-5

parallelogram

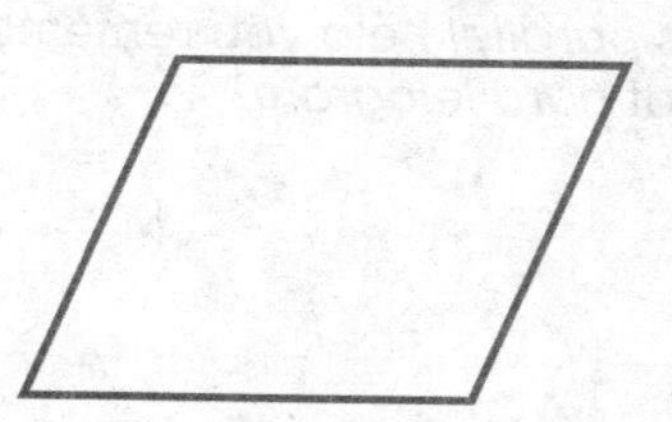

Lesson 12-1

pentagon

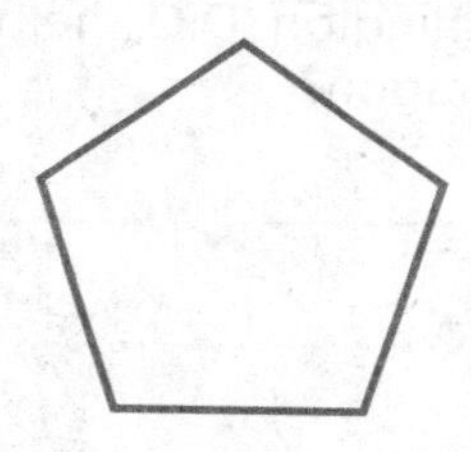

Lesson 12-1

polygon

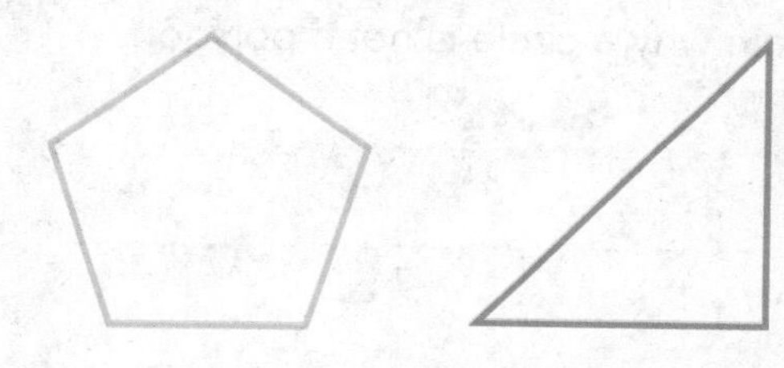

Lesson 12-7

prism

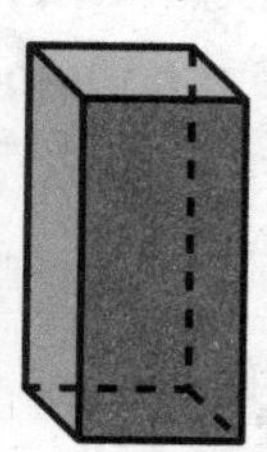

Lesson 12-5

rectangle

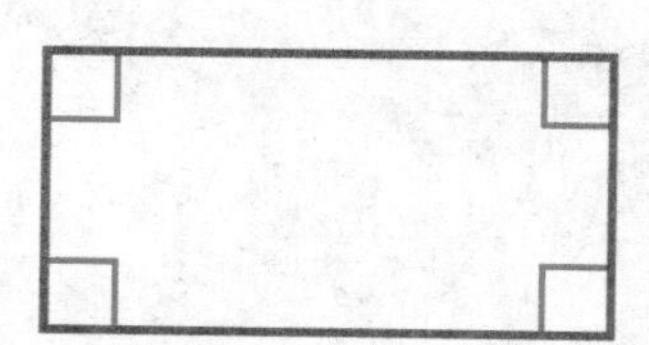

Lesson 12-6

rectangular prism

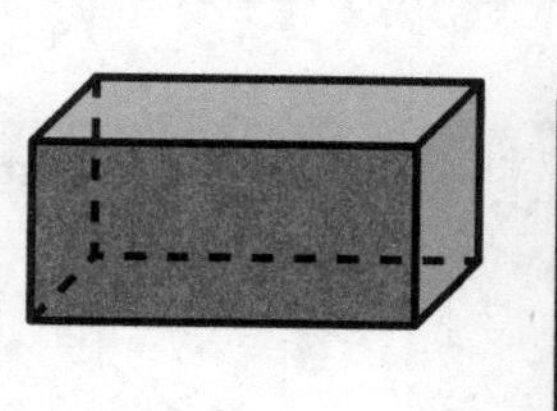

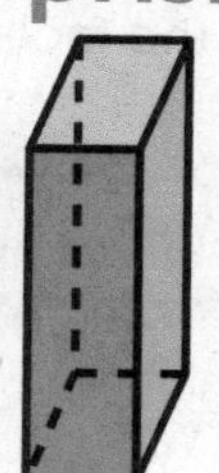

Lesson 12-1

regular polygon

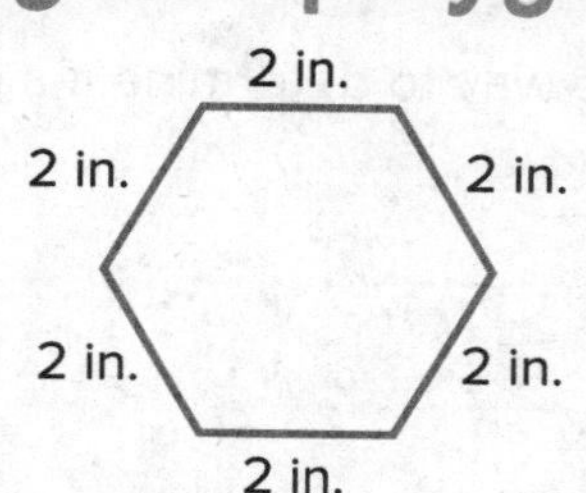

Ideas for Use

- Develop categories for the words. Sort them by category. Ask another student to guess each category.
- Draw or write examples for each card. Be sure your examples are different from what is shown on each card.

A quadrilateral in which each pair of opposite sides is parallel and congruent.

How does *parallel* help you remember the meaning of *parallelogram?*

A polygon with eight sides.

Okto is a Greek root meaning "eight." How can this help you remember this vocabulary word?

A closed figure made up of line segments that do not cross each other.

Explain why a circle is not a polygon.

A polygon with five sides.

How can the Pentagon, a government building in Washington, D.C., help you remember pentagon?

A quadrilateral with four right angles; opposite sides are equal and parallel.

Compare a rectangle with a square.

A three-dimensional figure with two parallel, congruent faces, called bases. At least three faces are rectangles.

Look at the orange prism on the front of the card. What shape are its bases?

A polygon in which all sides and angles are congruent.

What is one way to determine if a polygon is regular?

A prism that has rectangular bases.

Describe the faces of any rectangular prism.

MY Vocabulary Cards

Lesson 12-5

rhombus

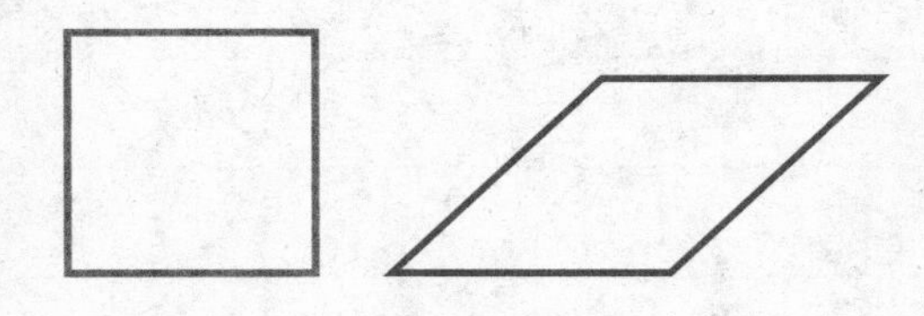

Lesson 12-3

right triangle

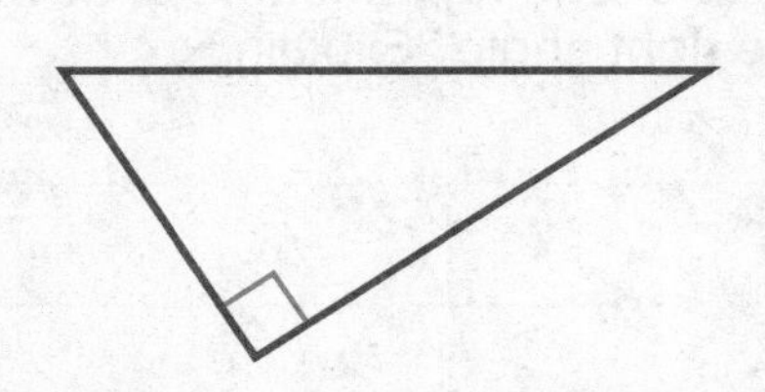

Lesson 12-3

scalene triangle

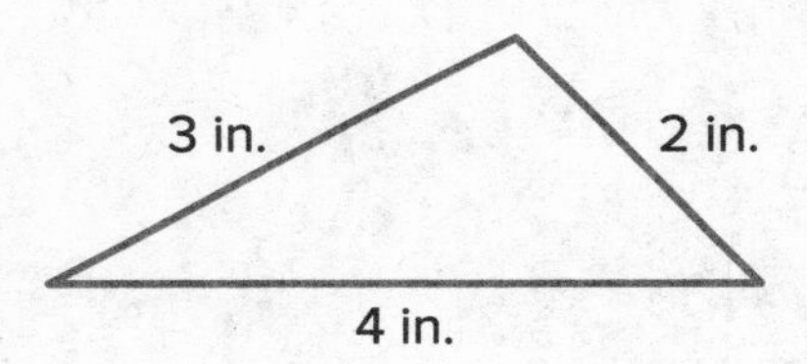

Lesson 12-5

square

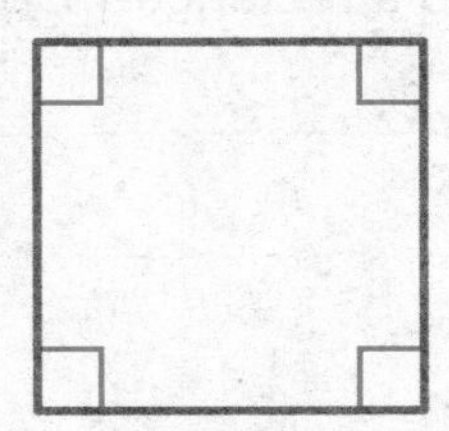

Lesson 12-6

three-dimensional figure

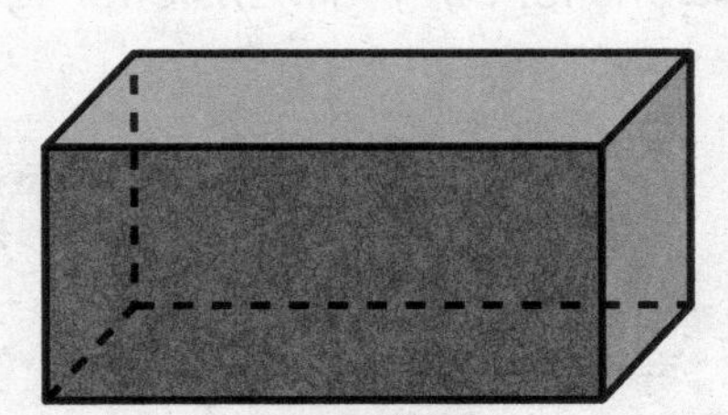

Lesson 12-5

trapezoid

Lesson 12-7

triangular prism

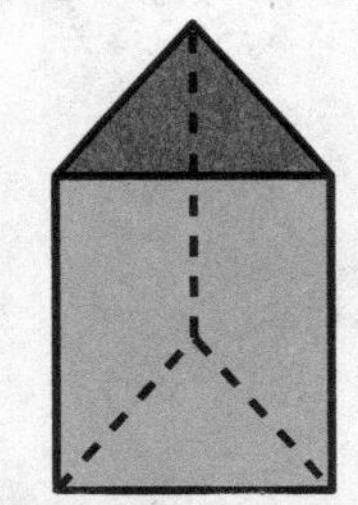

Lesson 12-8

unit cube

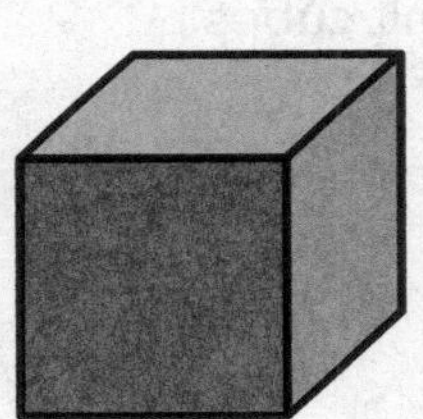

Ideas for Use

- Practice your penmanship! Write each word in cursive.
- Sort cards so that only polygons are displayed. Explain your sorting to a partner.

A triangle with 1 right angle and 2 acute angles.

Is it possible for a right triangle to have more than one right angle? Explain.

A parallelogram with four congruent sides.

Explain whether a rectangle is a rhombus.

A parallelogram with four congruent sides and four right angles.

Is a square also a rectangle? Explain.

A triangle with no congruent sides.

Draw a scalene triangle below.

A quadrilateral with exactly one pair of opposite sides parallel.

Draw a picture of two quadrilaterals—one that is a trapezoid and one that is not.

A figure that has length, width, and height.

Write a tip to help you remember the number of dimensions for *three-dimensional figures.*

A cube with a side length of one unit.

Draw a rectangular prism below that has a volume of 8 unit cubes.

A prism that has triangular bases.

Describe the bases of any triangular prism.

MY Vocabulary Cards

Lesson 12-7

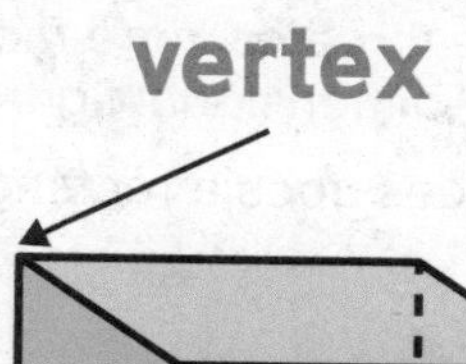

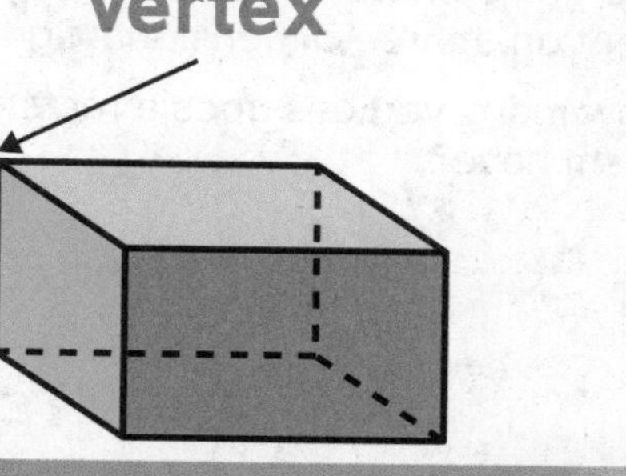

Lesson 12-8

volume

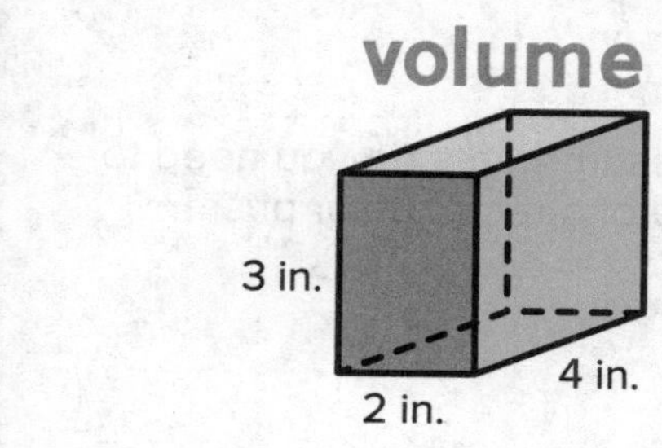

$V = \ell wh$

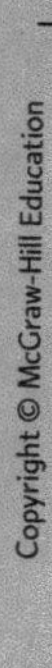

Ideas for Use

- Write key concepts from some lessons on the front of blank cards. Write a few examples on the back of each card to help you study.
- Use a blank card to write this chapter's essential question. Use the back of the card to write or draw examples that help you answer the question.

The amount of space inside a three-dimensional figure.

How many measurements do you need to find the volume of a rectangular prism? What are they?

The point where three or more faces meet on a three-dimensional figure.

How many vertices does a rectangular prism have?

MY Foldable

FOLDABLES® Follow the steps on the back to make your Foldable.

Pentagon

Regular

Not Regular

Hexagon

Regular

Not Regular

Octagon

Regular

Not Regular

Polygons

Regular	Not Regular

Triangle

Regular	Not Regular

Quadrilateral

Regular	Not Regular

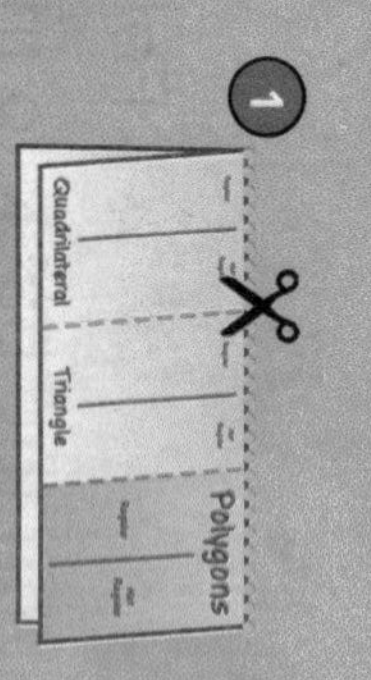

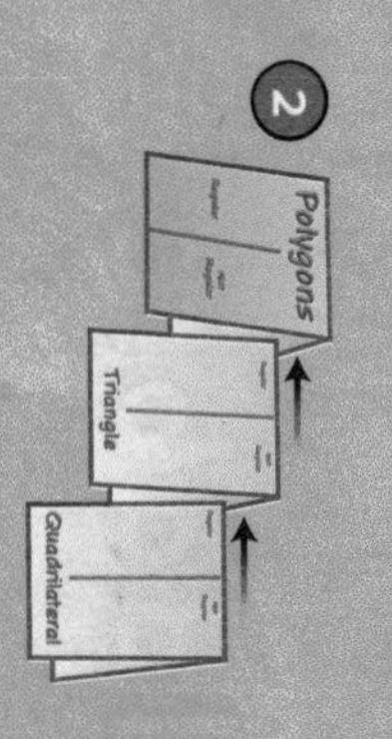

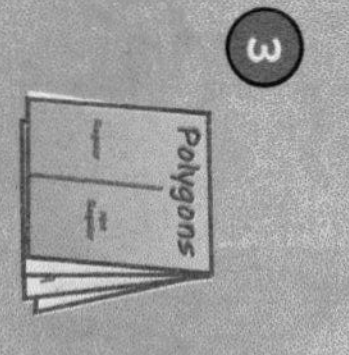

Name ....

Lesson 1
Polygons

ESSENTIAL QUESTION
How does geometry help me solve problems in everyday life?

A **polygon** is a closed figure made up of line segments that do not cross each other.

Polygons	Not Polygons
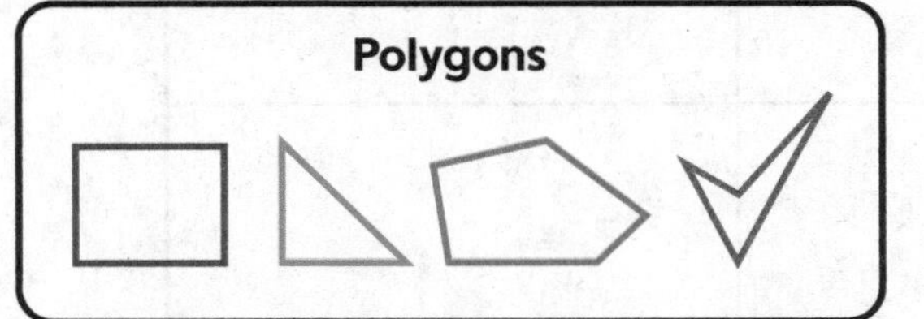	

Math in My World

Example 1

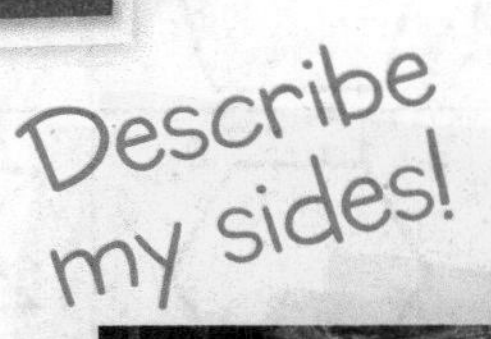

The building shown is the Pentagon in Washington, D.C. Describe the sides of the figure formed by the red outline. Does the red outline form a polygon?

The figure has ________ sides.

Do the sides ever cross each other? ________
The figure is a polygon.

A **regular polygon** is a polygon with congruent sides and congruent angles. **Congruent sides** are equal in length. **Congruent angles** have the same degree measure.

Example 2

Determine if the polygon appears to be *regular* or *not regular*.

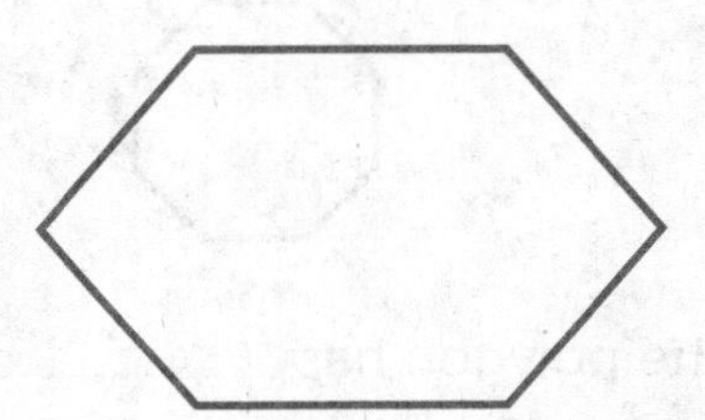

The top and bottom sides appear ________ than the other sides.

Are all six sides of the polygon congruent? ________

It is ________ regular.

Polygons are a subcategory of two-dimensional figures. A *subcategory* is a subdivision that has common characteristics within a larger category.

Example 3

Complete the table below.

Polygon	Regular	Not Regular	Number of Sides	Draw another polygon that is not regular.
Triangle				
Quadrilateral				
Pentagon				
Hexagon				
Octagon				

Guided Practice

1. Name the polygon. Determine if it appears to be *regular* or *not regular.*

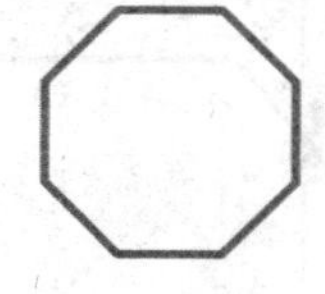

The polygon has ________ sides.

The sides appear to be ____________.

It is a ________________.

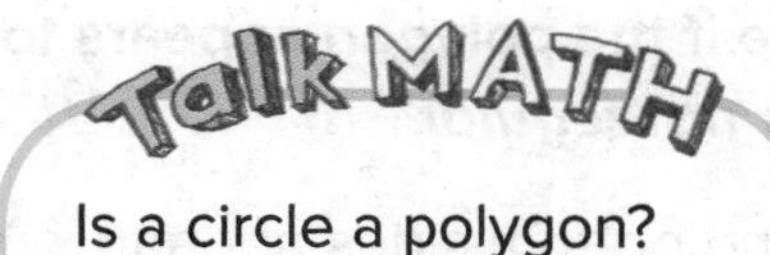

Is a circle a polygon? Explain.

Name

Independent Practice

Processes & Practices 7 **Identify Structure Name each polygon. Determine if it appears to be *regular* or *not regular*.**

2.

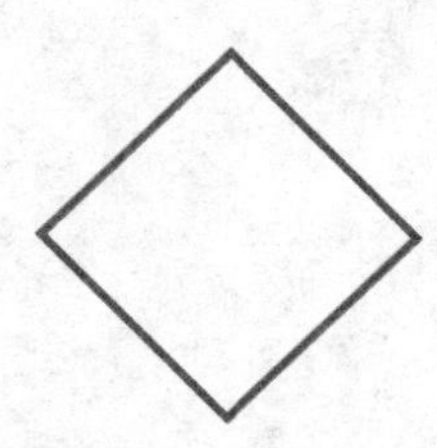

3.

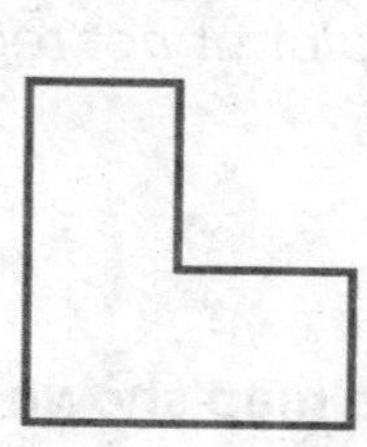

4.

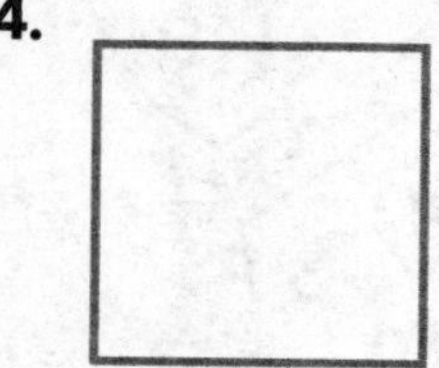

5.

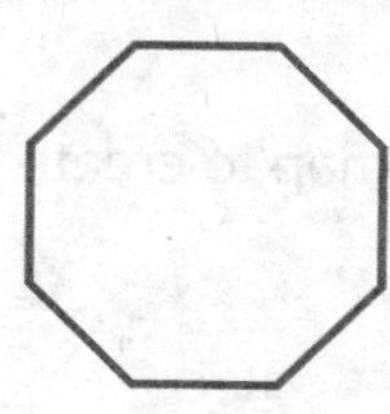

Draw each polygon.

6. triangle; not regular

7. pentagon; not regular

8. quadrilateral; not regular

9. triangle; regular

Problem Solving

10. What polygons make up the design?

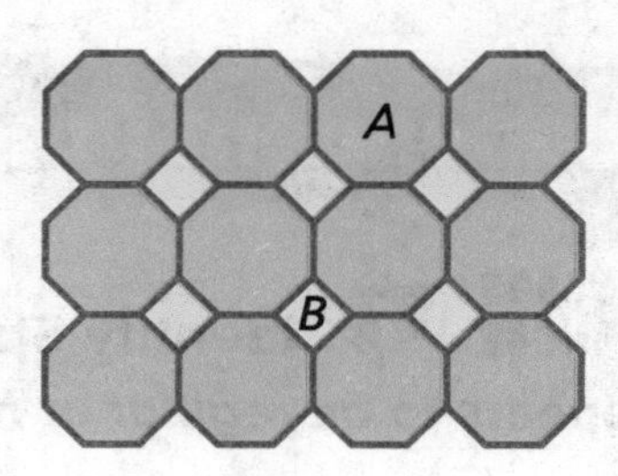

11. Describe polygon B as *regular* or *not regular.*

For Exercises 12 and 13, use the map shown at the right.

12. Circle the polygon that is a quadrilateral.

Brain Builders

13. Use the city streets on the map to create a regular polygon.

14. Processes & Practices 1 **Make Sense of Problems** Explain why every square is a regular polygon. Draw and label a model to support your answer.

15. **Building on the Essential Question** How can polygons be considered a subcategory of two-dimensional figures? Name another subcategory of two-dimensional figures.

Name ....

MY Homework

Lesson 1

Polygons

Homework Helper

Need help? **connectED.mcgraw-hill.com**

Name the polygon used to form the greeting card shown. Does the red outline appear to be a regular polygon?

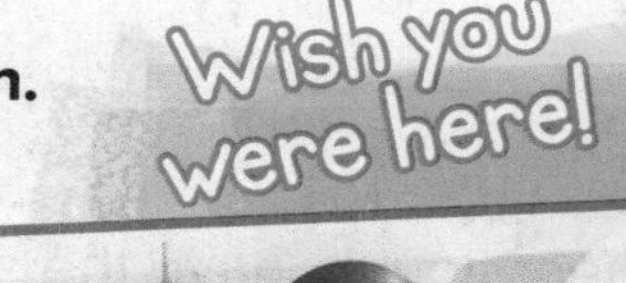

The polygon has four sides.

The top and bottom sides appear to be slightly longer than the other sides.

It is a quadrilateral.

It is not regular.

Practice

Name each polygon. Determine if it appears to be *regular* or *not regular*.

1.

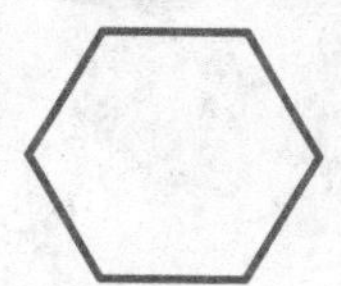

____

2.

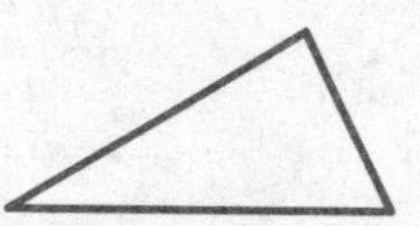

____

Vocabulary Check

Fill in each blank with the correct word(s) to complete each sentence.

3. A polygon is a ____________ figure made up of line segments that do not cross each other.

4. A regular polygon is a polygon with ____________ sides and ____________ angles.

Problem Solving

For Exercises 5–7, use the tangram pieces shown at the right.

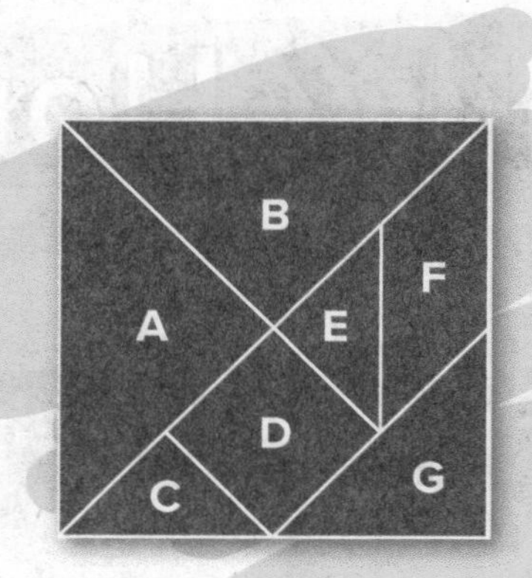

5. Which of the polygon(s) appear to be regular?

6. What polygons are represented in the tangrams?

7. Congruent figures have the same size and shape. Which polygons appear to be congruent?

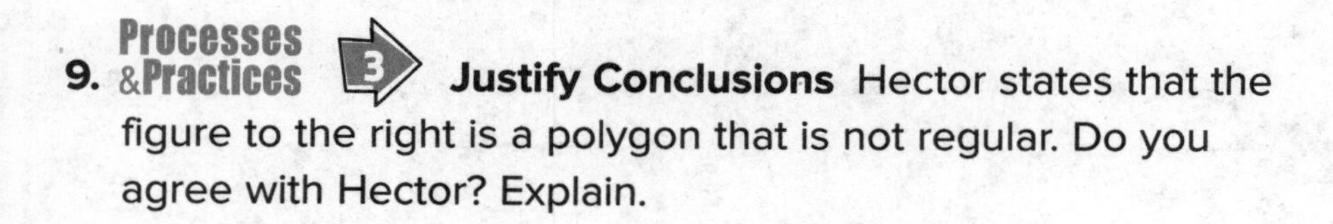

8. Name the polygons used to form the front and sides of the tent shown. Determine if they appear to be *regular* or *not regular.*

9. **Processes & Practices** 3 **Justify Conclusions** Hector states that the figure to the right is a polygon that is not regular. Do you agree with Hector? Explain.

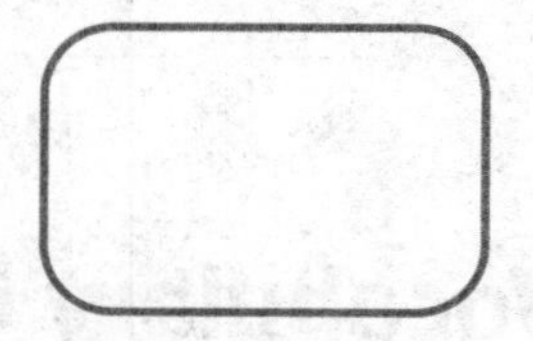

10. **Test Practice** Which of the following figures is a polygon?

Ⓐ 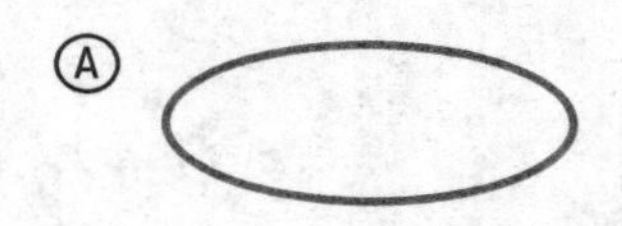

Ⓑ 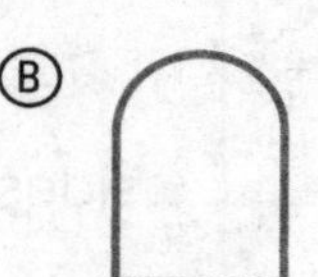

Ⓒ 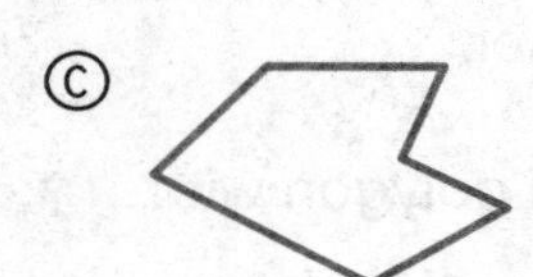

Ⓓ 

Name

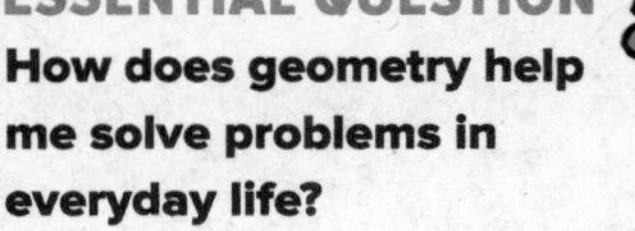

Lesson 2 Hands On

Sides and Angles of Triangles

ESSENTIAL QUESTION
How does geometry help me solve problems in everyday life?

A triangle is a polygon with three sides and three angles.

Measure It

Measure the sides of each pair of triangles below to the nearest tenth of a centimeter. Then record the measures.

Pair A

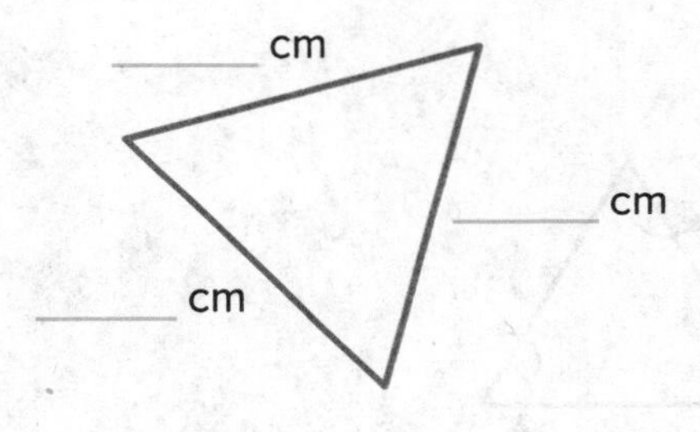

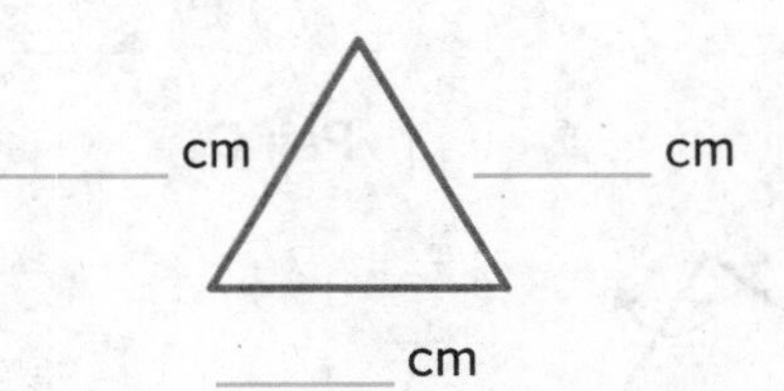

Pair B

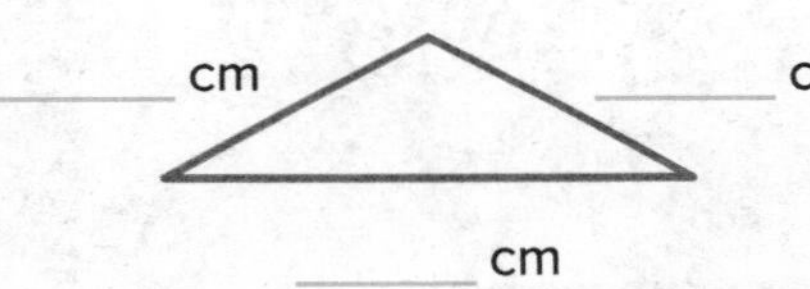

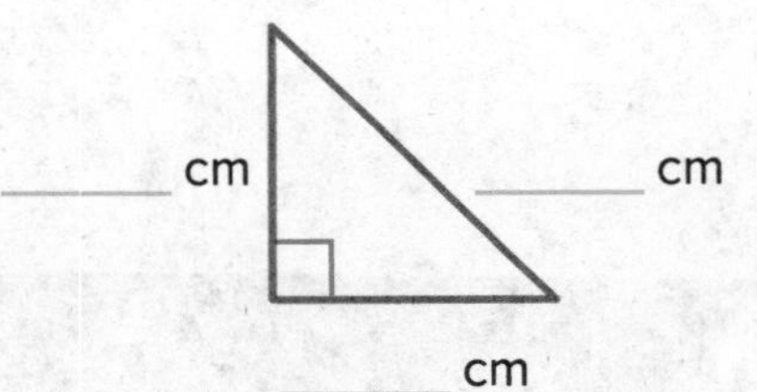

Pair C

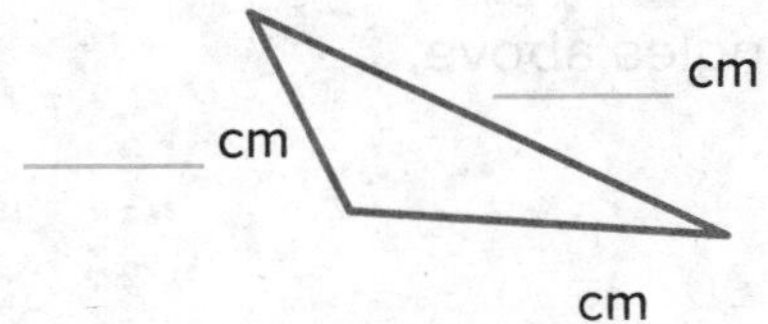

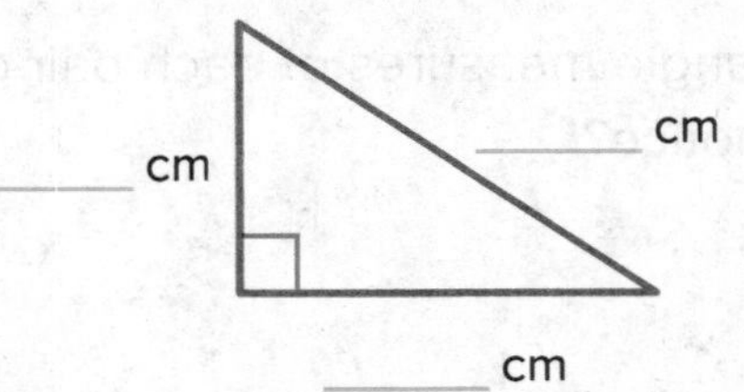

Cut out and use centimeter ruler.

Talk About It

1. Compare the side lengths of each pair of triangles above. What do you notice?

____

____

Try It

Measure the angles of each pair of triangles below to the nearest degree. Then record the measures.

Pair A

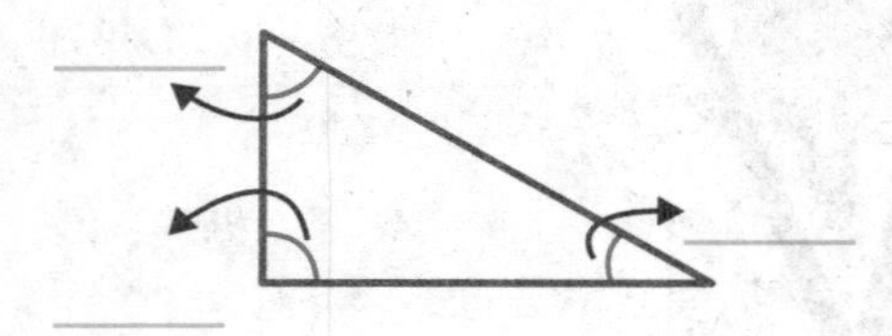

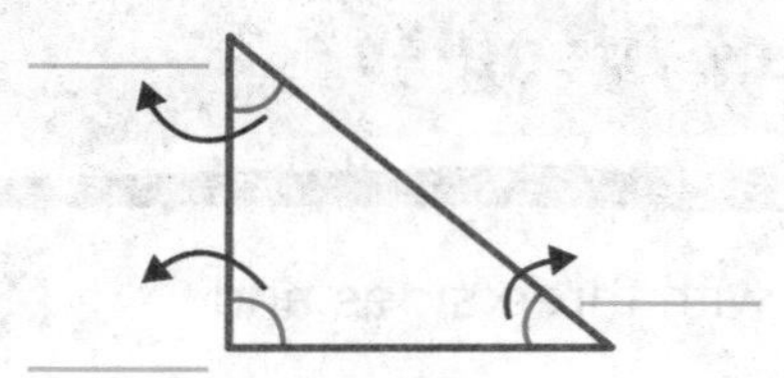

Pair B

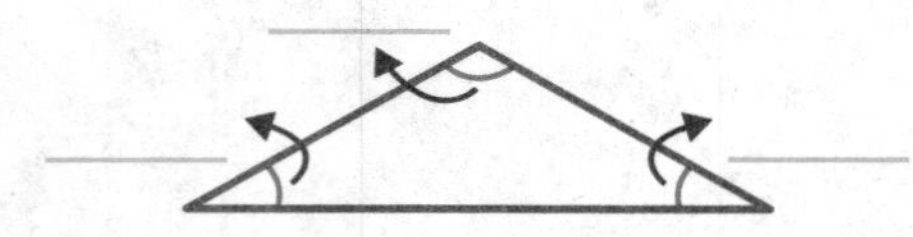

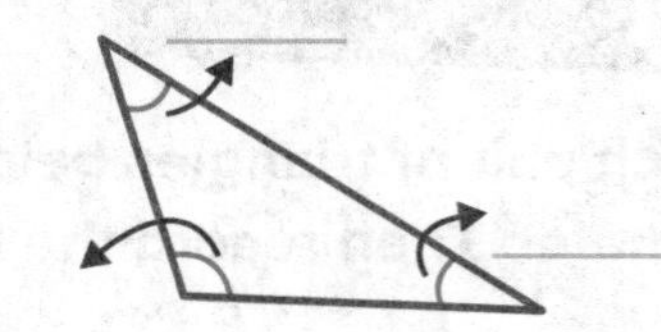

Pair C

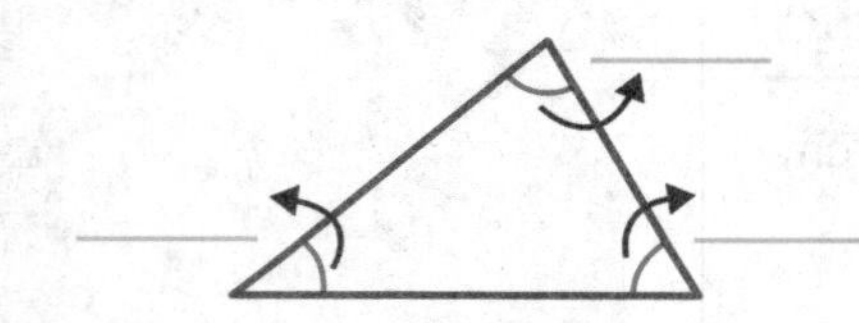

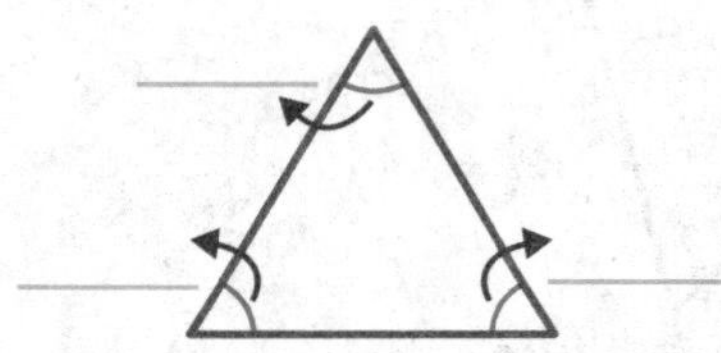

Talk About It

2. Compare the angle measures of each pair of triangles above. What do you notice?

3. Processes & Practices 1 **Make Sense of Problems** Explain how a triangle is a special kind of polygon.

Name

Practice It

Measure the sides of each triangle to the nearest tenth of a centimeter. Then describe the number of congruent sides.

4.

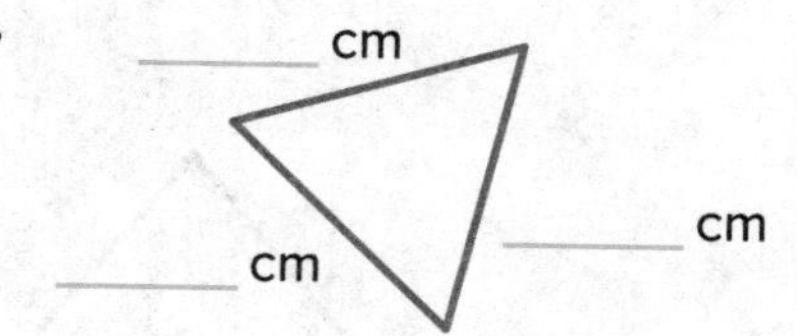

5.

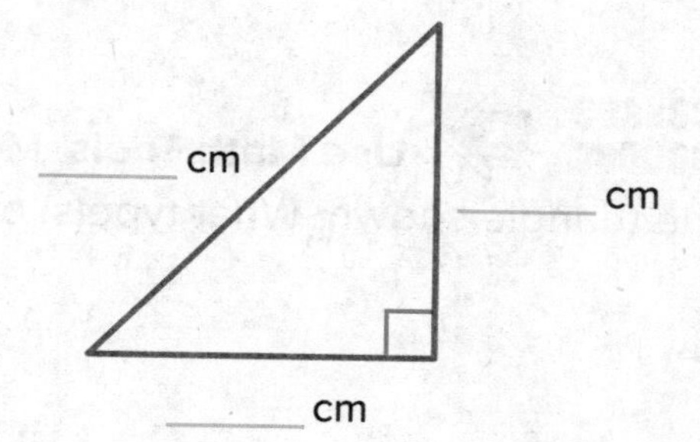

6.

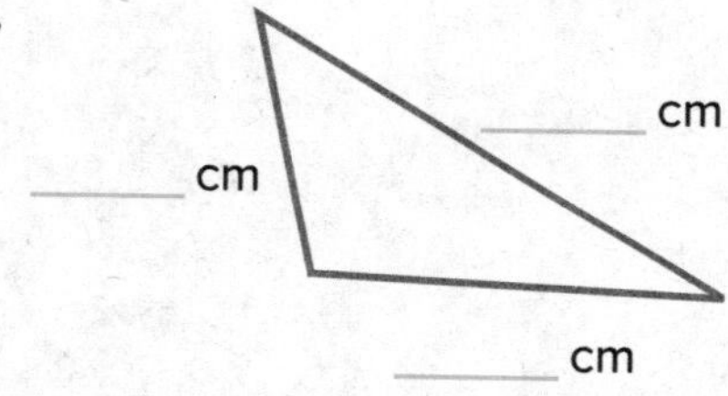

7.

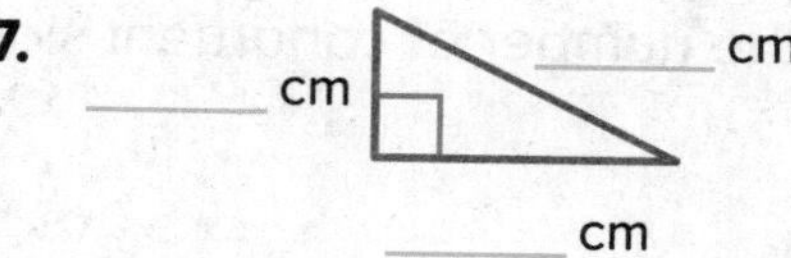

Measure the angles of each triangle to the nearest degree. Then describe the number of acute, right, or obtuse angles.

8.

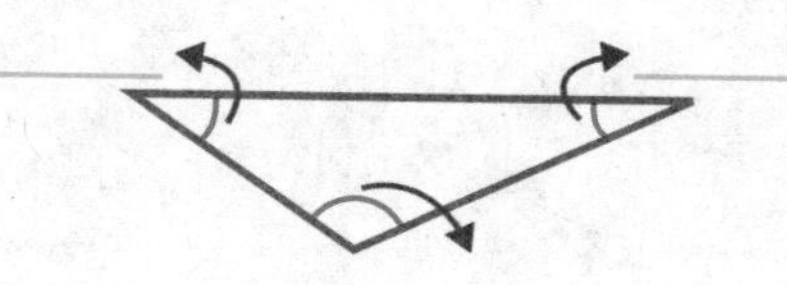

9.

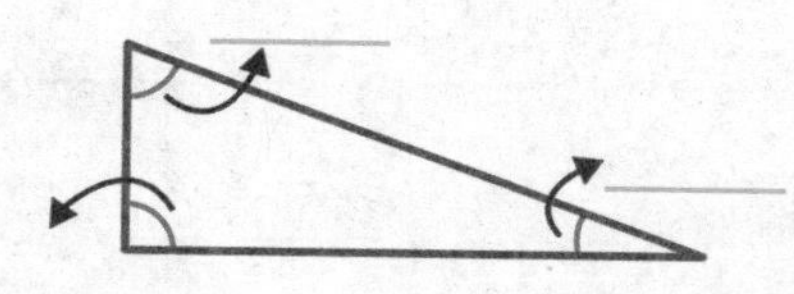

10.

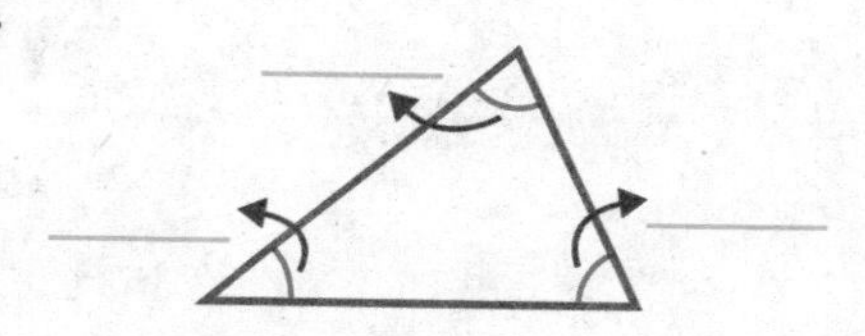

11.

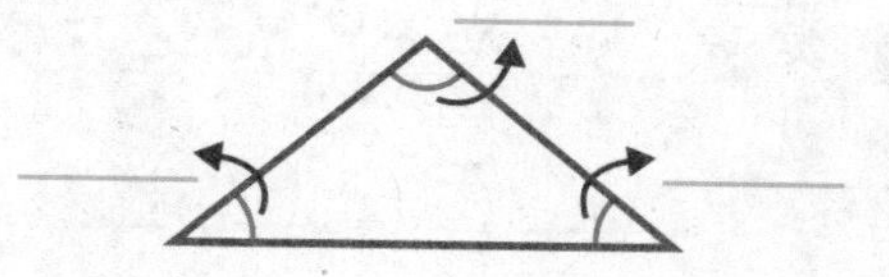

Apply It

12. In music, the "triangle" is an instrument with three congruent sides. If you know that the perimeter of the triangle is 18 inches, what is the measure of one side?

____

13. Processes & Practices 5 **Use Math Tools** Measure the angles in the triangle shown. What type(s) of angles are they?

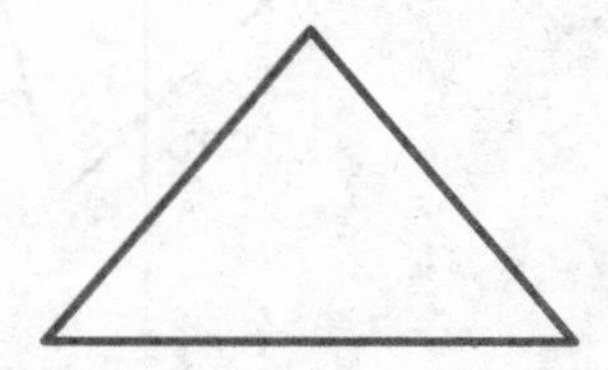

____

14. Refer to Exercise 13. Measure the sides of the triangle. Then describe the number of congruent sides.

____

15. Processes & Practices 3 **Which One Doesn't Belong?** Circle the triangle that does not belong with the other three. Explain your reasoning.

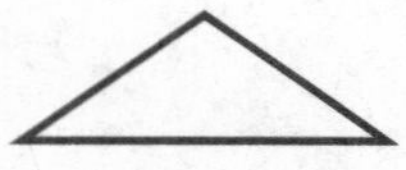 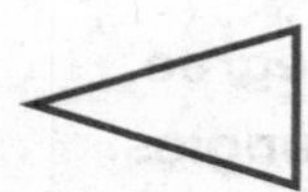 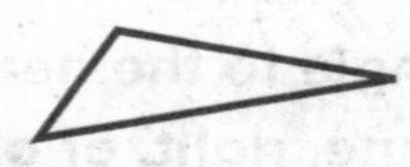 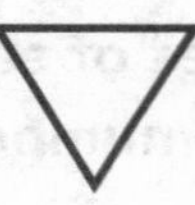

____

____

Write About It

16. How are all triangles the same, and how can they be different?

____

____

____

Name

MY Homework

Lesson 2

Hands On: Sides and Angles of Triangles

Homework Helper

eHelp

Need help? connectED.mcgraw-hill.com

Measure the sides of each triangle to the nearest tenth of a centimeter. Then describe the number of congruent sides.

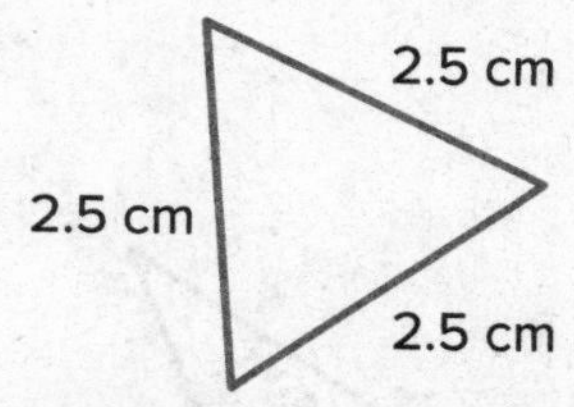

The triangle has 3 congruent sides.

The triangle has 2 congruent sides.

Measure the angles of each triangle to the nearest degree. Then describe the number of acute, right, or obtuse angles.

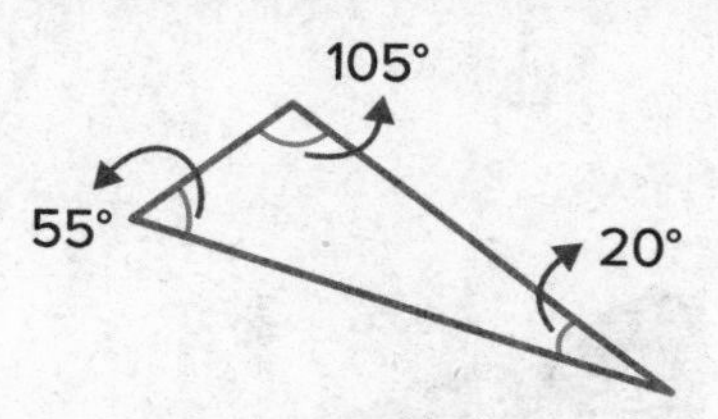

The triangle has 1 obtuse angle and 2 acute angles.

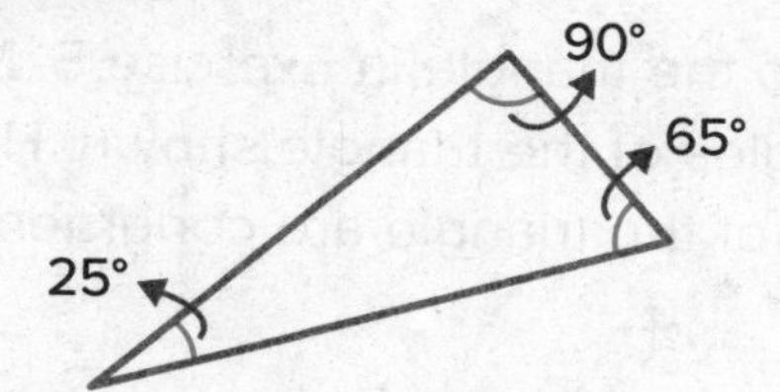

The triangle has 1 right angle and 2 acute angles.

Practice

Measure the sides of each triangle to the nearest tenth of a centimeter. Then describe the number of congruent sides.

1. ______ cm ______ cm ______ cm

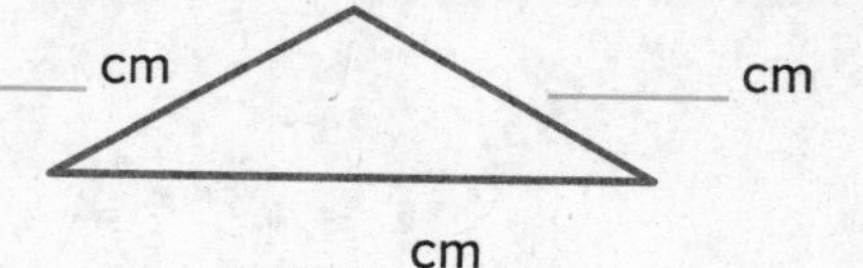

____

____

2. ______ cm ______ cm ______ cm

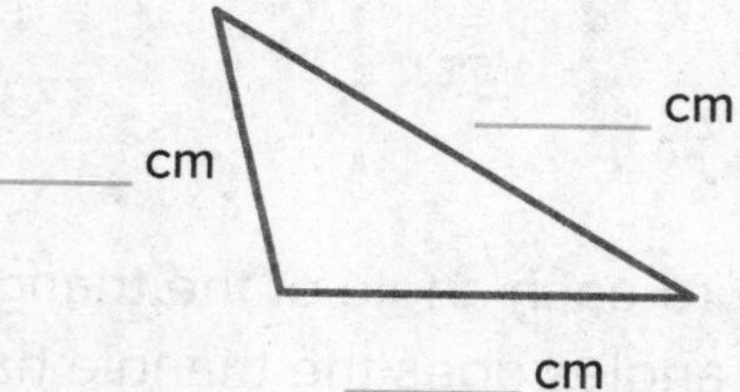

____

____

Cut out and use this centimeter ruler. →

cm 0 1 2 3 4 5

Measure the angles of each triangle to the nearest degree. Then describe the number of acute, right, or obtuse angles.

3.

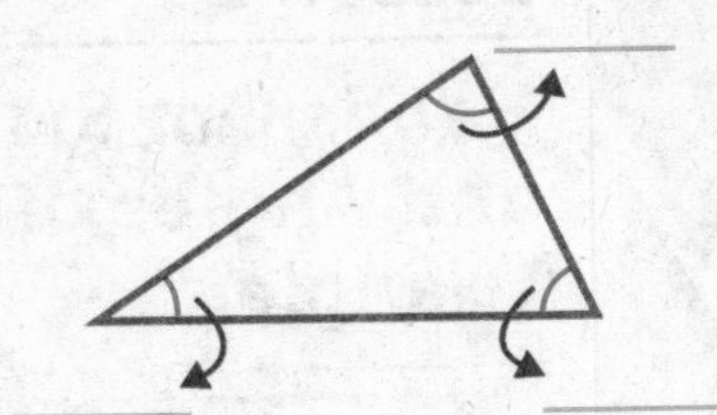

4.

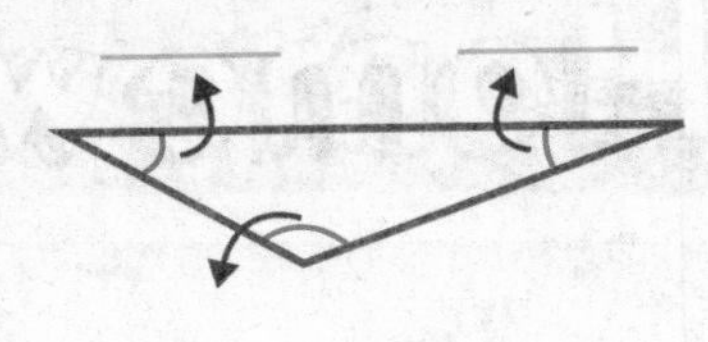

Problem Solving

5. Measure the sides of the triangle shown. How many sides of the triangle are congruent?

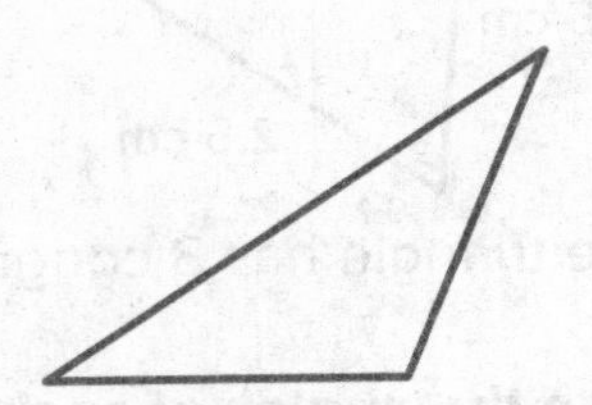

6. Refer to the triangle in Exercise 5. Measure the angles of the triangle shown. How many angles of the triangle are congruent?

7. In billiards, a rack is used to organize billiard balls at the beginning of the game. Jason is making the wood rack and found that each angle is congruent and that the sum of the angles is 180°. What is the measure of each angle?

8. Measure each angle of the triangle. How many acute angles does the triangle have?

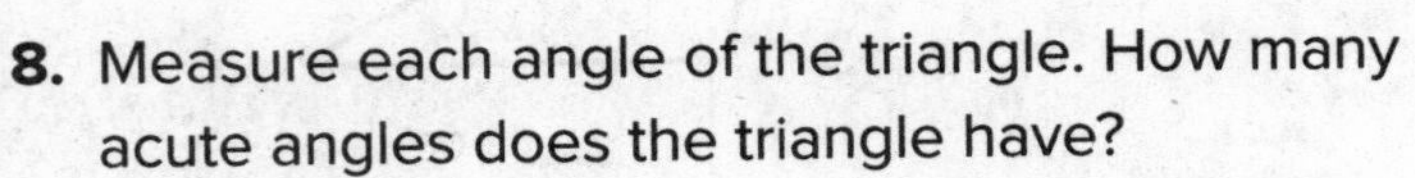

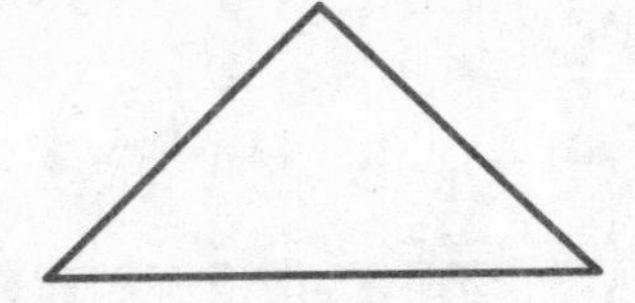

Name ....

Lesson 3
Classify Triangles

ESSENTIAL QUESTION
How does geometry help me solve problems in everyday life?

You can classify triangles using one or more attributes. An **attribute** is a characteristic of a figure, like side measures and angle measures.

Math in My World

Example 1

The Hammond family traveled from Columbus, Ohio, to Dallas, Texas, and then to Atlanta, Georgia, before returning home. The distance of each flight is shown on the map. Find the number of congruent sides.

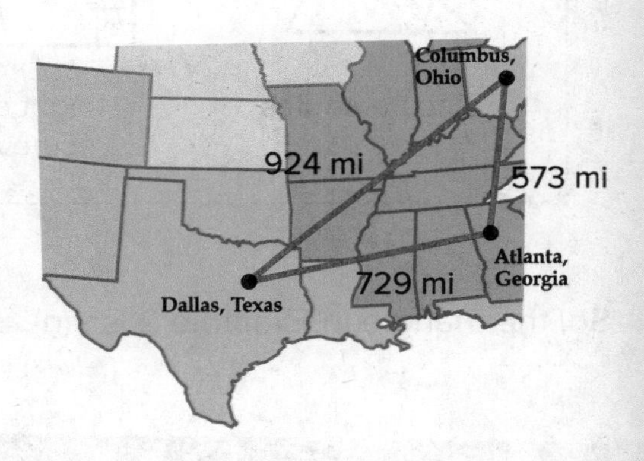

The lengths of the sides of the triangle are

924 miles, 573 miles, and ________ miles.

How many sides of the triangle are congruent? ________

Key Concept Classify Triangles by Sides

Isosceles Triangle	Equilateral Triangle	Scalene Triangle
3 in. 3 in. 2 in.	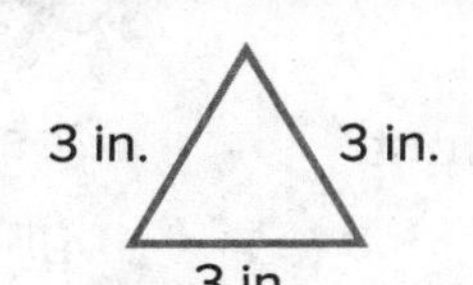	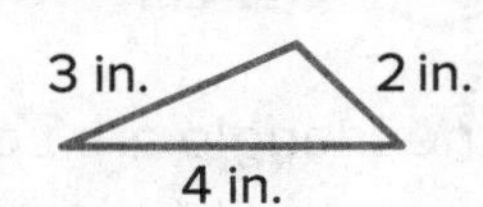
at least two sides congruent	all sides congruent	no sides congruent

So, the triangle formed on the map in Example 1 is a

____________ triangle.

Example 2

Triangles form the sides of the Khafre Pyramid in Egypt. Determine the number of acute, obtuse, or right angles in the triangle.

How many angles of the triangle are acute? ________

How many angles of the triangle are obtuse? ________

How many angles of the triangle are right? ________

Key Concept Classify Triangles by Angles

Acute Triangle	Right Triangle	Obtuse Triangle
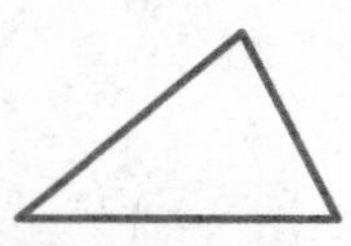	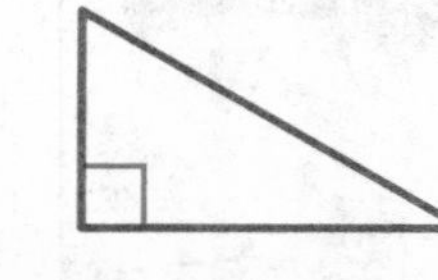	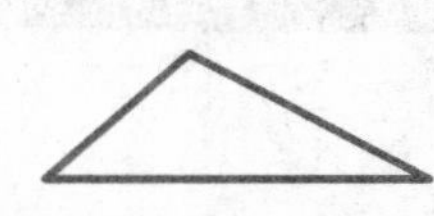
3 acute angles	1 right angle, 2 acute angles	1 obtuse angle, 2 acute angles

So, the triangle in Example 2 is a(n) ____________________.

Guided Practice

1. Classify the triangle based on its sides.

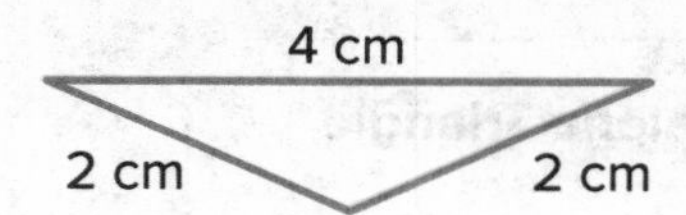

How many sides of the triangle are congruent?

____

The triangle is a(n) ____________________.

Talk MATH

Describe an isosceles right triangle.

2. Classify the triangle based on its angles.

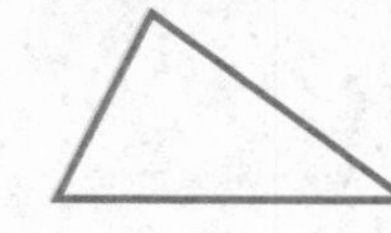

The triangle is a(n) ____________________.

Name ....

Independent Practice

Determine the number of congruent sides for each triangle. Then classify the triangle based on its sides.

3. 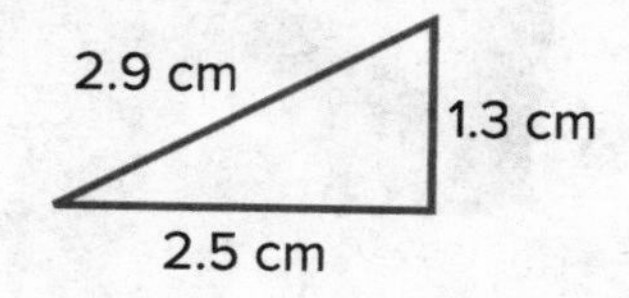

____

4. 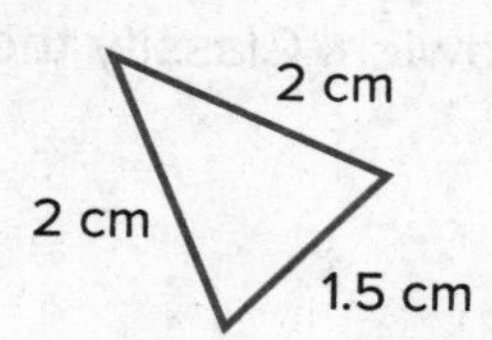

____

Classify each triangle based on its angles.

5. 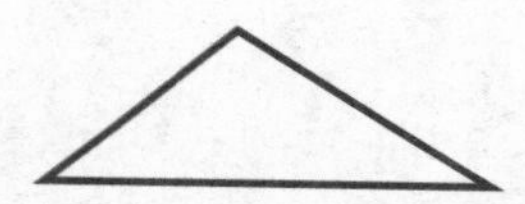

____

6. 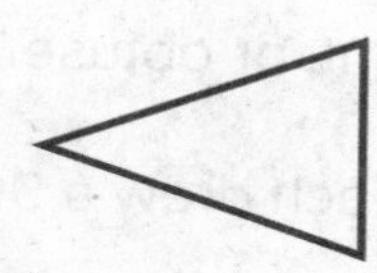

____

7. 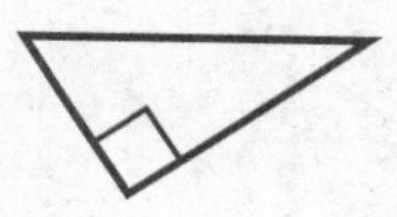

____

8. 

____

Draw each triangle.

9. equilateral triangle

10. right triangle

Problem Solving

11. Half of a rectangular sandwich looks like a triangle. Classify it based on its angles.

12. Processes & Practices 7 **Identify Structure** Measure the sides of the sandwich. Classify the triangle based on its sides.

Brain Builders

13. Processes & Practices 3 **Draw a Conclusion** Emma, Gabriel, Jorge, and Makayla each drew a different triangle. Use the clues below to describe each person's triangle as isosceles, equilateral, or scalene and also as acute, right, or obtuse.

- Gabriel and Jorge each drew a 90° angle in their triangles.
- Gabriel's triangle does not have any congruent sides.
- One angle in Emma's triangle measures greater than 90°.
- Each side of Makayla's triangle and two sides of Emma's and Jorge's triangles are four centimeters long.

14. **Building on the Essential Question** How do I classify triangles using their attributes? Include a sketch of each type of triangle.

Name

MY Homework

Lesson 3

Classify Triangles

Homework Helper

Need help? connectED.mcgraw-hill.com

There is a large pyramid standing in front of the Louvre museum in Paris, France. The sides of the pyramid are shaped like triangles. Classify the red triangle based on its angles.

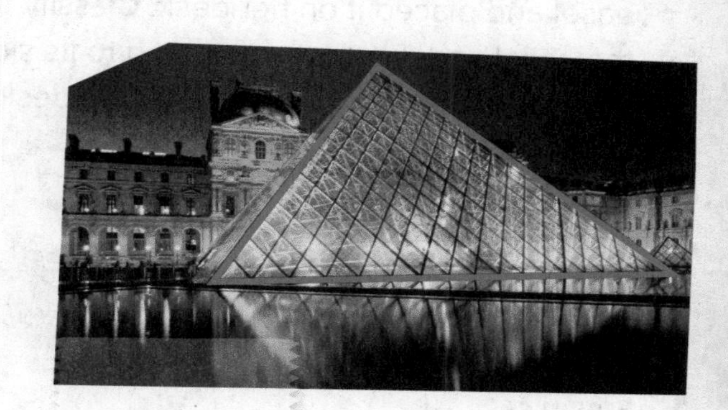

There are three acute angles.

So, the triangle formed by the side of the pyramid is an acute triangle.

Practice

1. Determine the number of congruent sides. Then classify the triangle based on its sides.

1.4 cm, 3.1 cm, 2.7 cm

How many sides of the triangle are congruent?

______

The triangle is a ______________.

Vocabulary Check

Fill in each blank with the correct term(s) or number(s) to complete each sentence.

2. An equilateral triangle is a triangle with ______ congruent sides.
3. An acute triangle is a triangle with ______ angles each less than ______.
4. An obtuse triangle is a triangle with one angle that is greater than ______.

Problem Solving

5. Look at the triangle on the top of the White House in the photo. Describe the sides and angles of the triangle.

6. Serena has an art easel with sides of equal length. She opened the easel and placed it on her desk. Classify the type of triangle formed by the easel and the desk according to its sides. Next, classify the type of triangle formed by the easel and the desk according to its angles.

Brain Builders

7. **Processes & Practices 7** **Identify Structure** The image shown at the right contains many triangles. Describe the different types of triangles found in the image. Explain how you classified each type.

8. **Processes & Practices 3** **Justify Conclusions** A triangle has two sides that are perpendicular. Could the triangle be isosceles, equilateral, or scalene? Explain. Include a drawing to support your answer.

9. **Test Practice** Which of the following figures is an obtuse triangle?

 Ⓐ 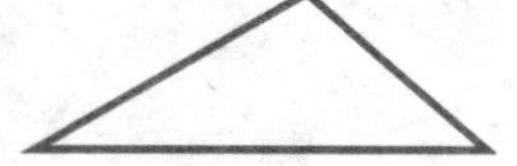

 Ⓑ 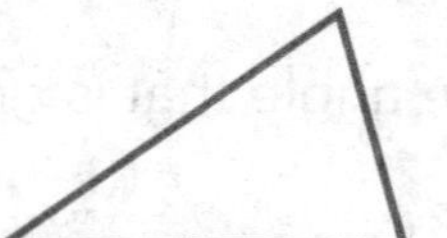

 Ⓒ 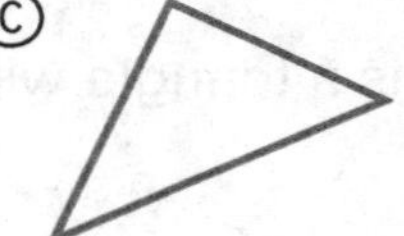

Ⓓ 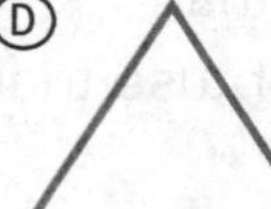

Check My Progress

Vocabulary Check

State whether each sentence is *true* or *false*.

1. A triangle with no congruent sides is a **scalene triangle**. ______

2. A polygon that has 4 sides and 4 angles is a **pentagon**. ______

3. Sides or angles with the same measure are **congruent**. ______

4. A **right triangle** is a triangle with two right angles. ______

Concept Check

Name each polygon. Determine if it appears to be *regular* or *not regular*.

5. 

______

6. 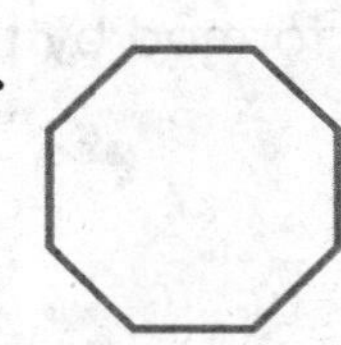

______

Measure the sides of each triangle to the nearest tenth of a centimeter. Then describe the number of congruent sides.

7. 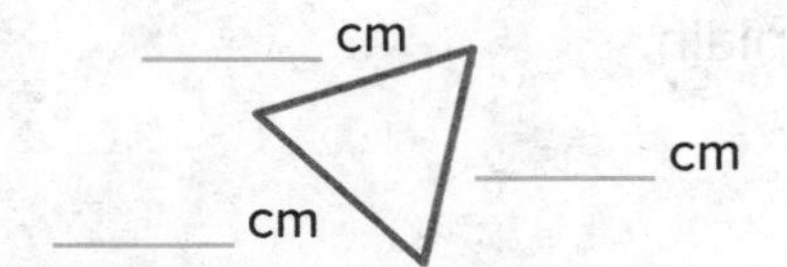

______

______

8. 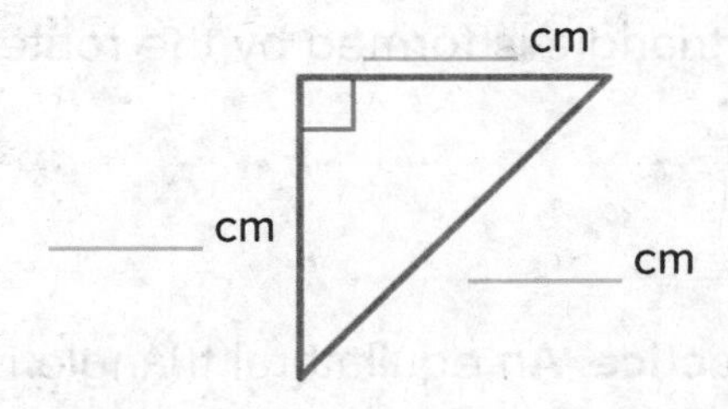

______

______

Problem Solving

9. Name the polygon shown by the video game screen at the right. Determine if it appears to be *regular* or *not regular*.

____

10. Steve has three lengths of fence. He connects them to make a triangular pen for his dog. If the lengths are 5 meters, 6 meters, and 10 meters, what type of triangle is formed by the dog pen?

____

11. Name the polygon shown by the banner at the right. Determine if it appears to be *regular* or *not regular*.

____

Brain Builders

12. What type of triangles are formed by the diagonal of a square and its sides? Explain.

____

____

____

13. Lindsay was going to visit her grandmother, shop at the mall, and then return home. The route she took was in the shape of a triangle. The distance between each place she visited was 10 miles. What type of triangle is formed by the route she traveled? Explain.

____

____

14. **Test Practice** An equilateral triangle is folded in half. What type of triangle is formed?

Ⓐ scalene triangle
Ⓑ isosceles triangle
Ⓒ equilateral triangle
Ⓓ obtuse triangle

Name ....

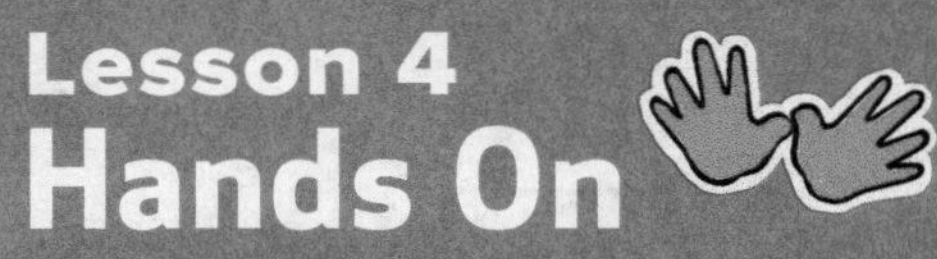

Lesson 4
Hands On
Sides and Angles of Quadrilaterals

ESSENTIAL QUESTION
How does geometry help me solve problems in everyday life?

A quadrilateral is a polygon with four sides and four angles.

Measure It

Measure the sides and angles of each figure to determine if any are congruent. Then determine if any sides are parallel. Complete the table.

Figure 1   Figure 2   Figure 3   Figure 4

Attribute	Figure(s)
Opposite sides are congruent.	
Opposite sides are parallel.	
Opposite angles are congruent.	

Each figure has ________ sides and ________ angles.

Talk About It

1. What common attributes do all of the figures have?

____

2. Does Figure 3 have all the attributes of Figure 2? Explain.

____

____

Try It

Measure the sides and angles of each figure to determine if any are congruent. Then determine if any sides are parallel. Complete the table.

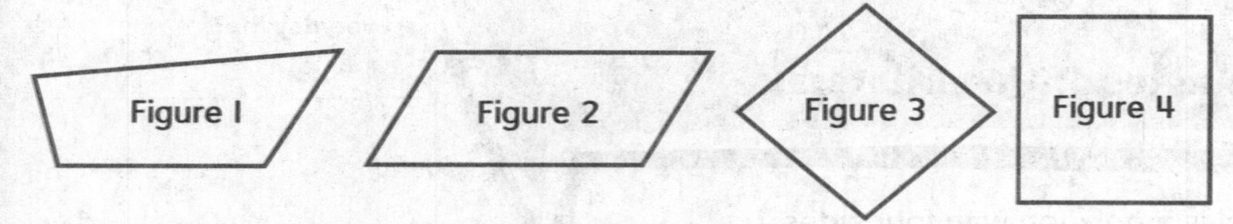

Attribute	Figure(s)
Opposite sides are congruent.	
Opposite sides are parallel.	
Opposite angles are congruent.	

Talk About It

3. Does Figure 3 have all the attributes of Figure 2? Explain.

4. What are some additional attributes that Figure 4 has that Figure 3 doesn't have?

5. Processes &Practices 1 **Make Sense of Problems** Explain how Figure 2 is a special kind of polygon.

6. Which figure does not have any of the attributes listed in the table?

Name ....

Practice It

Measure the sides and angles of each figure to determine if any are congruent or parallel. Then answer Exercises 7–13.

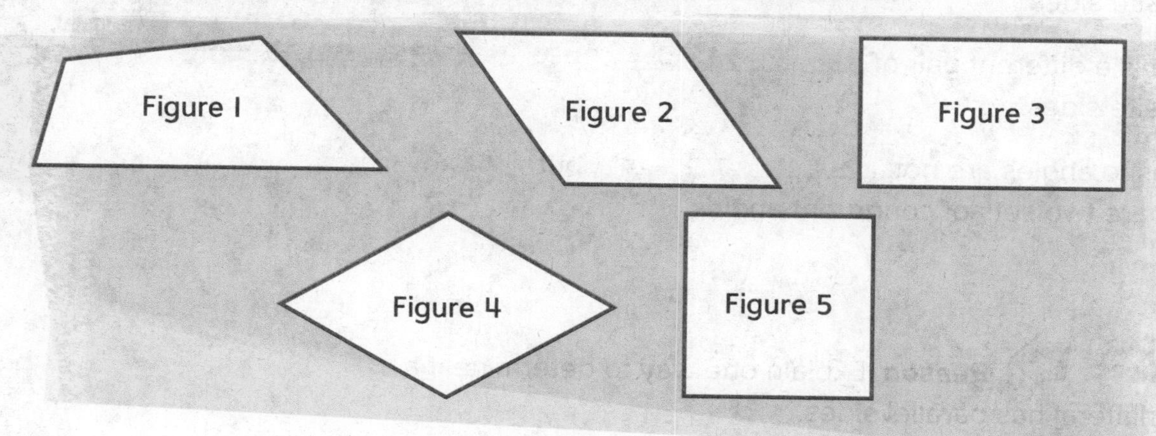

7. Complete the attributes of Figure 1.

Opposite sides are ______________ and ______________.

Opposite angles are ______________.

The figure has ______ sides and ______ angles.

8. Complete the attributes of Figure 2.

Opposite sides are ______________ and ______________.

Opposite angles are ______________.

The figure has ______ sides and ______ angles.

9. Which figures have all the attributes of Figure 1? ______________

10. Which figures have all the attributes of Figure 2? ______________

11. Which figures have all the attributes of Figure 3? ______________

12. Which figures have four right angles? ______________

13. Which figures have four equal sides? ______________

Apply It

14. Complete the attributes of the red quadrilateral outlining one side of the Chichen Itza pyramid in Mexico.

There is one pair of ______________ opposite sides.

There is a different pair of ______________ opposite sides.

Opposite angles are not ______________, but there are two sets of congruent angles.

15. Processes & Practices 2 **Reason** Explain one way to determine if a quadrilateral has parallel sides.

____

____

16. Processes & Practices 3 **Which One Doesn't Belong?** Circle the quadrilateral that does not belong with the other three. Explain your reasoning.

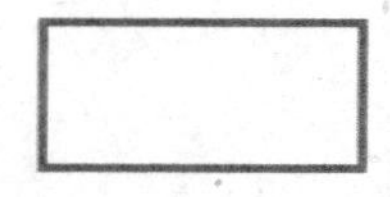
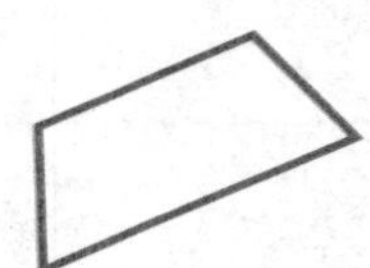

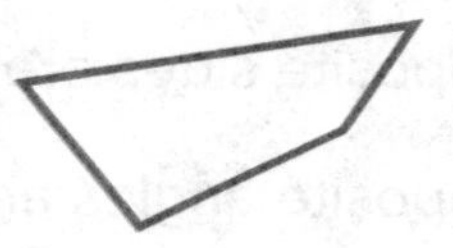

____

____

Write About It

17. How are all quadrilaterals alike and how can they be different?

____

____

Name

Lesson 4

Hands On: Sides and Angles of Quadrilaterals

Homework Helper

Need help? connectED.mcgraw-hill.com

Measure the sides and angles of each figure to determine if any are congruent. Then determine if any sides are parallel. Complete the table.

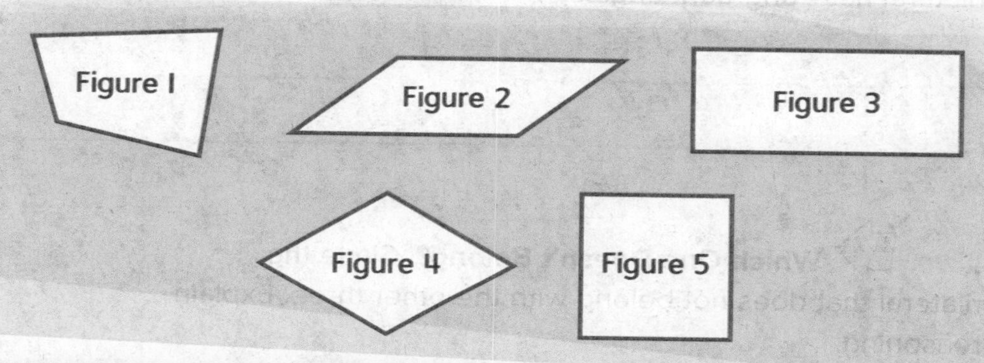

Attribute	Figure(s)
Opposite sides are congruent.	2, 3, 4, 5
Opposite sides are parallel.	2, 3, 4, 5
Opposite angles are congruent.	2, 3, 4, 5

Each figure has 4 sides and 4 angles.

Practice

Refer to the figures above in the Homework Helper to solve Exercises 1–3.

1. Complete the attributes of Figure 2.

Opposite sides are ______________ and ______________.

Opposite angles are ______________.

The figure has ______ sides and ______ angles.

2. Which figures have all the attributes of Figure 2? ____________

3. Which figures have four right angles? ____________

Problem Solving

4. The state of Nevada is in the shape of a quadrilateral. Complete the attributes of the outline of the state of Nevada.

There is one set of ______________ opposite sides.

Opposite sides are not ______________.

Opposite angles are not ______________, but there are two right angles.

5. **Processes & Practices 2** **Reason** Explain one way to determine if a quadrilateral has congruent angles.

____

____

6. **Processes & Practices 3** **Which One Doesn't Belong?** Circle the quadrilateral that does not belong with the other three. Explain your reasoning.

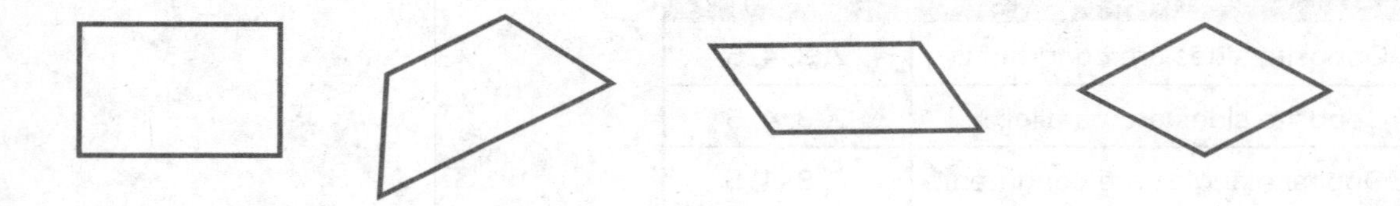

____

____

Vocabulary Check

Fill in each blank with the correct term or number to complete the sentence.

7. A quadrilateral is a polygon with ______ sides and ______ angles.

Name ....

Lesson 5
Classify Quadrilaterals

ESSENTIAL QUESTION
How does geometry help me solve problems in everyday life?

You can classify quadrilaterals using one or more attributes like congruent sides, parallel sides, and right angles.

Math in My World

Example 1

Trina cut out polygon mats to use for her travel photos. Use the figures below to determine the missing attribute(s) of each type of quadrilateral.

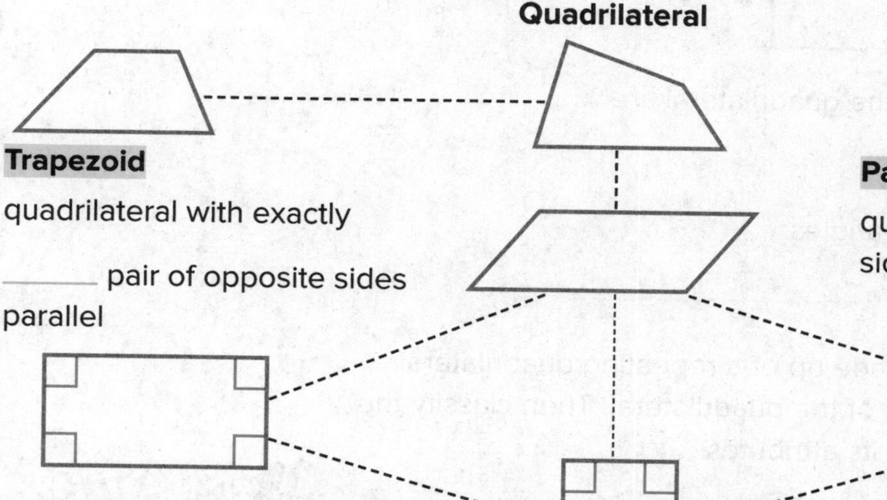

Trapezoid
quadrilateral with exactly

______ pair of opposite sides parallel

Parallelogram
quadrilateral with opposite sides congruent and

______

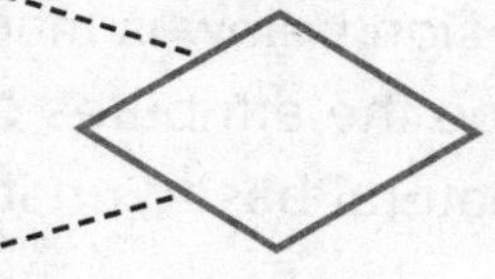

Rectangle
parallelogram with

______ right angles

Square
parallelogram with

______ sides congruent

and ______ right angles

Rhombus
parallelogram with

______ sides congruent

A square has all the attributes of a rectangle and a ____________.

Example 2

One side of the Realia building in Madrid, Spain, is shown at the right. Describe the attributes of the quadrilateral. Then classify it based on its attributes.

The quadrilateral has opposite sides ______________

and ______________.

So, it is a ______________.

Guided Practice

1. Describe the attributes of the quadrilateral below. Then classify the quadrilateral based on its attributes.

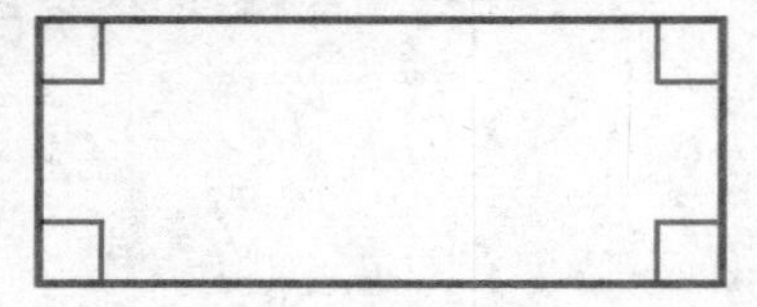

The opposite sides of the quadrilateral are ______________

and ______________.

There are ______ right angles.

So, the quadrilateral is a ______________.

2. The design below is made up of a repeating quadrilateral. Describe the attributes of the quadrilateral. Then classify the quadrilateral based on its attributes.

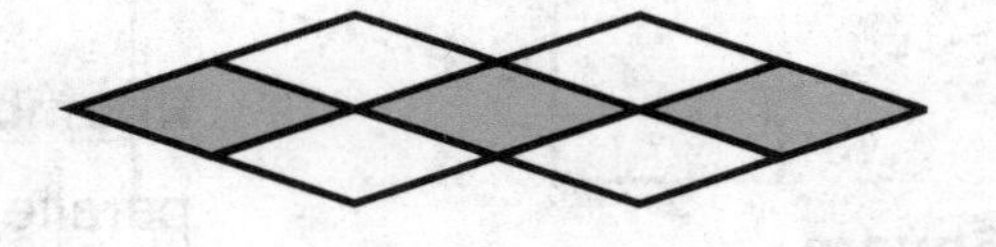

The quadrilateral has ______ congruent sides.

Opposite sides are ______________.

So, the quadrilateral is a ______________.

Talk MATH

Tell why a square is a special kind of rectangle.

Name

Independent Practice

Describe the attributes of each quadrilateral. Then classify the quadrilateral.

3.

4. 

5. Circle the quadrilateral(s) that have all the attributes of a parallelogram.

rectangle rhombus square trapezoid

6. Circle the quadrilateral(s) that have all the attributes of a rhombus.

rectangle square trapezoid parallelogram

State whether the following statements are *true* or *false*. If *false*, explain why.

7. All parallelograms have opposite sides congruent and parallel. Since rectangles are parallelograms, all rectangles have opposite sides congruent and parallel.

8. All squares have four congruent sides. Since rectangles are squares, all rectangles have four congruent sides.

Problem Solving

9. Processes & Practices 7 **Identify Structure** Many aircraft display the shape of the American flag as shown below to indicate motion. Classify the quadrilateral.

____

10. Adena used a quadrilateral in her art design. The quadrilateral has no sides congruent and only one pair of opposite sides parallel. Classify the shape of the quadrilateral she used.

____

Brain Builders

11. Traci planted two tomato gardens. One garden is rectangular. The shape of the second garden has all the attributes of the rectangular garden. In addition, it has four congruent sides. Classify the shape of the second tomato garden. Explain.

____

____

12. Processes & Practices 4 **Model Math** Draw a parallelogram that is neither a square, rhombus, nor rectangle. Explain how you determined your parallelogram.

____

____

13. **Building on the Essential Question** How do I classify quadrilaterals using their attributes? Include a real-world example.

____

____

Name

Lesson 5

Classify Quadrilaterals

Homework Helper

Need help? connectED.mcgraw-hill.com

Describe the attributes of the quadrilateral. Then classify it based on its attributes.

The quadrilateral has all sides congruent and opposite sides parallel.

It has four right angles.

So, the quadrilateral is a square.

Practice

Describe the attributes of each quadrilateral. Then classify the quadrilateral.

1.

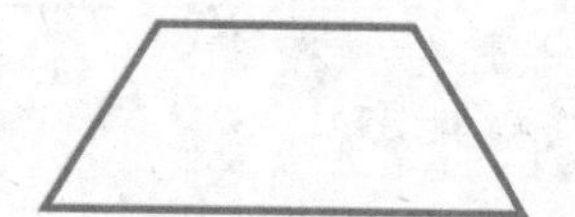

2.

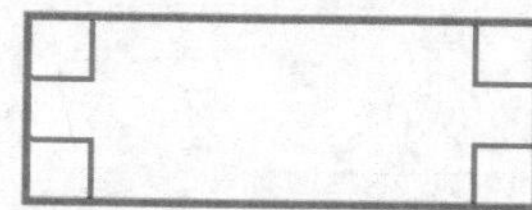

3. Circle the quadrilateral(s) that have all the attributes of a rectangle.

trapezoid parallelogram square rhombus

Problem Solving

Name all the quadrilaterals that have the given attributes.

4. opposite sides parallel ______

5. four right angles ______

6. exactly one pair of opposite sides parallel ______

Brain Builders

7. Is it possible to draw a quadrilateral with 4 congruent sides that is *not* a parallelogram? Explain.

______

______

8. Processes & Practices  **Model Math** Write a real-world problem that involves classifying a quadrilateral. Then solve the problem and explain the solution.

______

______

______

Vocabulary Check

Fill in each blank with the correct term or number to complete each sentence.

9. A rectangle is a parallelogram with ______ right angles.

10. A trapezoid is a quadrilateral with exactly ______ pair of parallel sides.

11. **Test Practice** Which statement about the figures shown below is true?

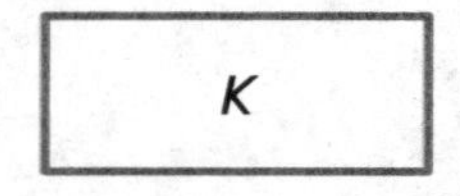

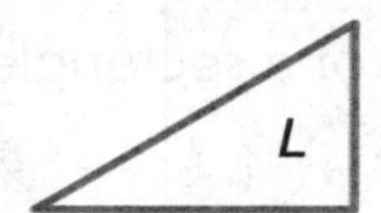

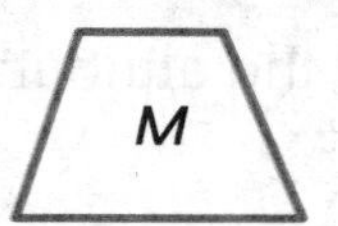

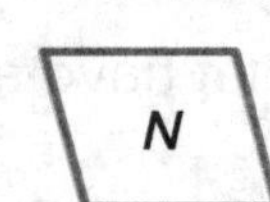

Ⓐ Figures *K* and *N* are rectangles.

Ⓑ Figures *L* and *N* are quadrilaterals.

Ⓒ Figures *K* and *N* are parallelograms.

Ⓓ Figures *M* and *N* are parallelograms.

Name ________________________________

Lesson 6
Hands On
Build Three-Dimensional Figures

ESSENTIAL QUESTION
How does geometry help me solve problems in everyday life?

A **three-dimensional figure** has length, width, and height. A **net** is a two-dimensional pattern of a three-dimensional figure. You can use a net to build a three-dimensional figure.

A **cube** is a three-dimensional figure with six faces that are congruent squares. **Congruent figures** have the same size and shape.

A **rectangular prism** is a three-dimensional figure with six rectangular faces. Opposite faces are parallel and congruent.

A **face** is a flat surface.

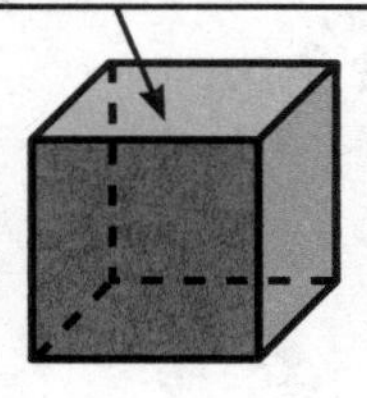

Cube

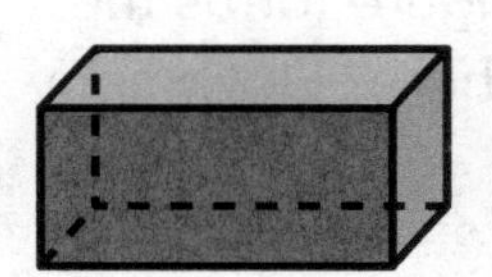

Rectangular Prism

Build It

1 **Copy the net shown onto grid paper.**

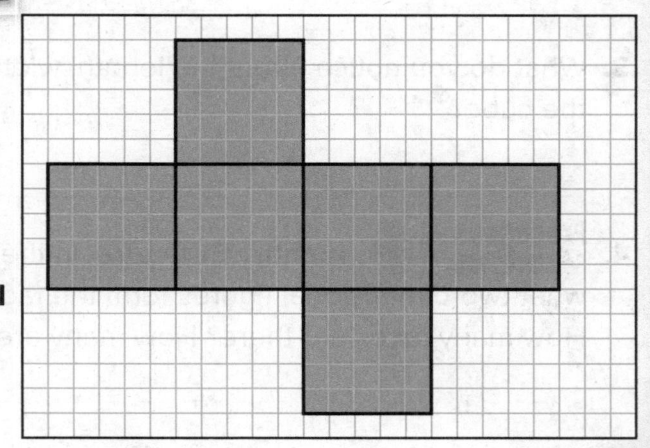

2 **Cut out the net. Fold along the lines to form a three-dimensional figure. What figure did you form?**

____

Try It

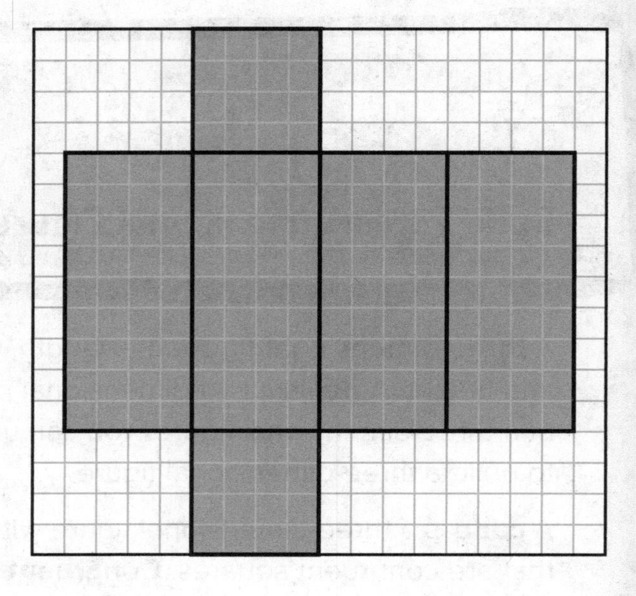

1. Copy the net shown onto grid paper.

2. Cut out the net. Fold along the lines to form a three-dimensional figure. What figure did you form?

____

How are the two figures you just built alike?

____

____

____

How are the two figures you just built different?

____

Talk About It

1. In the first activity, what two-dimensional figure forms the faces of the figure? How many faces are there? How many are congruent?

____

2. Identify the length, width, and height of the cube you formed in the first activity.

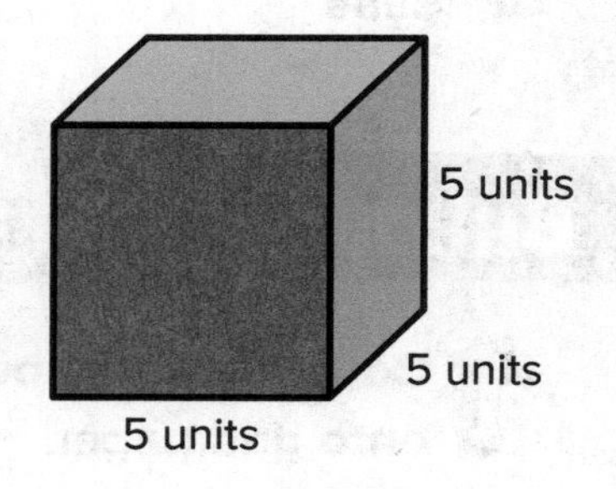

____

3. What do you notice about the length, width, and height of the cube?

____

4. **Processes & Practices** 7 **Identify Structure** In the second activity, what two-dimensional figures form the faces of the figure? How many faces are there? How many are congruent?

____

____

Name

Practice It

For Exercises 5 and 6, refer to the grid at the right.

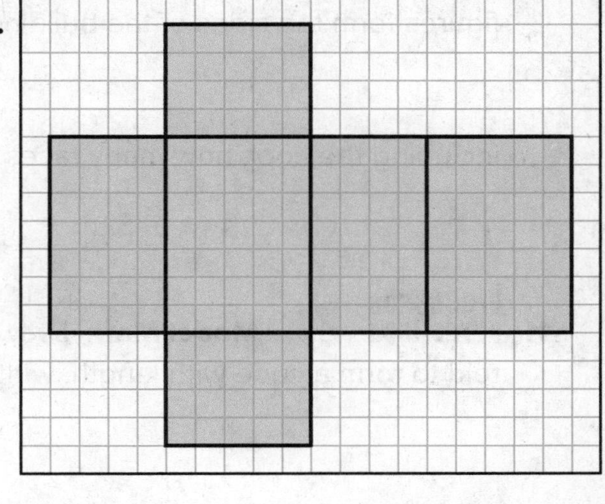

5. Copy the net onto grid paper. Cut out the net and fold along the lines to form a three-dimensional figure. What figure did you form?

____

6. What two-dimensional figure forms the faces of the figure?

____

How many faces are there? ________ Describe the congruent faces.

____

For Exercises 7–9, refer to the grid at the right.

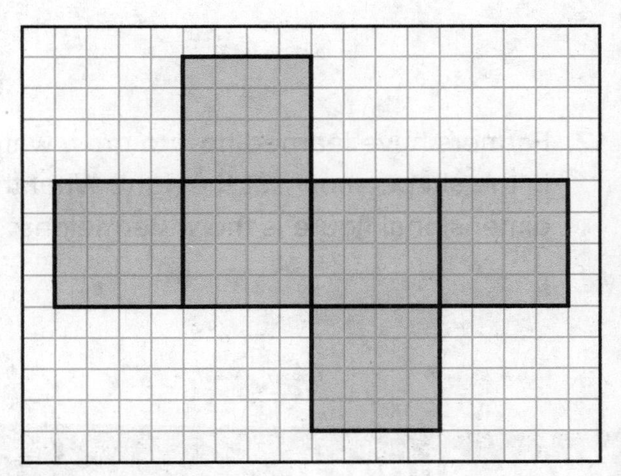

7. Copy the net onto grid paper. Cut out the net and fold along the lines to form a three-dimensional figure. What figure did you form?

____

8. What two-dimensional figure forms the faces of the figure?

____

How many faces are there?

____

Describe the congruent faces.

____

9. Identify the length, width, and height of the figure you formed.

____

Apply It

10. The rectangular prism-shaped building shown at the right was used for the 2008 Olympics in Beijing, China. What two-dimensional figures form the sides of the building?

____

Including the floor, how many faces are there?

____

11. **Processes & Practices** 4 **Model Math** Draw two different nets that would fold to form a cube with length, width, and height each 4 units.

12. Farmers have learned how to grow watermelons in the shape shown at the right. What three-dimensional figure is the watermelon?

____

Write About It

13. How are nets used to build three-dimensional figures?

____

____

____

Name ____________________

Lesson 6

Hands On: Build Three-Dimensional Figures

Homework Helper

Need help? connectED.mcgraw-hill.com

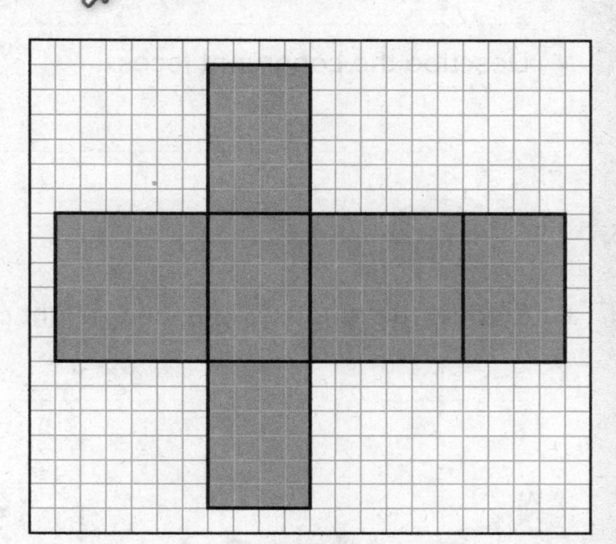

The net shown was used to form the three-dimensional figure below.

The three-dimensional figure formed from the net is a rectangular prism.

The faces of the rectangular prism are rectangles.

The figure has 6 faces.

The four rectangles are congruent, and the two squares are congruent.

The figure formed has a length of 4 units, a width of 6 units, and a height of 6 units.

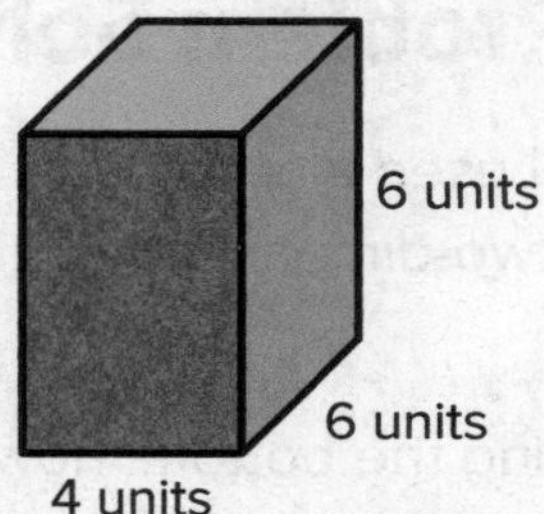

Vocabulary Check

Fill in each blank with the correct word(s) to complete each sentence.

1. A three-dimensional figure has ____________, width, and ____________.
2. A net is a two-dimensional ____________ of a three-dimensional figure.
3. A cube is a three-dimensional figure with six square faces that are ____________.

Practice

For Exercises 4–6, refer to the grid at the right.

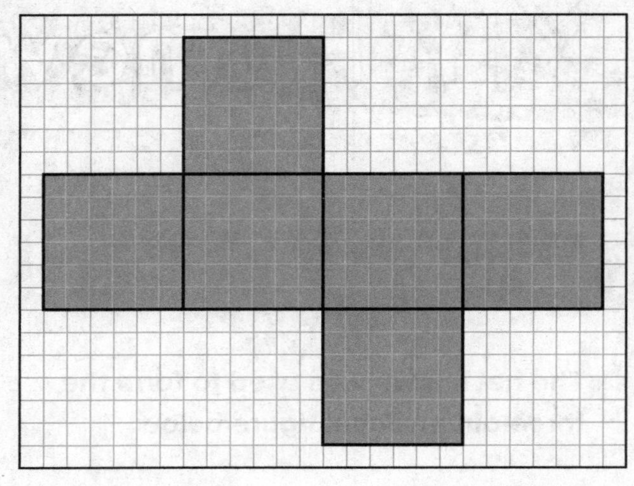

4. What three-dimensional figure is formed using the net shown?

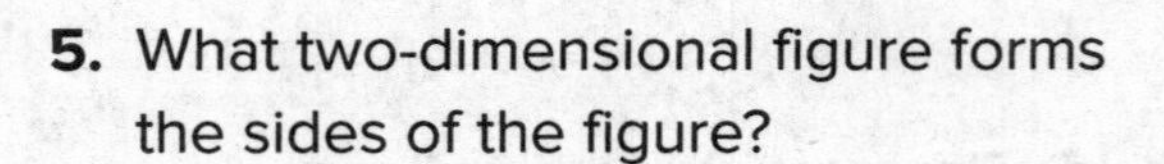

____

5. What two-dimensional figure forms the sides of the figure?

____

Describe the congruent faces.

____

____

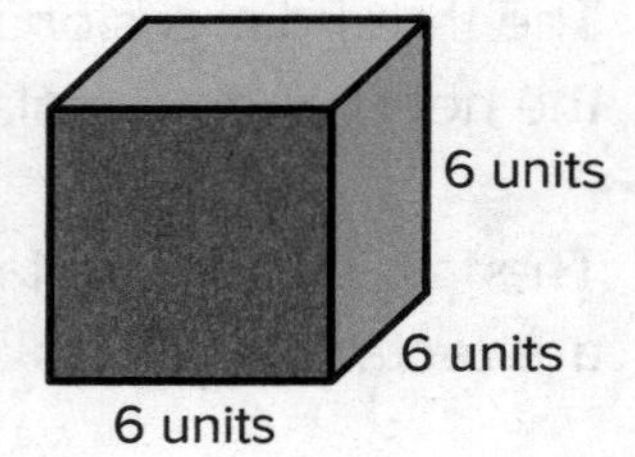

6. Identify the length, width, and height of the figure formed.

____

Problem Solving

7. Rachel used a rectangular prism-shaped box to ship a package to her friend. What two-dimensional figure forms the faces of the box?

____

Including the bottom, how many faces are there?

____

Describe the faces.

____

____

8. Joseph is forming a three-dimensional figure using a net. The figure has six congruent square faces. What type of figure did he make?

____

Name ....

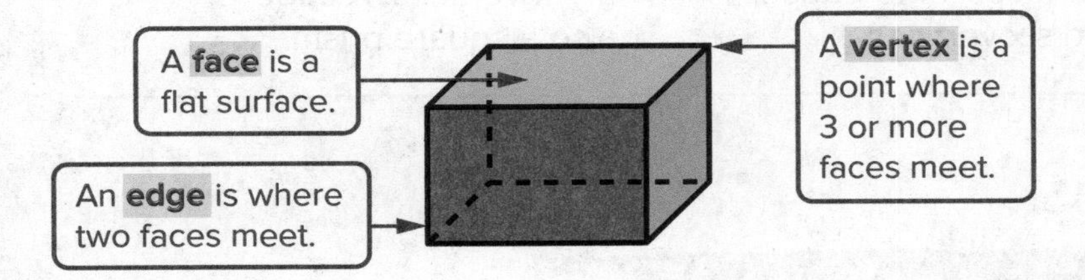

Lesson 7
Three-Dimensional Figures

ESSENTIAL QUESTION
How does geometry help me solve problems in everyday life?

A **three-dimensional figure** has length, width, and height.

A **face** is a flat surface.

An **edge** is where two faces meet.

A **vertex** is a point where 3 or more faces meet.

Math in My World

Describe the faces, edges, and vertices of the figure outlined on the luggage bag. Then identify the shape of the figure.

faces The figure has ________ faces. Each face appears to be a rectangle.

edges There are ________ edges. The opposite edges are parallel and congruent.

vertices The figure has ________ vertices.

Prisms are three-dimensional figures. A **prism** has at least three faces that are rectangles. The top and bottom faces, called the **bases**, are congruent parallel polygons.

The figure above is a rectangular prism. In a **rectangular prism**, the bases are congruent rectangles. A rectangular prism has six rectangular faces, twelve edges, and eight vertices.

Key Concept Prisms

Rectangular Prism	Triangular Prism	Cube
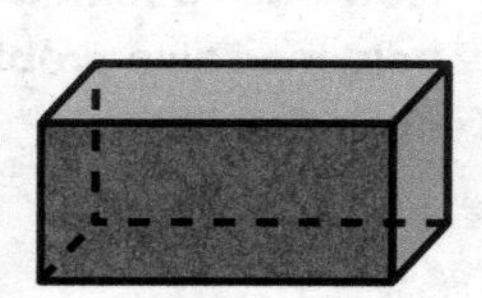	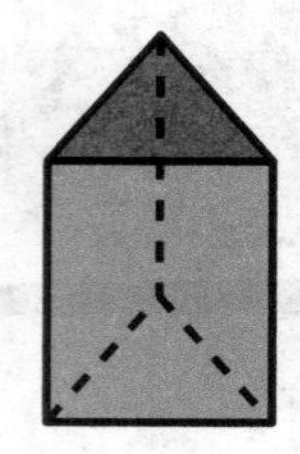	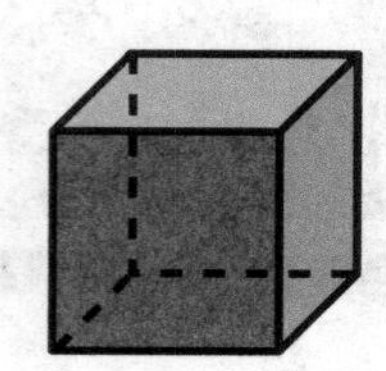
A rectangular prism has six rectangular faces, twelve edges, and eight vertices.	A triangular prism has triangular bases. It has five faces, nine edges, and six vertices.	A cube has six square faces, twelve edges, and eight vertices. A cube is also a square prism.

Guided Practice

1. Describe the faces, edges, and vertices of the three-dimensional figure. Then identify it.

faces This figure has ________ faces. The ______________________ bases are congruent and parallel. The other faces are ______________________.

edges There are ________ edges. The edges that form the vertical sides of the rectangles are parallel and ______________________.

vertices This figure has ________ vertices.

The figure is a ______________________.

Talk MATH

Describe the differences between a triangular prism and a rectangular prism.

Name

Independent Practice

Describe the faces, edges, and vertices of each three-dimensional figure. Then identify it.

2. 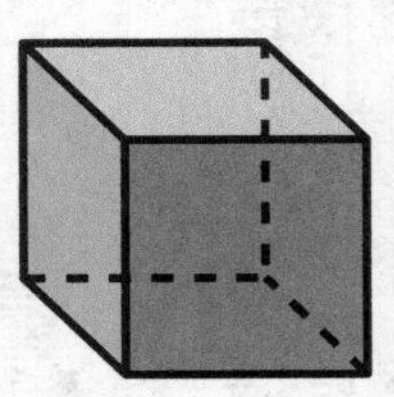

3. 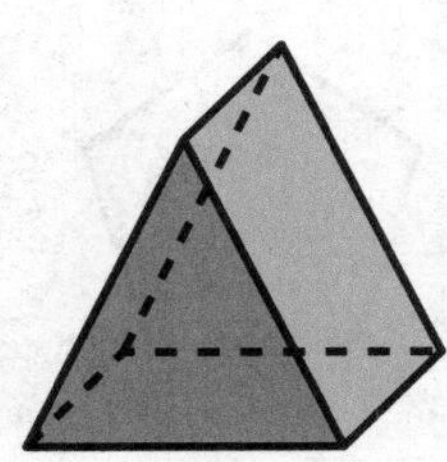

4. 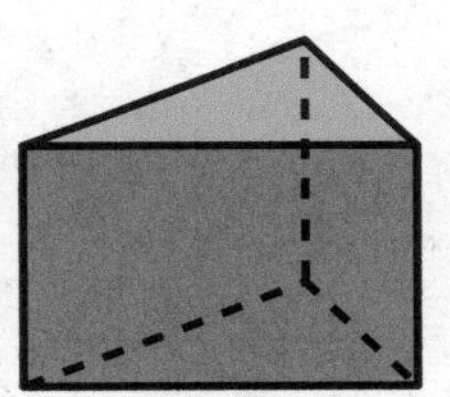

5. 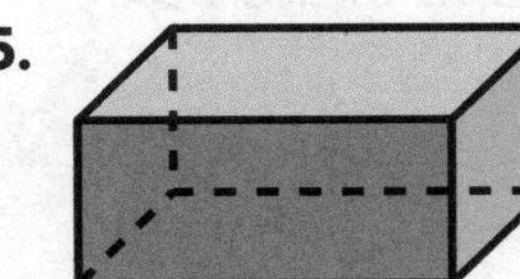

6. 

7. 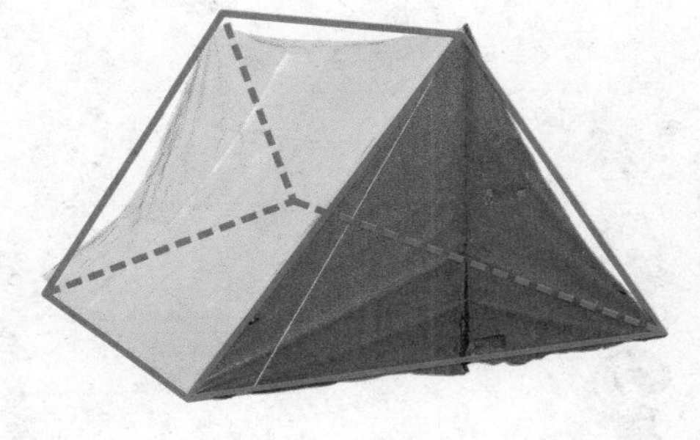

Problem Solving

8. **Processes & Practices 7** **Identify Structure** The Aon Center in Chicago is in the shape of a rectangular prism. Circle the two-dimensional figures that make up the faces of the prism.

 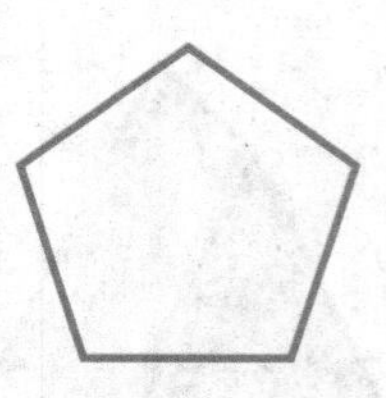 

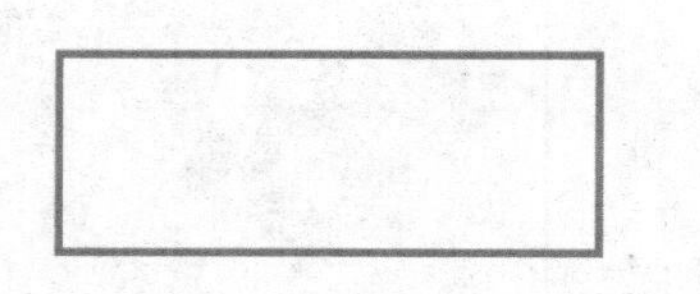 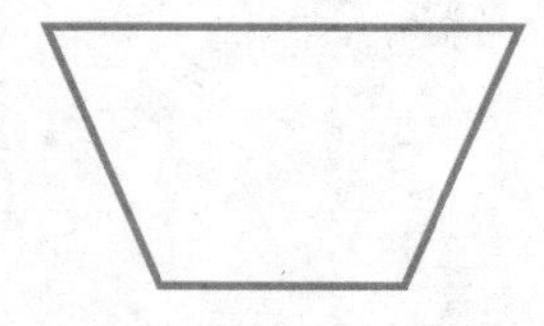

9. Describe the number of vertices and edges in an unopened cereal box. Identify the shape of the box.

____

Brain Builders

10. **Processes & Practices 4** **Model Math** What figure is formed if only the height of a cube is increased? Draw the figure to support your answer.

____

____

____

11. **Building on the Essential Question** How are rectangular prisms, triangular prisms, and cubes different? How are they the same?

____

____

Name

Lesson 7

Three-Dimensional Figures

Homework Helper

Need help? connectED.mcgraw-hill.com

Describe the faces, edges, and vertices of the ramp. Then identify the shape of the ramp.

faces This figure has 5 faces. The triangular bases are congruent and parallel. The other faces are rectangles.

edges There are 9 edges. The edges that form the horizontal sides of the rectangles are parallel and congruent.

vertices This figure has 6 vertices.

The ramp is a triangular prism.

Practice

Describe the faces, edges, and vertices of each three-dimensional figure. Then identify it.

1. 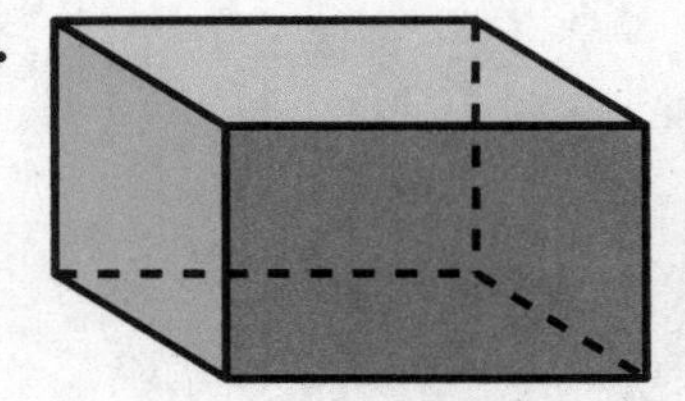

____

____

____

2. 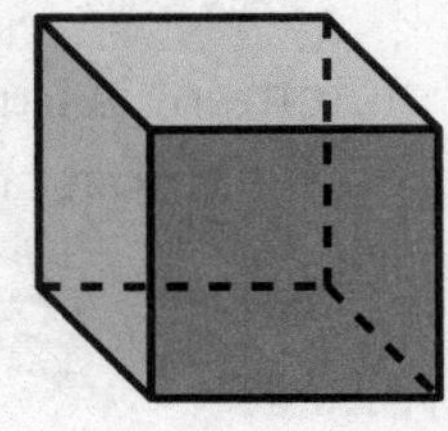

____

____

____

Problem Solving

3. Rhett made a simple drawing of his house. It is a three-dimensional figure with four faces that are rectangular and two that are square. What kind of figure is it?

____

4. A toy box has 6 faces that are squares. There are 12 edges and 8 vertices. Identify the shape of the toy box.

____

Brain Builders

5. Processes & Practices  **Make Sense of Problems** Gabriel is playing a board game. When it is his turn, he tosses a three-dimensional figure that has 6 square faces. What kind of figure is it? How many edges and vertices does it have? Draw the figure to support your answer.

____

Vocabulary Check

Fill in the blank with the correct term or number to complete the sentence.

6. A vertex is a point where ________ or more edges meet.

7. Test Practice Which statement is true about the three-dimensional figure that most closely represents the slice of pie?

Easy as pie

Ⓐ The figure has 4 vertices.

Ⓑ The figure has 6 vertices.

Ⓒ The figure has 10 edges.

Ⓓ The figure has 12 edges.

Check My Progress

Vocabulary Check 

Circle the correct term or terms to complete each sentence.

1. A (**rectangular prism**, **triangular prism**) is a three-dimensional figure that has six rectangular faces and eight vertices.
2. A (**rectangle**, **rhombus**) is a parallelogram with all sides congruent.
3. A(n) (**vertex**, **edge**) of a three-dimensional figure is where two faces meet.
4. A (**prism**, **trapezoid**) has at least three faces that are rectangles.

Concept Check

Describe the attributes of each quadrilateral. Then classify the quadrilateral.

5. 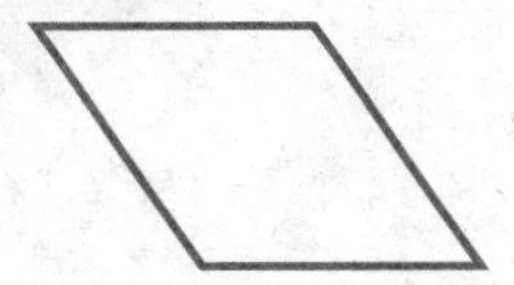

6. 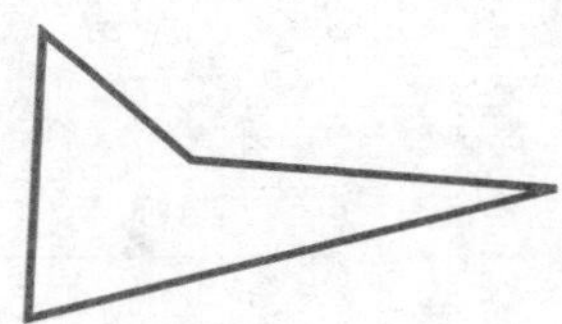

7. Circle the quadrilateral(s) that have all the attributes of a rhombus.

   trapezoid   square   parallelogram   rectangle

8. Circle the quadrilateral(s) that have all the attributes of a rectangle.

   parallelogram   square   trapezoid   rhombus

Describe the faces, edges, and vertices of each three-dimensional figure. Then identify it.

9.

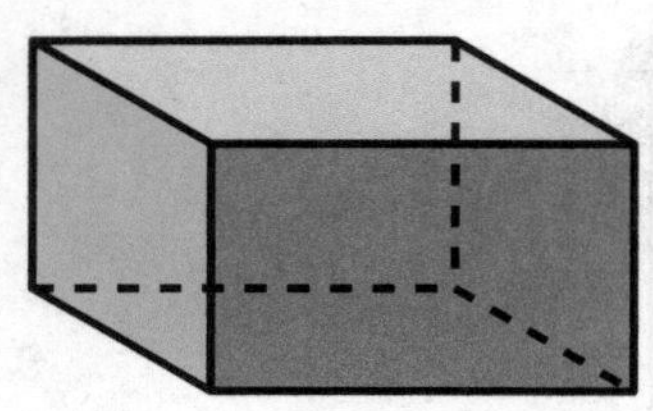

10.

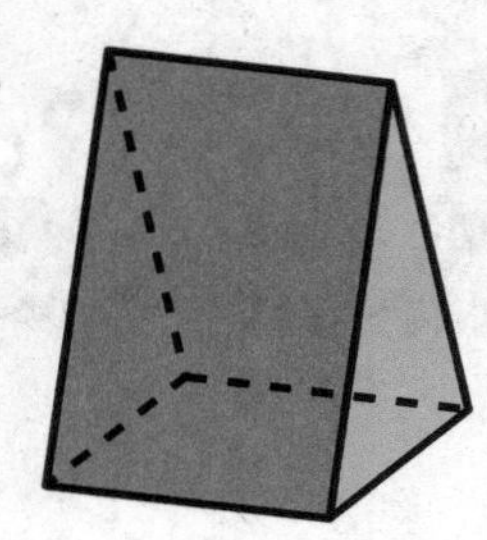

Brain Builders

11. Kendra claims that the shape of the xylophone shown at the right is a rectangle since there are right angles and parallel sides. Is Kendra correct? Explain.

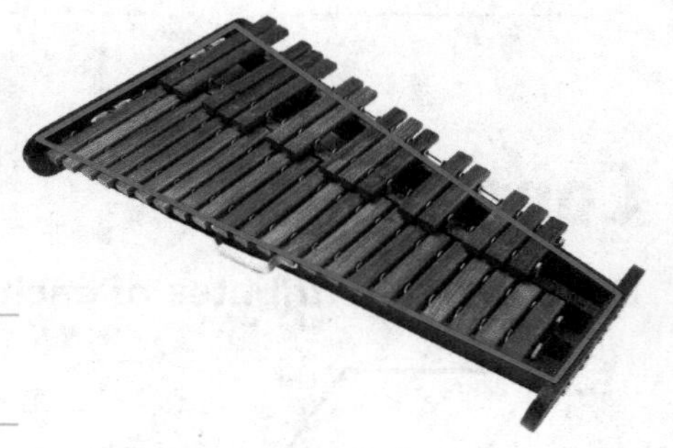

12. Garrett cut a piece of cheese to eat for a snack. The cheese is a prism with 3 faces that are rectangular and 2 that are triangular. What kind of figure is it? Draw the figure to support your answer.

13. Test Practice Which is *not* a true statement?

Ⓐ All squares are parallelograms.

Ⓑ Some rhombi are squares.

Ⓒ All rectangles are squares.

Ⓓ All rectangles are parallelograms.

Name

Lesson 8
Hands On
Use Models to Find Volume

ESSENTIAL QUESTION
How does geometry help me solve problems in everyday life?

Volume is the amount of space inside a three-dimensional figure. Centimeter cubes can help you find the volume of a prism.

Use centimeter cubes to build four different rectangular prisms. Complete the fourth and fifth columns of the table below for each prism.

Prism	Length (cm)	Width (cm)	Height (cm)	Number of Cubes	Volume (cubic cm)
A	1	2	1		
B	2	2	1		
C	3	2	2		
D	4	2	2		

A prism built from cubes has no gaps or overlaps.

A cube with a side length of one unit is called a **unit cube.**
A unit cube has a volume of 1 cubic unit, or 1 unit^3.
A **cubic unit** is a unit for measuring volume.

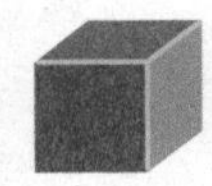
1 cubic unit

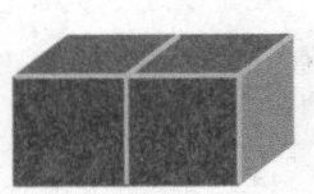
2 cubic units

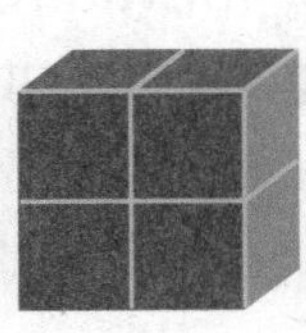
4 cubic units

So, if you use 12 centimeter cubes to build a rectangular prism, the prism has a volume of ________ cubic centimeters, or ________ cm^3.

Try It

Use centimeter cubes to build the rectangular prism shown. Complete the table for each layer.

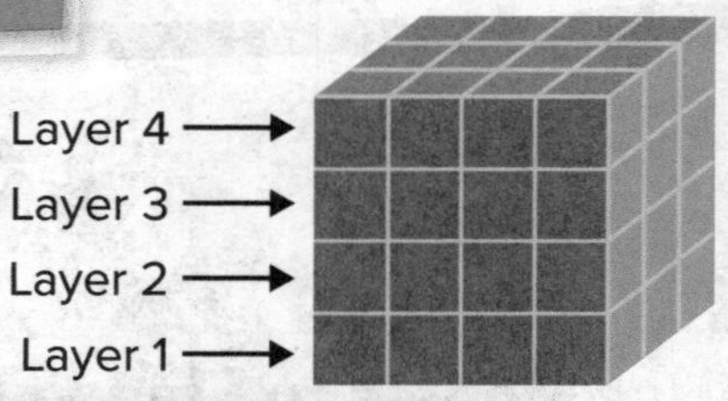

Layer	Length (cm)	Width (cm)	Height (cm)	Number of Cubes	Volume (cubic cm)
1					
2					
3					
4					

How many cubes were used to build the prism? ______________

What is the volume? ______________

Talk About It

1. Describe the relationship between the number of cubes needed to build a rectangular prism and its volume, in cubic units.

____

____

2. Describe the pattern in the table between the length, width, height, and volume of each prism.

____

____

3. Use ℓ for length, w for width, and h for height to write a formula for the volume V of a rectangular prism.

____

4. Processes & Practices 5 **Use Math Tools** Use your formula to find the volume of the prism at the right in appropriate units. Verify your solution by counting the number of cubes.

____

Name

Practice It

Processes &Practices 5 Use Math Tools Use centimeter cubes to build the rectangular prism shown.

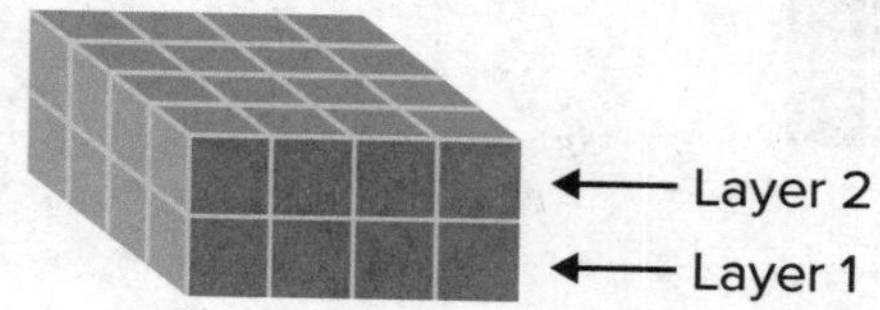

5. Complete the table below.

Layer	Length (cm)	Width (cm)	Height (cm)	Number of Cubes
1				
2				

6. How many cubes were used to build the prism? ______

What is the volume? ______ cm^3

Use centimeter cubes to build the rectangular prism shown.

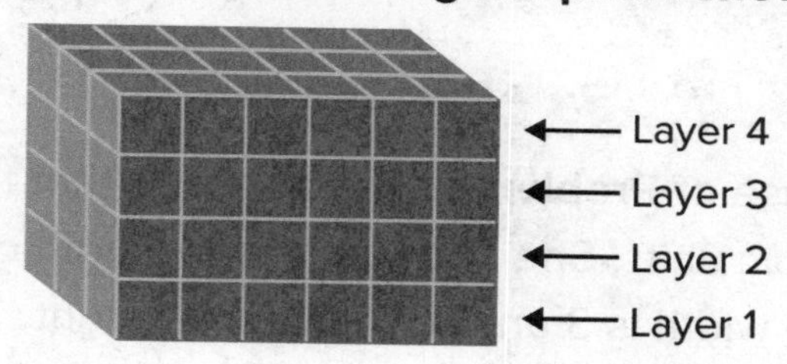

7. Complete the table below.

Layer	Length (cm)	Width (cm)	Height (cm)	Number of Cubes	Volume (cubic cm)
1					
2					
3					
4					

8. How many cubes were used to build the prism? ______

What is the volume? ______ cm^3

Apply It

Use the prism shown for Exercises 9–11.

9. What shape is the base of the prism?

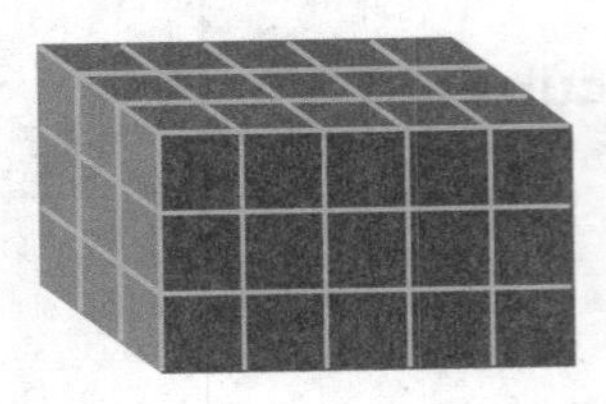

10. Processes & Practices 6 **Explain to a Friend** Explain to a friend how to find the area of the base of the prism.

11. Find the volume of the prism above by multiplying the area of the base by the height. Verify your solution by counting the number of centimeter cubes.

12. Processes & Practices 1 **Make Sense of Problems** Valerie knows that the volume of a prism is 36 cubic units. She knows that the length of the prism is 4 units and the width is 3 units. What is the height of the prism?

Write About It

13. Describe a way to find the volume of a rectangular prism without using models.

Name

MY Homework

Lesson 8
Hands On: Use Models to Find Volume

Homework Helper

Need help? connectED.mcgraw-hill.com

Centimeter cubes were used to build the rectangular prism shown. The table shows the number of cubes that were used to build each layer.

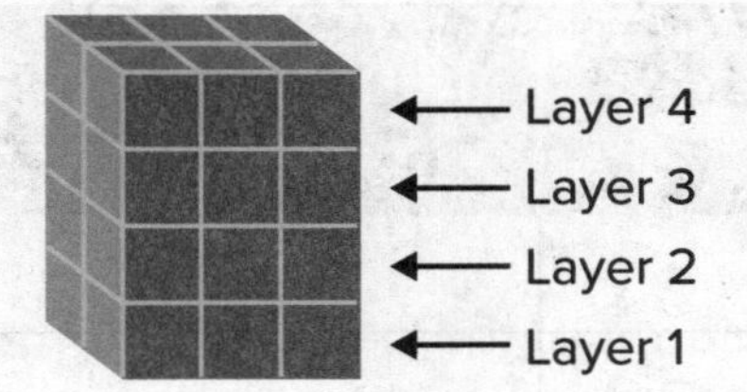

Layer	Length (cm)	Width (cm)	Height (cm)	Number of Cubes	Volume (cubic cm)
1	3	2	1	6	6
2	3	2	1	6	6
3	3	2	1	6	6
4	3	2	1	6	6

So, 24 cubes were used to build the prism.

The volume of the prism is 24 cubic centimeters, or 24 cm^3.

Vocabulary Check

Fill in each blank with the correct term or number to complete each sentence.

1. Volume is the amount of ________ inside a three-dimensional figure.

2. A cube with a side length of ________ unit is called a unit cube.

3. The volume of a rectangular prism can be found by multiplying the length by the ________ by the height.

Practice

For Exercises 4–7, centimeter cubes were used to build the rectangular prism shown.

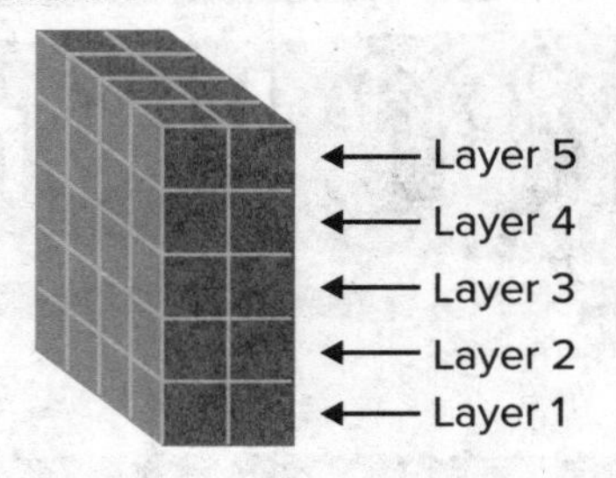

4. How many cubes were needed to build Layer 1?

____

5. Complete the table below.

Layer	Length (cm)	Width (cm)	Height (cm)	Number of Cubes	Volume (cubic cm)
1					
2					
3					
4					
5					

6. How many cubes were used to build the prism?

____

7. What is the volume of the prism?

____

Problem Solving

8. **Processes & Practices** 1 **Make Sense of Problems** Samir knows that the volume of a prism is 40 cubic units. He also knows that the width of the prism is 2 units and the height is 5 units. What is the length of the prism?

____

9. Centimeter cubes were used to build the prism. What is the volume of the prism?

____

Name ____________________

Lesson 9
Volume of Prisms

How does geometry help me solve problems in everyday life?

Volume is the amount of space inside a three-dimensional figure. You can use either formula below to find the volume of a prism.

$V = \ell \times w \times h$ $\quad$ V = volume, ℓ = length, w = width, and h = height

$B = \ell w$

$V = B \times h$ $\quad$ V = volume, B = area of the base, and h = height

Common units of volume are cubic inches, cubic feet, cubic yards, cubic centimeters, and cubic meters.

Math in My World

Example 1

On his family vacation to the beach, Armando filled a cooler with water and snacks. Find the volume of the cooler.

30 in.

One Way **Use $V = \ell \times w \times h$.**

$V = \ell \times w \times h$ — Volume formula

$V =$ ______ $\times$ ______ $\times$ ______ — $\ell = 30$, $w = 15$, $h = 20$

$V =$ ______ — Multiply.

Another Way **Use $V = B \times h$.**

$V = B \times h$ — Volume formula

$V =$ ______ $\times$ ______ — $B = 30 \times 15$, $h = 20$

$V =$ ______ — Multiply.

The volume of the cooler is ______ cubic inches.

Remember that the Associative Property of Multiplication tells you that the way in which factors are grouped does not change the product. You can group the factors to make the multiplication easier.

Example 2

Find the volume of the prism.

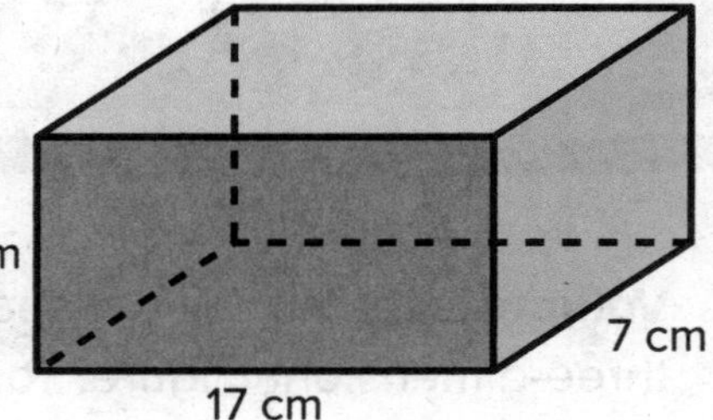

$V = \ell \times w \times h$ Volume formula

$V = 17 \times 7 \times 9$

$V =$ ______ $\times$ (______ $\times$ ______) Associative Property

$V =$ ______ $\times$ ______ Multiply.

$V =$ ______ Multiply.

The volume of the prism is ______ cm^3.

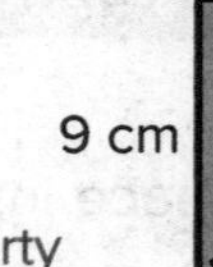

Guided Practice

Find the volume of each prism.

1.

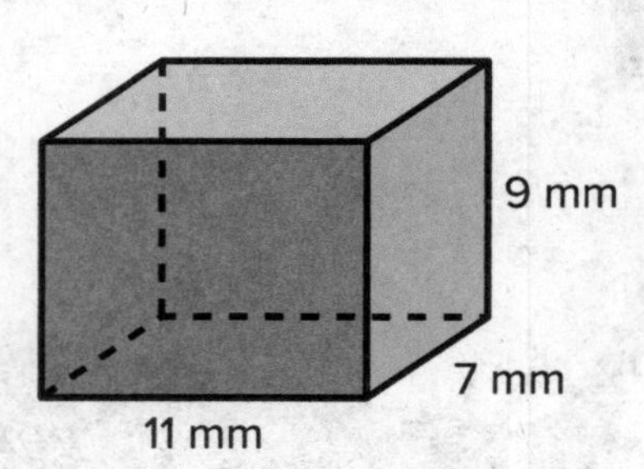

$V = \ell \times w \times h$

$V =$ ______ $\times$ ______ $\times$ ______

$V =$ ______ mm^3

2.

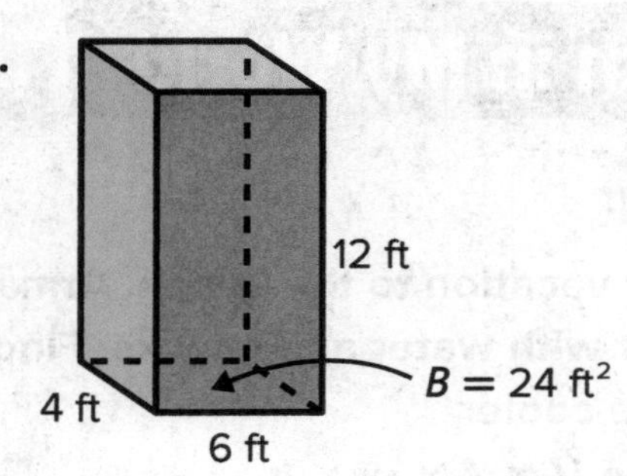

$V = B \times h$

$V =$ ______ $\times$ ______

$V =$ ______ ft^3

If you know the area of the base of a rectangular prism and the prism's height, which formula would you use? Why?

Name

Independent Practice

Processes & Practices 2 **Use Symbols Find the volume of each prism. Use the formula $V = \ell \times w \times h$ or $V = B \times h$.**

3.

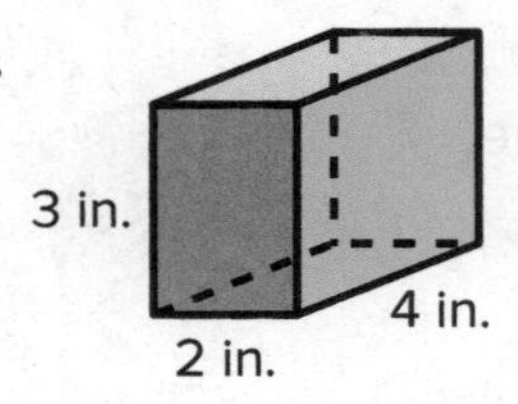

$V =$ ____________

4.

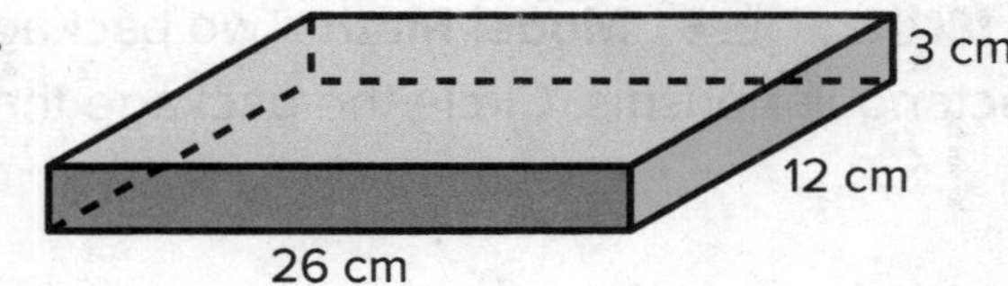

$V =$ ____________

5.

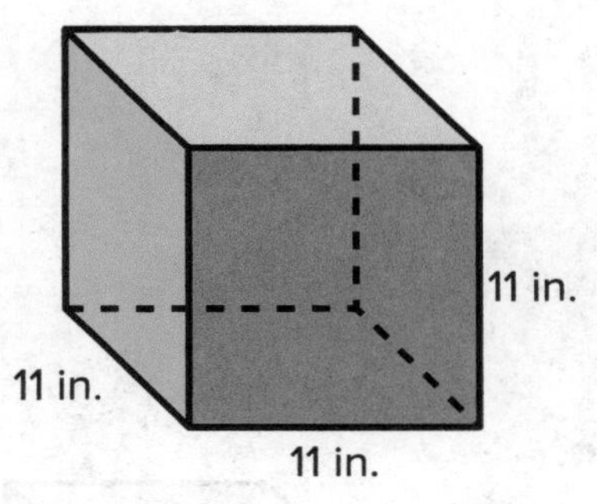

$V =$ ____________

6.

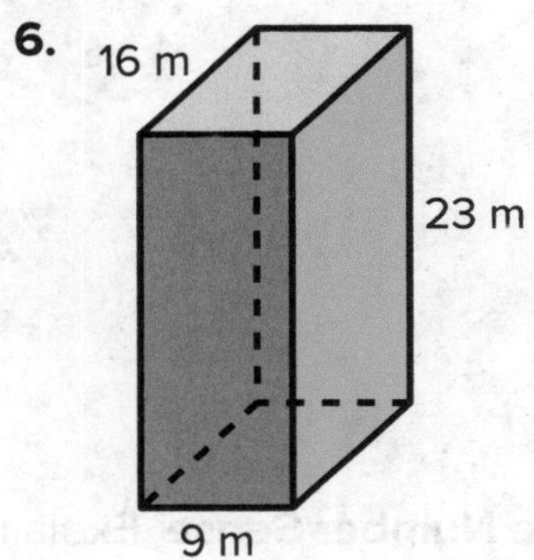

$V =$ ____________

7.

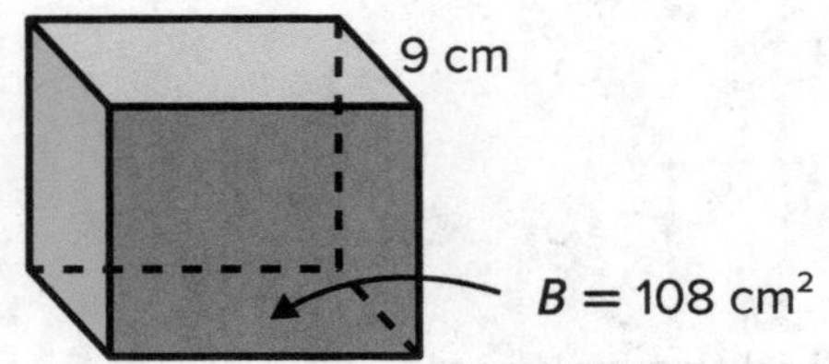

$V =$ ____________

8.

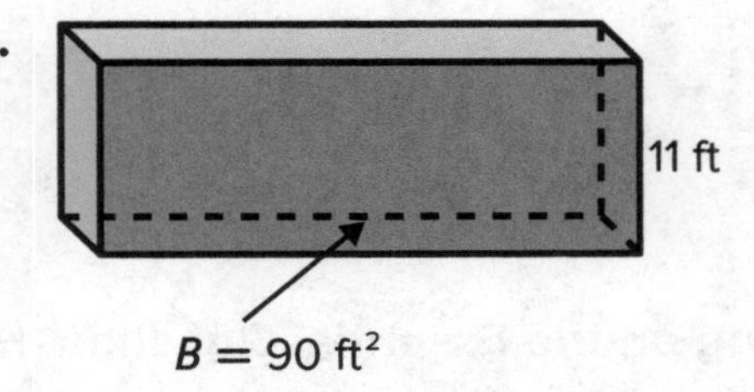

$V =$ ____________

Problem Solving

9. Find the volume of the Frog Queen building in Graz, Austria. The building is 18 meters long, 17 meters tall, and 18 meters wide.

10. Processes & Practices 4 **Model Math** Two packages are in the shape of rectangular prisms. Circle the package that has the smaller volume.

Brain Builders

11. Processes & Practices 2 **Use Number Sense** Explain how the Associative Property can be used to mentally find the volume of the prism shown. Then state the volume.

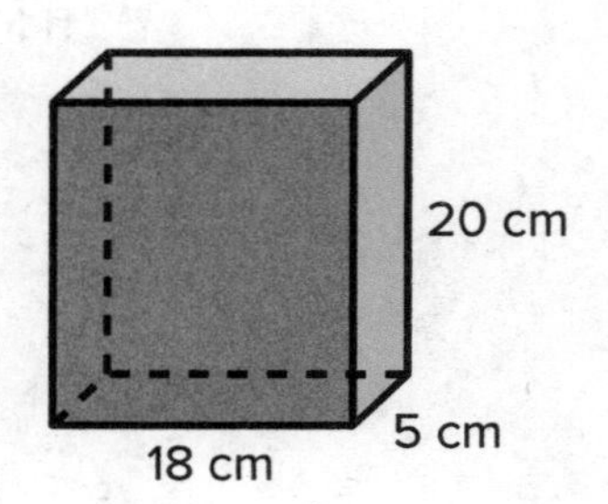

12. **Building on the Essential Question** How do I find the volume of rectangular prisms?

Name ______________________

MY Homework

Lesson 9

Volume of Prisms

Homework Helper

Need help? connectED.mcgraw-hill.com

Find the volume of the prism.

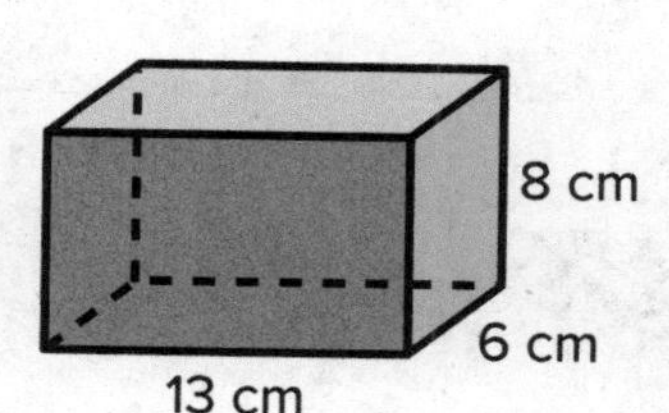

$V = \ell \times w \times h$ Volume formula

$V = 13 \times 6 \times 8$ $\ell = 13$, $w = 6$, $h = 8$

$V = 624$ Multiply.

The volume of the prism is 624 cm^3.

Practice

Find the volume of each prism.

1.

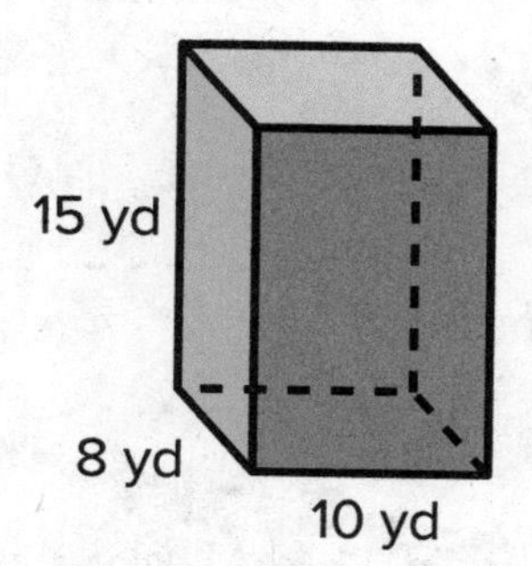

$V =$ ______________

2.

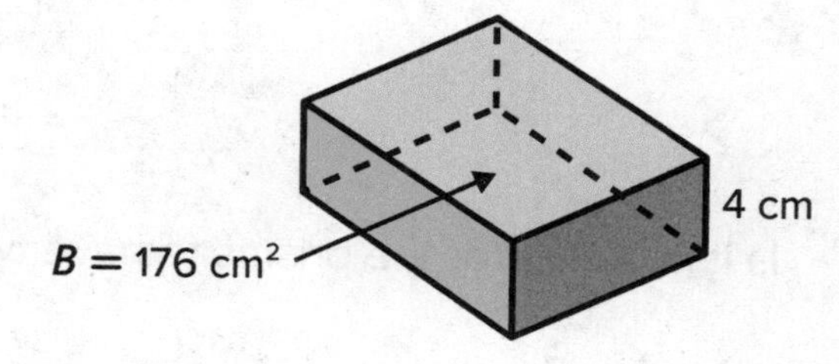

$V =$ ______________

Vocabulary Check

Fill in the blank with the correct term or number to complete the sentence.

3. Volume is measured in __________ units.

Problem Solving

4. The Donaldsons' swimming pool measures 15 meters long, 8 meters wide, and 3 meters deep. How many cubic meters of water will the pool hold?

____

5. The hotel that the Hutching family is staying at on vacation is shaped like a rectangular prism. It is 234 feet long, 158 feet wide, and 37 feet tall. What is the volume of the hotel?

____

Brain Builders

6. Jena has a small jewelry box in the shape of a cube with side lengths of 2 inches. How much greater is the volume of her large jewelry box with dimensions 7 inches, 5 inches, and 4 inches?

____

7. Processes & Practices 4 **Model Math** Describe the dimensions of two different prisms that each have a volume of 2,400 cubic centimeters. Then draw each prism.

____

____

Is it possible for the prisms to have two of the same dimensions?

____

8. **Test Practice** What is the volume of the prism formed by the luggage bag?

Ⓐ 6,000 in^3

Ⓑ 6,600 in^3

Ⓒ 7,200 in^3

Ⓓ 7,400 in^3

Name ....

Lesson 10
Hands On 
Build Composite Figures

ESSENTIAL QUESTION
How does geometry help me solve problems in everyday life?

A **composite figure** is made up of two or more three-dimensional figures.

Build It

Tools Tutor

A composite figure is shown below. Use centimeter cubes to build the figure.

1. Count the number of cubes needed to make the base layer.
   How many cubes did you use? → ☐

2. Count the number of cubes needed to make the top layer.
   How many cubes did you use? → + ☐

3. Add the number of cubes for the base and the top. → ☐

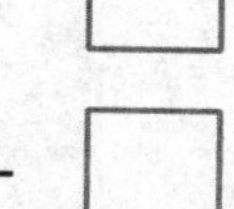

Talk About It

1. How many cubes did it take to build the figure? ______

2. What is the volume of the composite figure?

______ cubic centimeters

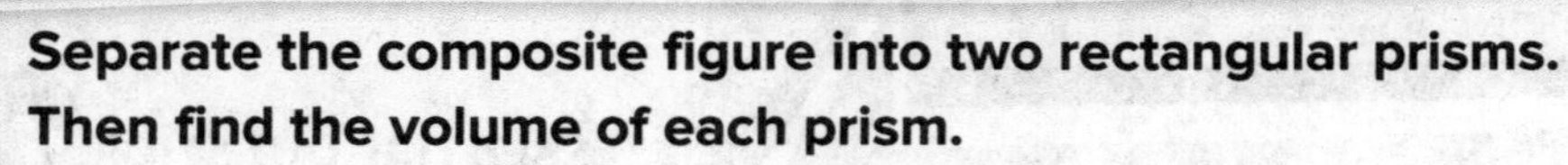

Try It

Separate the composite figure into two rectangular prisms. Then find the volume of each prism.

Find the volume of the top prism.

$V = \ell \times w \times h$

$V =$ ______ $\times$ ______ $\times$ ______

$V =$ ______

The volume of the top prism is ______ cubic centimeters.

Find the volume of the bottom prism.

$V = \ell \times w \times h$

$V =$ ______ $\times$ ______ $\times$ ______

$V =$ ______

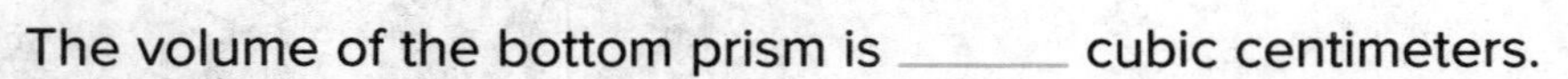

The volume of the bottom prism is ______ cubic centimeters.

Add the volumes to find the volume of the composite figure.

______ + ______ = ______

So, the volume of the composite figure is ______ cubic centimeters.

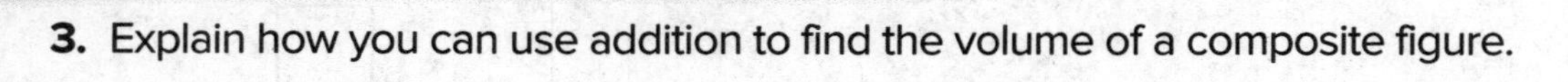

Talk About It

3. Explain how you can use addition to find the volume of a composite figure.

____

____

4.   **Make Sense of Problems** Explain how you would find the volume of the composite figure shown.

____

____

5. What is the volume of the figure in Exercise 4?

______ cubic centimeters

Name

Practice It

Use the model at the right to build the composite figure using centimeter cubes.

6. Separate the figure into prisms. Make a drawing of each prism used to build the composite figure.

7. How many cubes did it take to build the figure?

______

8. What is the volume of this figure?

______ cubic centimeters

Use the model at the right to build the composite figure using centimeter cubes.

9. Separate the figure into prisms. Make a drawing of each prism used to build the composite figure.

10. How many cubes did it take to build the figure?

______

11. What is the volume of this figure?

______ cubic centimeters

Apply It

Tanela arranged centimeter cubes into the composite figure shown. Use the composite figure for Exercises 12 and 13.

12. Processes & Practices 4 **Model Math** Separate the figure into prisms. Make a drawing of each prism used to build the composite figure.

13. What is the volume of the composite figure? Check your answer by building a model and counting the number of cubes. _______ cubic centimeters

14. Circle the composite figure that has a volume of 24 cubic centimeters.

15. Processes & Practices 1 **Make Sense of Problems** Explain how to use the formula of a rectangular prism to find the volume of a composite figure that is composed of rectangular prisms.

____

____

Write About It

16. How can you use models to find the volume of composite figures?

____

____

____

Name ____________________

Lesson 10

Hands On: Build Composite Figures

Homework Helper

Need help? connectED.mcgraw-hill.com

A composite figure is shown at the right. Centimeter cubes were used to build the figure. Find the volume.

1 Six cubes were used to make the base layer.

2 Four cubes were used to make the two top layers.

3 Add the number of cubes for the base and the top.
$6 + 4 = 10$

So, a total of 10 cubes were used to build the figure.
The volume is 10 cubic centimeters.

Practice

Refer to the composite figure at the right.

1. How many cubes are needed to build the bottom layer?

____

2. How many cubes are needed to build the top two layers?

____

3. Use addition to add the bottom and top layers.

____

4. What is the volume of the composite figure?

______ cubic centimeters

Problem Solving

Jared built the composite figure at the right using centimeter cubes.

5. Separate the figure into prisms. Make a drawing of each prism used to build the composite figure.

6. How many cubes did it take Jared to build the figure?

____

7. What is the volume of this figure? __________ cubic centimeters

8.  **Find the Error** Gabriele built a composite figure using 12 cubes for the bottom layer and 10 cubes for the top layer. She said that the volume of the composite figure was 12×10, or 120 cubic centimeters. Find and correct her error.

____

____

Vocabulary Check

Fill in the blank with the correct term or number to complete the sentence.

9. A composite figure is made up of two or more

__________ figures.

10. The composite figure was built using centimeter cubes. What is the volume of the composite figure shown?

$V =$ __________ cubic centimeters

Name ....

Lesson 11
Volume of Composite Figures

ESSENTIAL QUESTION
How does geometry help me solve problems in everyday life?

A **composite figure** is made up of two or more three-dimensional figures. To find the volume, separate the figure into figures with volumes you know how to find.

Math in My World

Example 1

The Arc de Triomphe in Paris, France, is roughly in the shape of the composite figure shown. Find the volume of the Arc de Triomphe.

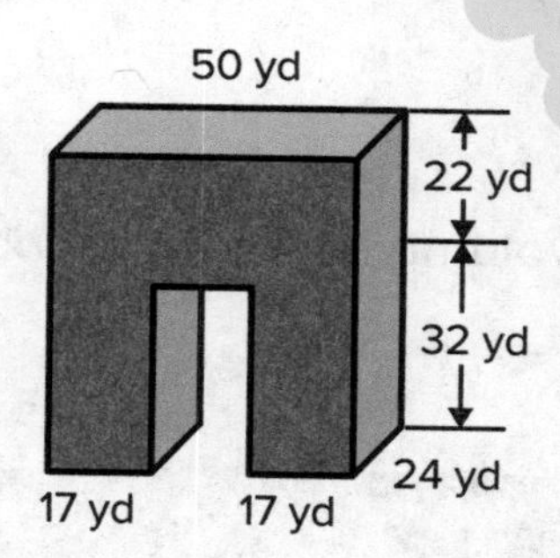

Separate the figure into three rectangular prisms. Find the volume of each prism.

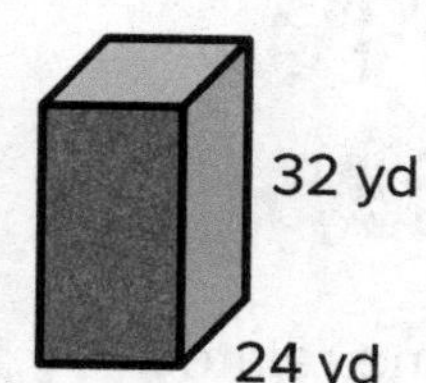

$V = \ell \times w \times h$ → $V =$ ______

$V = 17 \times 24 \times 32$

32 yd, 24 yd, 17 yd

$V = \ell \times w \times h$ → $V =$ ______

$V = 17 \times 24 \times 32$

22 yd, 24 yd, 50 yd

$V = \ell \times w \times h$ → $V =$ ______

$V = 50 \times 24 \times 22$

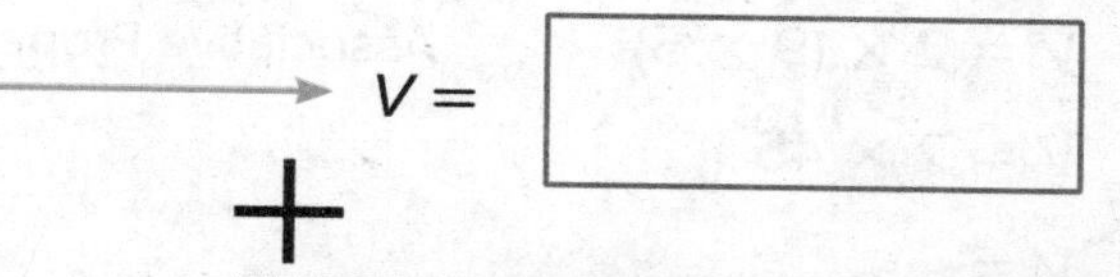

So, the total volume is ______ cubic yards, or ______ yd^3.

Example 2

Find the volume of the composite figure.

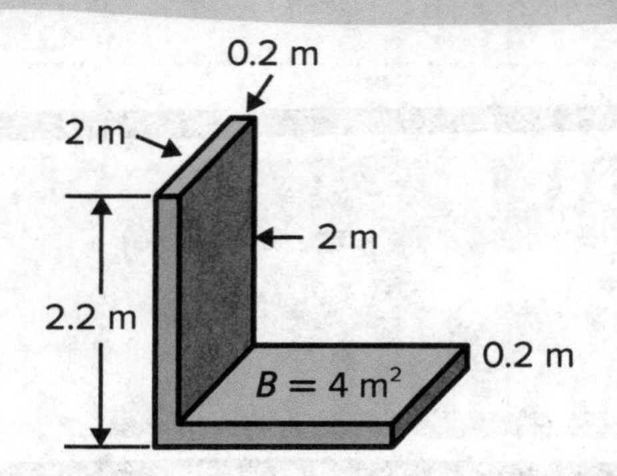

Separate the figure into two prisms.
Find the volume of each prism.

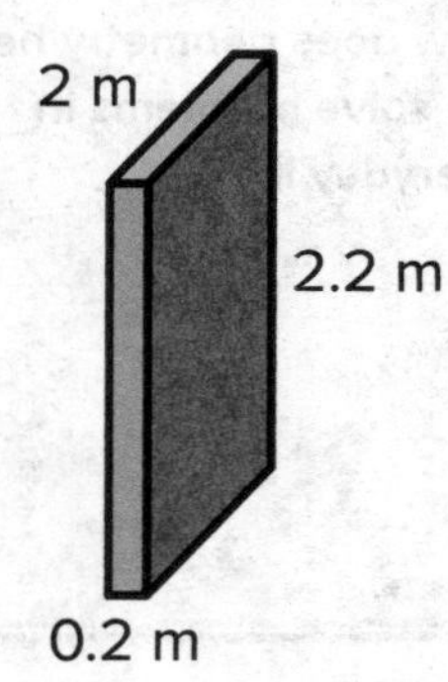

$V = \ell \times w \times h$

$V =$ ____ $\times$ ____ $\times$ ____ → $V =$ [ ]

0.2 m

$B = 4\text{ m}^2$

$V = B \times h$

$V =$ ____ $\times$ ____ → $V =$ [ ]

+

Add the volumes. The total volume is ______ cubic meters, or [ ] m^3.

Guided Practice

1. Find the volume of the composite figure.

Bottom Prism

$V = B \times h$

$V = 126 \times 11$

$V =$ ______

9 cm

2 cm

5 cm

11 cm

$B = 126\text{ cm}^2$

Top Prism

$V = \ell \times w \times h$

$V = 2 \times 9 \times 5$

$V = 2 \times (9 \times 5)$ Associative Property

$V = 2 \times 45$

$V =$ ______

The total volume is ______ + ______

or ______ cubic centimeters.

How is volume related to the operation of addition?

Name

Independent Practice

Find the volume of each composite figure.

2.

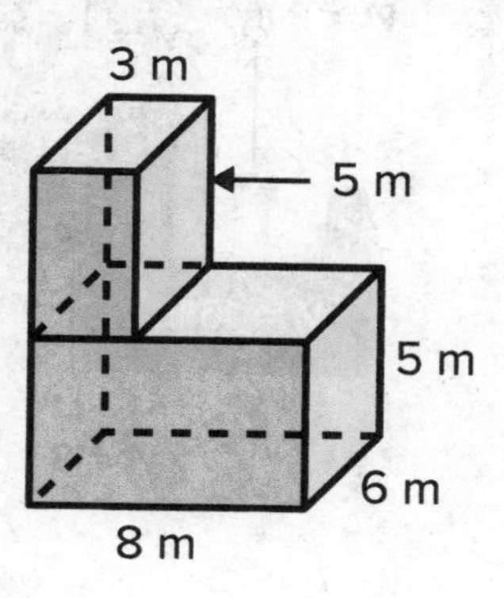

$V =$ ______

3.

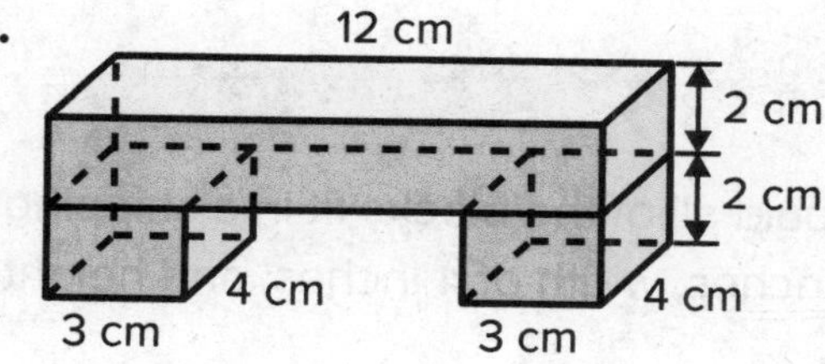

$V =$ ______

4.

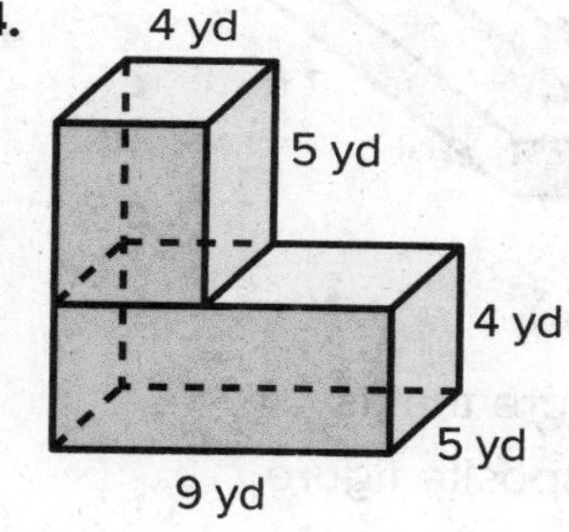

$V =$ ______

5.

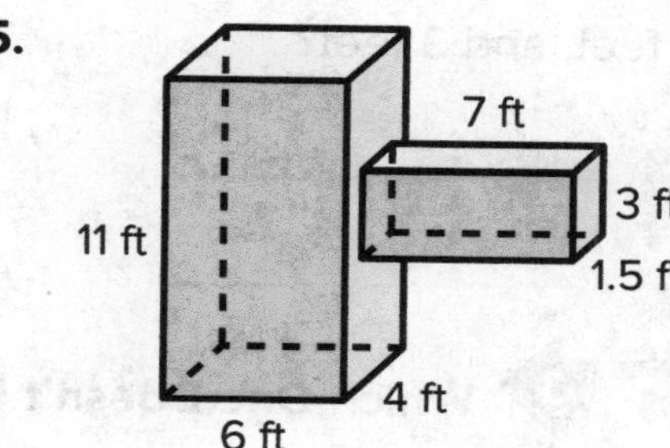

$V =$ ______

6.

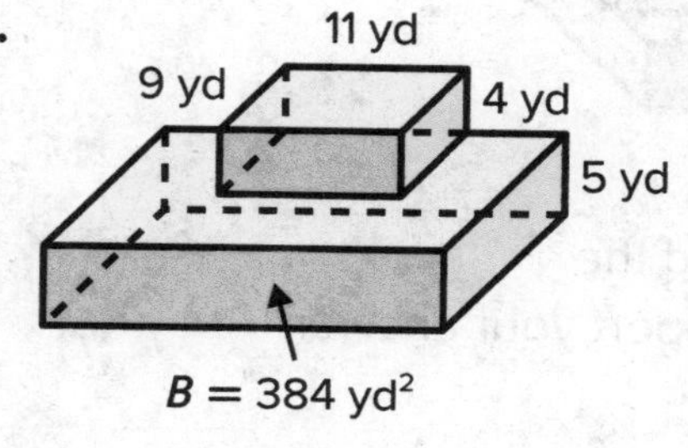

$V =$ ______

7.

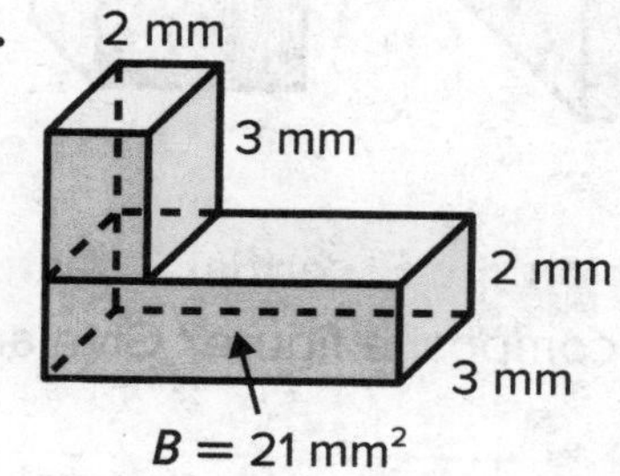

$V =$ ______

Brain Builders

8. Mrs. Stafford ordered the set of blocks shown at the right for her classroom. Find the total volume of all the blocks.

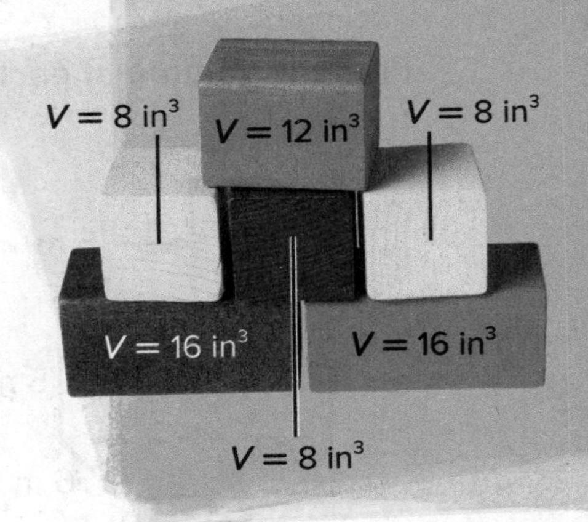

Would all of the blocks fit in a shipping box with a length of 4 inches, width of 4 inches, and height of 4 inches? Explain.

9. The figure represents a piece of foam packaging. How does the volume of the foam compare to the volume of a rectangular prism with dimensions 3 feet, 7 feet, and 3 feet?

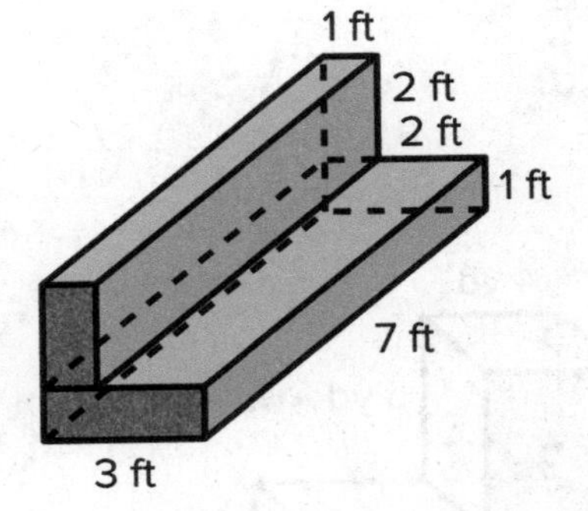

10. Processes &Practices 3 **Which One Doesn't Belong?** Circle the figure that is not a composite figure. Redraw the figure so that it is a composite figure.

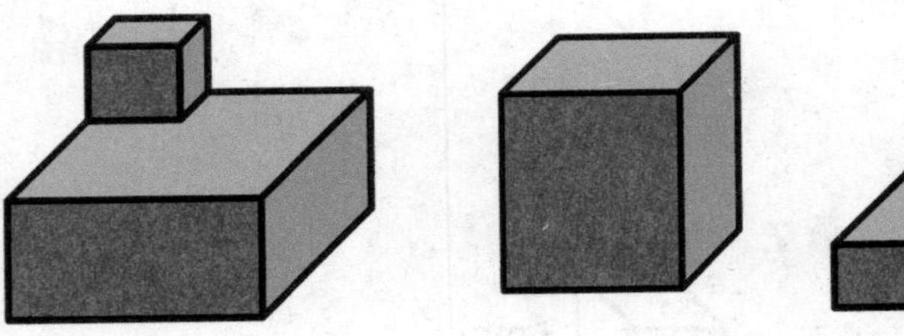

11. **Building on the Essential Question** How can I find the volume of a composite figure? Give an example to support your answer.

Name

MY Homework

Lesson 11

Volume of Composite Figures

Homework Helper

Need help? connectED.mcgraw-hill.com

Find the volume of the composite figure.

3 yd
4 yd
2 yd
4 yd
9 yd
10 yd

The figure was separated into two prisms. Find the volume of each rectangular prism.

Top prism

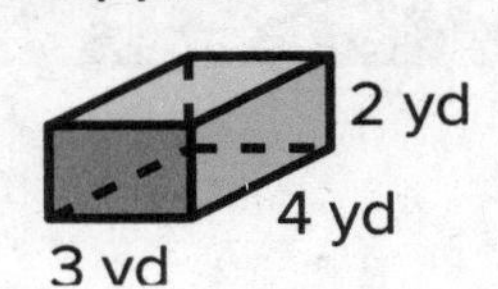

$V = \ell \times w \times h$

$V = 3 \times 4 \times 2$ → 24

Bottom prism

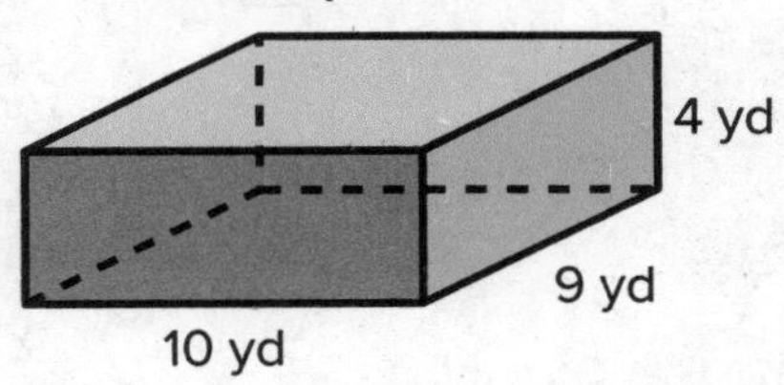

$V = \ell \times w \times h$

$V = 10 \times 9 \times 4$ → $+360$

384

The total volume of the composite figure is 24 + 360 or 384 cubic yards.

Practice

Find the volume of each composite figure.

1.

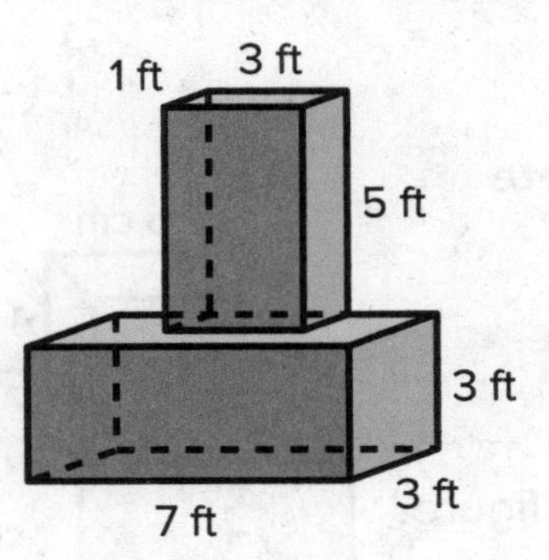

$V =$ __________

2.

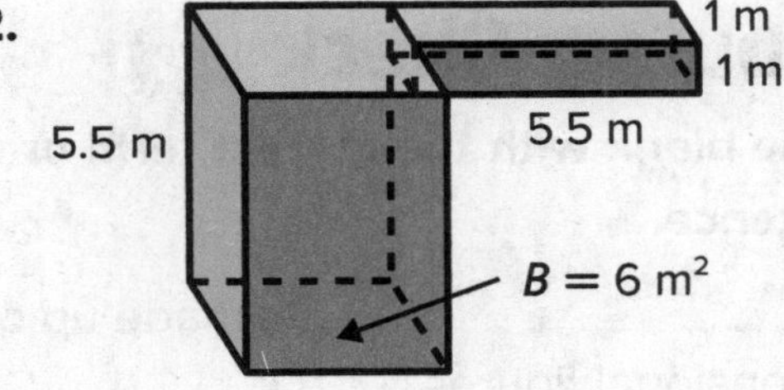

$V =$ __________

Problem Solving

3. Maci is decorating the cake shown. Find the volume of the cake.

Brain Builders

4. The building shown is in the shape of a composite figure. The building was remodeled to include an additional 2,600 cubic yards of space. How many cubic yards of space are in the remodeled building?

5. Processes & Practices 4 **Model Math** Draw an example of a composite figure that has a volume between 750 and 900 cubic units. Explain how you find the volume of your figure.

Vocabulary Check

Fill in the blank with the correct term or number to complete the sentence.

6. A ______________ is made up of two or more three-dimensional figures.

7. Test Practice What is the total volume of the composite figure?

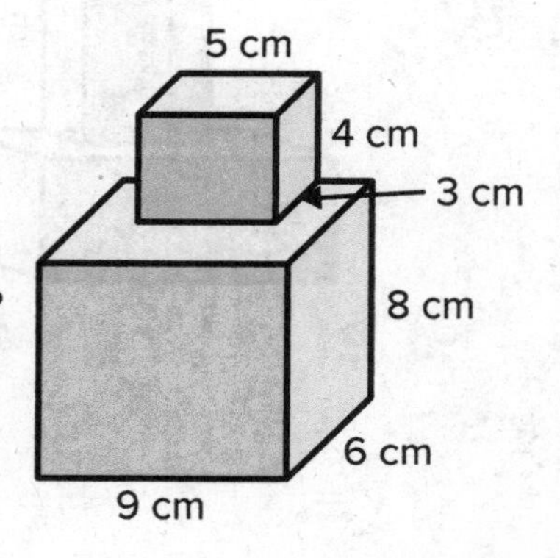

Ⓐ 282 cubic centimeters

Ⓑ 432 cubic centimeters

Ⓒ 492 cubic centimeters

Ⓓ 502 cubic centimeters

Name ....

Lesson 12

Problem-Solving Investigation

STRATEGY: Make a Model

ESSENTIAL QUESTION
How does geometry help me solve problems in everyday life?

Learn the Strategy

Nick is helping his sister put away her alphabet blocks. To fill one layer, it takes nine blocks. If there are six layers, how many blocks would be in the box?

1 Understand

What facts do you know?

There are ________ blocks in each layer and there are six layers.

What do you need to find?

The number of blocks in the box when there are ________ layers.

2 Plan

I can solve the problem by making a ________.

3 Solve

Arrange ________ cubes in a 3 × 3 array. Stack the cubes until there are ________ layers. There are a total of ________ cubes. So, the box would have ________ blocks.

4 Check

Is my answer reasonable? Explain.

Multiply.

6 × 9 = ________

Practice the Strategy

Evelyn wants to mail a package to her cousin. What is the volume of the package if it is 6 inches long, 4 inches wide, and 4 inches tall?

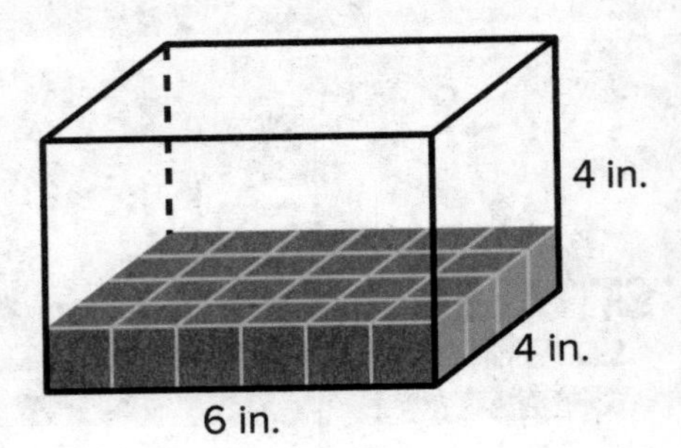

 Understand

What facts do you know?

What do you need to find?

2 Plan

3 Solve

4 Check

Is my answer reasonable? Explain.

Name

Apply the Strategy

Solve each problem by making a model.

1. On an assembly line that is 150 feet long, there is a work station every 15 feet. The first station is at the beginning of the line. How many work stations are there?

2. Processes & Practices 5 **Use Math Tools** A store is stacking cans of food into a rectangular prism display. The bottom layer has 8 cans by 5 cans. There are 5 layers. How many cans are in the display?

Brain Builders

3. Paul is making sand castles by stacking different sized rectangular prisms. The base of his castle is 4 feet by 4 feet by 2 feet. He places 4 more prisms on top of the base. If each dimension of the prism he places on top of the base is one-half the dimension of the base, how many cubic feet of sand are in Paul's sand castle?

4. Alice is creating a pyramid with toy rectangular-prism-shaped blocks that measure 1 inch by 2 inches by 2 inches. She places 8 blocks in the first row. Then she places 6 blocks on top of the first row and 4 blocks on top of the second row. If she continues the pattern until the top row has 2 blocks, what is the volume of the final pyramid?

Review the Strategies

Use any strategy to solve each problem.

- Make a model.
- Guess, check, and revise.
- Look for a pattern.
- Make a table.

5. Five friends are standing in a circle and playing a game where they toss a ball of yarn to one another. If each person is connected by the yarn to each other person only once, how many lines of yarn will connect the group?

____

6. Processes & Practices 8 **Look for a Pattern** In the figure below, there are 22 marbles in Box A. To go from Box A to Box B, exactly four marbles must pass through the triangular machine at a time. Exactly five marbles must pass through the square machine at a time. Describe how to move all the marbles from Box A to Box B in the fewest moves possible.

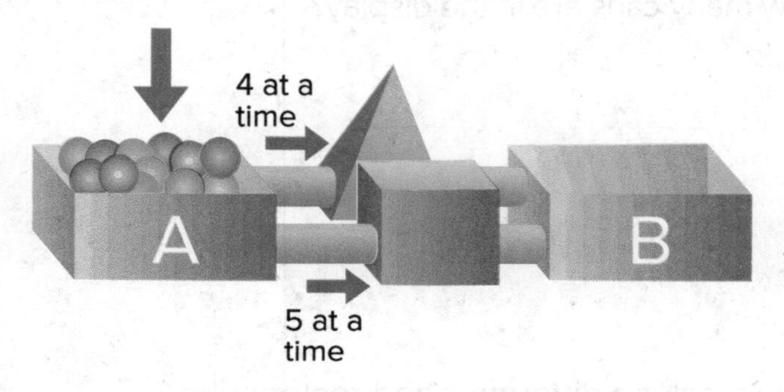

____

____

____

7. The volume of a rectangular prism is 5,376 cubic inches. The prism is 14 inches long and 16 inches wide. How tall is the prism?

____

8. The table at the right shows the number of minutes Danielle spent practicing the trumpet over the last 7 days. If she continues this pattern of practicing, in how many days will she have practiced 340 minutes?

Day	Time (min)
1	20
2	20
3	35
4	20
5	20
6	35
7	20

____

Name

Lesson 12

Problem Solving: Make a Model

Homework Helper

Need help? connectED.mcgraw-hill.com

Victor wants to build a brick wall. Each brick layer is 3 inches thick, and the wall will be 18 inches tall. How many layers will it have?

1 Understand

What facts do you know?

Each brick layer is 3 inches thick. The wall will be 18 inches tall.

What do you need to find?

the number of layers the wall will have

2 Plan

Solve the problem by making a model.

3 Solve

Make a model of the wall by using cubes.

Each cube represents a 3-inch brick.

He needs 6 cubes to build the 18-inch wall.

So, the wall would have 6 layers.

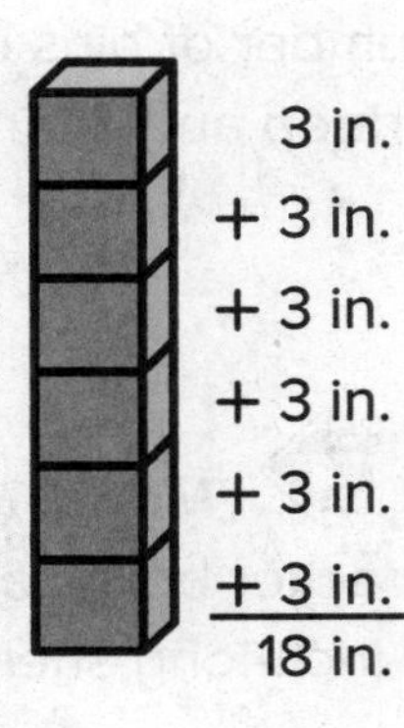

4 Check

Is my answer reasonable? Explain.

Multiply. $6 \times 3 = 18$ inches

Problem Solving

Solve each problem by making a model.

1. Nan and Sato are designing a coffee table using 4-inch tiles. Nan uses 30 tiles and Sato uses half as many. How many total tiles did they use?

If the area of the table is 36 inches by 24 inches, will they have enough tiles to cover the table? If not, how many more will they need?

2. The Jones family is landscaping their rectangular yard. Their yard is 160 square feet and one side is 10 feet long. What is the length of the other side of the yard?

They want to plant 3 1 foot by 1 foot trees. If they plant 3 trees that need to be 3 feet apart and 3 feet away from the fence around the yard, will they have the space?

Brain Builders

3. Billy is organizing his closet. He has clothing bins that measure 20 inches high, 18 inches wide, and 14 inches deep. What is the greatest number of bins he can fit in a 60-inch high closet that is 36 inches deep and 72 inches wide? Explain.

4. Processes &Practices 4 **Model Math** Bob is organizing his pantry. If he has cracker boxes as shown, how many boxes can he fit on a 20-inch-long shelf that is 14 inches deep? Explain.

Review

Chapter 12
Geometry

Vocabulary Check

Match each word to its definition. Write your answers on the lines provided.

1. **equilateral triangle** _______
2. **composite figure** _______
3. **parallelogram** _______
4. **volume** _______
5. **rectangular prism** _______
6. **regular polygon** _______
7. **triangular prism** _______
8. **obtuse triangle** _______
9. **face** _______
10. **polygon** _______
11. **square** _______
12. **pentagon** _______

A. a three-dimensional figure with six rectangular faces, twelve edges, and eight vertices

B. a flat surface of a three-dimensional figure

C. a triangle with one obtuse angle

D. a closed figure made up of line segments that do not cross each other

E. a figure made up of two or more three-dimensional figures

F. a polygon with five sides

G. a prism with two congruent triangular bases

H. a polygon with congruent sides and all congruent angles

I. a quadrilateral with opposite sides both parallel and congruent

J. a triangle with three congruent sides

K. the amount of space within a three-dimensional figure

L. a rectangle with four congruent sides

Concept Check

Name each polygon. Determine if it appears to be *regular* or *not regular.*

13.

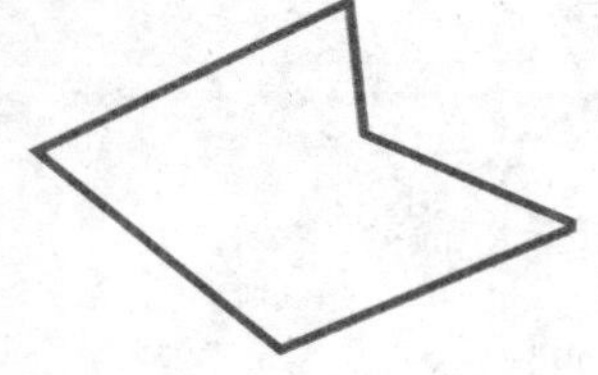

14.

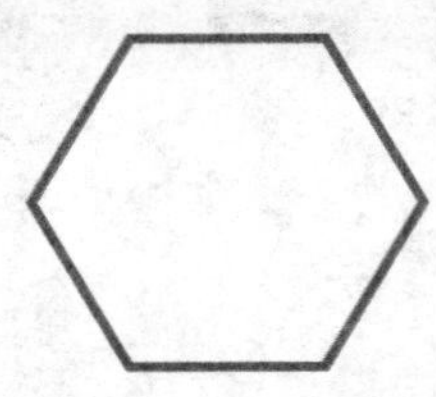

Describe the attributes of each quadrilateral. Then classify the quadrilateral.

15.

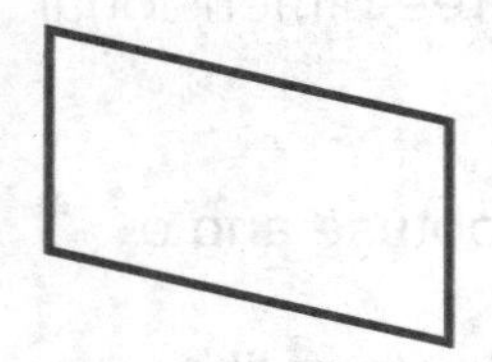

16.

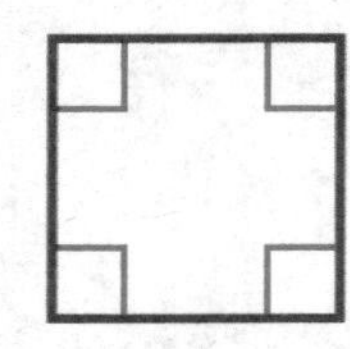

Describe the faces, edges, and vertices of each three-dimensional figure. Then identify it.

17.

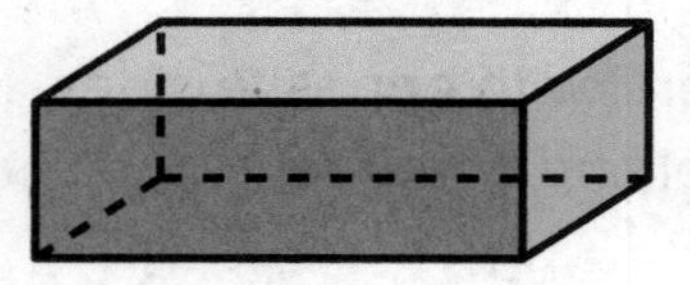

18.

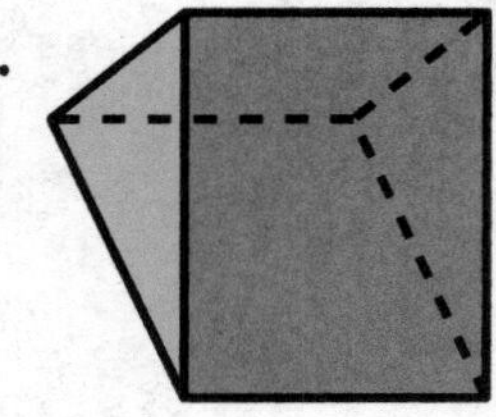

Problem Solving

19. A triangle forms the front of the Pantheon in Rome, Italy. Classify the triangle based on its sides. Then classify it based on its angles.

____

____

20. Shawn keeps his photos in a box like the one shown.

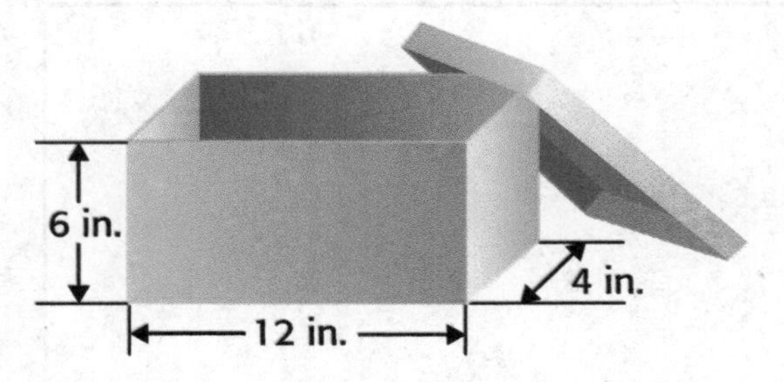

What is the volume of the box?

____

Brain Builders

21. Ishmail jogs around a track that is in the shape of a regular pentagon. One side of the pentagon is $\frac{1}{4}$ mile long. How far does Ishmail jog if he completes 6 laps around the track?

____

If the track was a hexagon, how far would Ishmail jog in 6 laps?

____

22. Test Practice Find the volume of the composite figure.

Ⓐ 2,700 in^3 Ⓒ 3,420 in^3

Ⓑ 2,780 in^3 Ⓓ 3,660 in^3

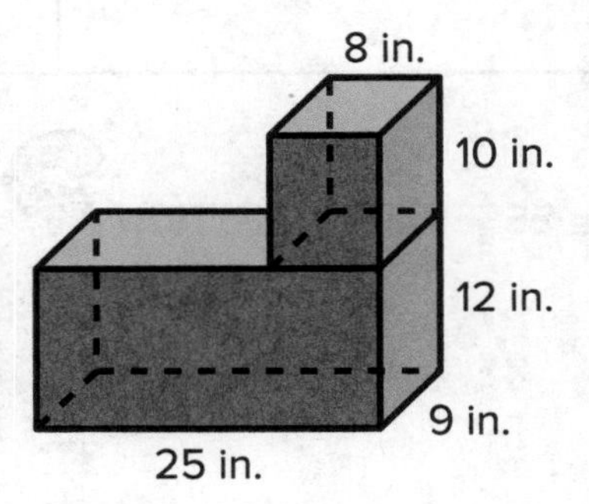

Reflect

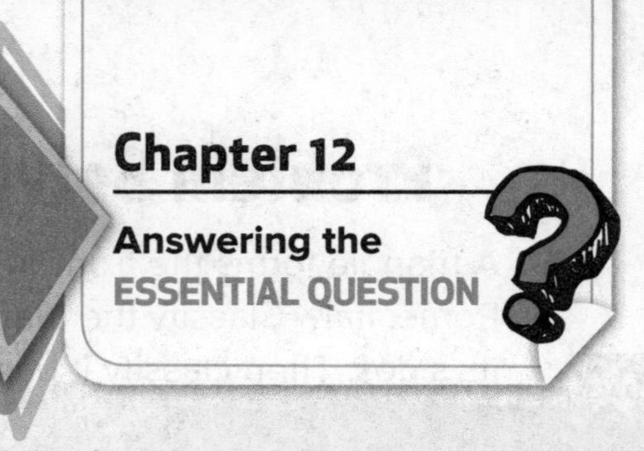

Use what you learned about geometry to complete the graphic organizer.

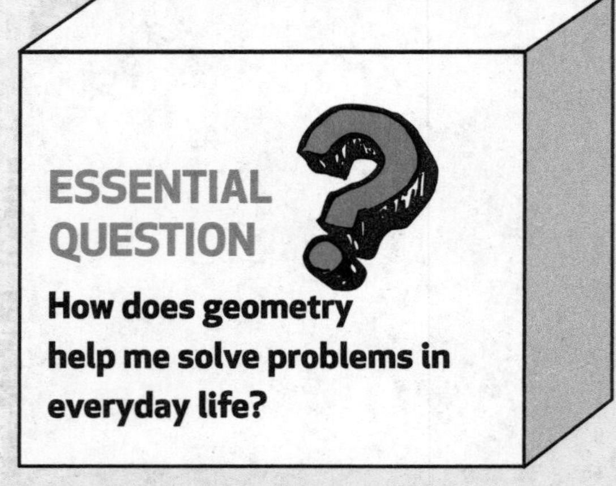

Faces, Edges, and Vertices

Vocabulary	Real-World Example

Now reflect on the ESSENTIAL QUESTION  Write your answer below.

Name

Date

Score

Performance Task

Constructing Macaroni Boxes

A pasta company is looking to construct a box for their new gluten-free macaroni. The dimensions of the box are shown below. Two overlapping flaps each cover the entire top and bottom of the box.

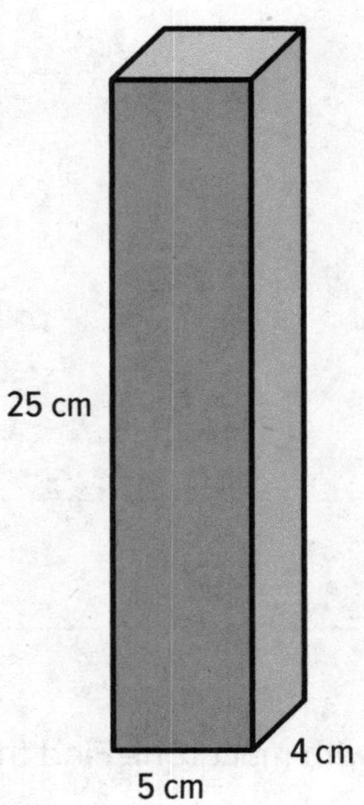

Show all your work to receive full credit.

Part A

Each box needs to be cut from a flat piece of cardboard. Draw a net for the box and label the length, width, and height.

Part B

The amount of cardboard needed is measured in square centimeters. Find the area of the net you drew in **Part A** in order to find the area of cardboard needed to construct a box. Explain your answer.

Part C

The box will be filled $\frac{4}{5}$ of the way with macaroni. Find the volume of macaroni that can be put in each box. Explain your answer.

Glossary/Glosario

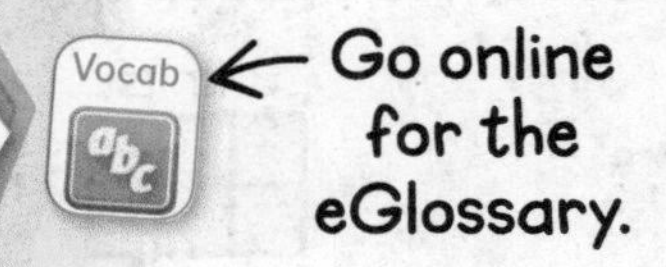

Go to the eGlossary to find out more about these words in the following 13 languages:

Arabic • Bengali • Brazilian Portuguese • Cantonese • English • Haitian Creole
Hmong • Korean • Russian • Spanish • Tagalog • Urdu • Vietnamese

English	Spanish/Español
acute angle An angle with a measure between 0° and 90°.	**ángulo agudo** Ángulo que mide entre 0° y 90°.

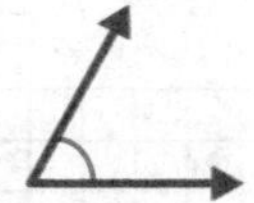

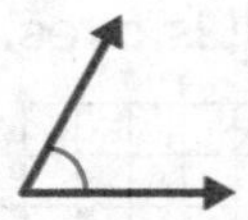

English	Spanish/Español
acute triangle A triangle with three acute angles.	**triángulo acutángulo** Triángulo con tres ángulos agudos.
algebra A branch of mathematics that uses symbols, usually letters, to explore relationships between quantities.	**álgebra** Rama de las matemáticas que usa símbolos, generalmente letras, para explorar relaciones entre cantidades.
angle Two rays with a common endpoint.	**ángulo** Dos semirrectas con un extremo común.

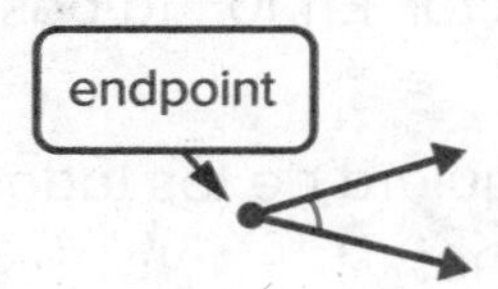

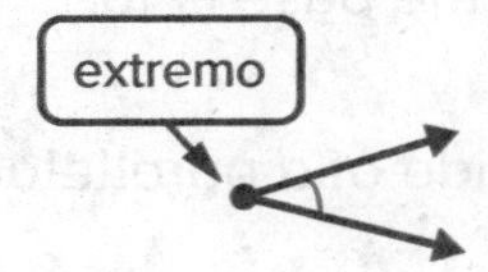

English	Spanish/Español
annex To place a zero to the right of a decimal without changing a number's value.	**agregar** Poner un cero a la derecha de un decimal sin cambiar el valor de un número.

Aa

area The number of square units needed to cover the surface of a closed figure.

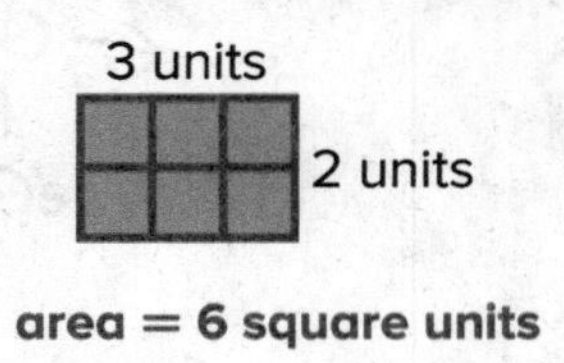

area = 6 square units

área Cantidad de unidades cuadradas necesarias para cubrir la superficie de una figura cerrada.

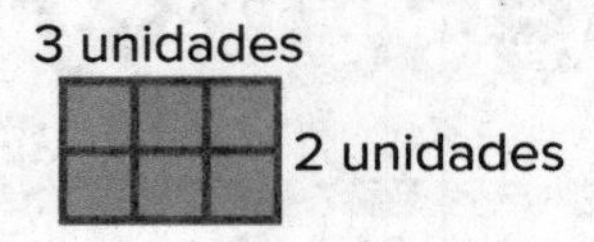

área = 6 unidades cuadradas

Associative Property Property that states that the way in which numbers are grouped does not change the sum or product.

propiedad asociativa Propiedad que establece que la manera en que se agrupan los números no altera la suma o el producto.

attribute A characteristic of a figure.

atributo Característica de una figura.

axis A horizontal or vertical number line on a graph. Plural is axes.

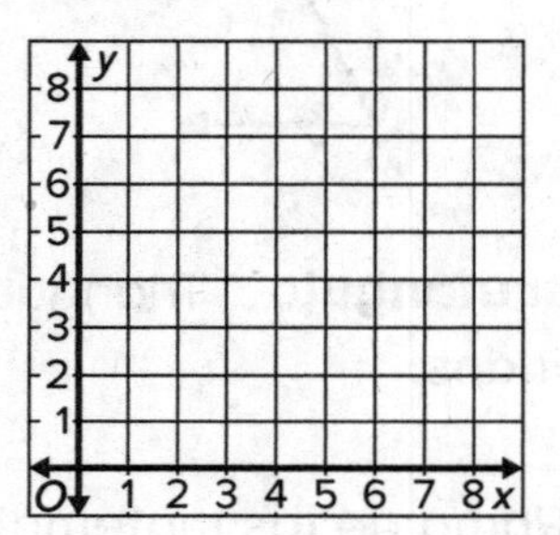

eje Recta numérica horizontal o vertical en una gráfica.

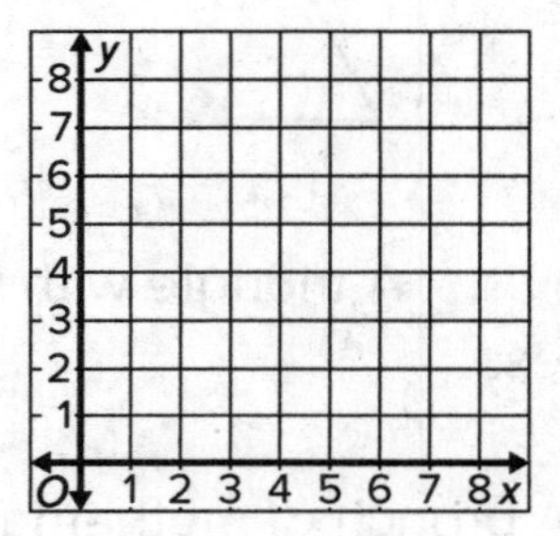

Bb

base In a power, the number used as a factor. In 10^3, the base is 10.

base En una potencia, el número que se usa como factor. En 10^3, la base es 10.

base Any side of a parallelogram.

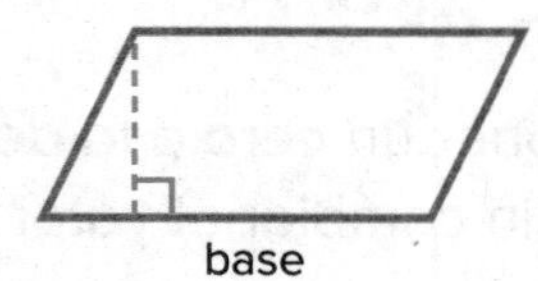

base Cualquiera de los lados paralelogramo.

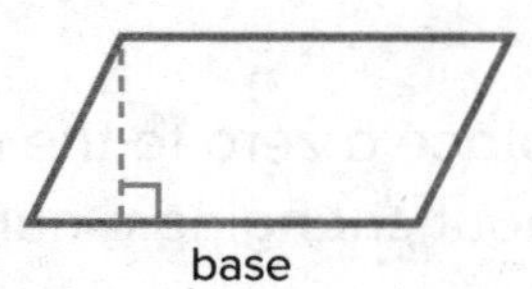

base One of the two parallel congruent faces in a prism.

base Una de las dos caras congruentes paralelas en un prisma.

capacity The amount a container can hold.

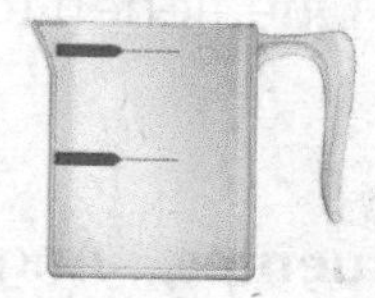

centimeter (cm) A metric unit for measuring length.

100 centimeters = 1 meter

common denominator A number that is a multiple of the denominators of two or more fractions.

common factor A number that is a factor of two or more numbers.

3 is a common factor of 6 and 12.

common multiple A whole number that is a multiple of two or more numbers.

24 is a common multiple of 6 and 4.

Commutative Property Property that states that the order in which numbers are added does not change the sum and that the order in which factors are multiplied does not change the product.

compatible numbers Numbers in a problem that are easy to work with mentally.

720 and 90 are compatible numbers for division because $72 \div 9 = 8$.

composite figures A figure made up of two or more three-dimensional figures.

capacidad Cantidad que puede contener un recipiente.

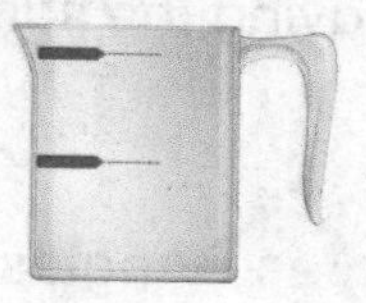

centímetro (cm) Unidad métrica de longitud.

100 centímetros = 1 metro

denominador común Número que es múltiplo de los denominadores de dos o más fracciones.

factor común Número que es un factor de dos o más números.

3 es factor común de 6 y 12.

múltiplo común Número natural múltiplo de dos o más números.

24 es un múltiplo común de 6 y 4.

propiedad conmutativa Propiedad que establece que el orden en que se suman los números no altera la suma y que el orden en que se multiplican los factores no altera el producto.

números compatibles Números en un problema con los cuales es fácil trabajar mentalmente.

$720 \div 90$ es una divisíon que usa son números compatibles porque $72 \div 9 = 8$.

figura compuesta Figura conformada por dos o más figuras tridimensionales.

Cc

composite number A whole number that has more than two factors.

12 has the factors 1, 2, 3, 4, 6, and 12.

congruent Having the same measure.

congruent angles Angles of a figure that are equal in measure.

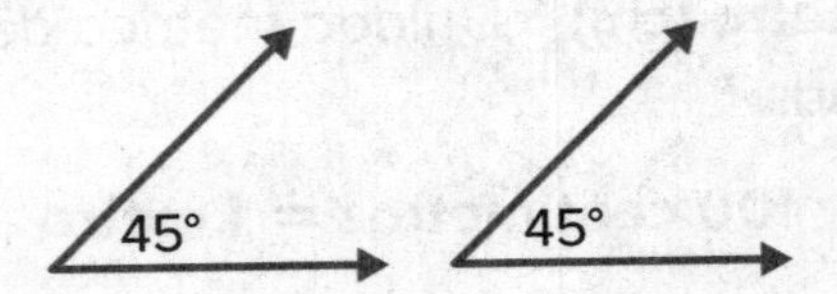

congruent figures Two figures having the same size and the same shape.

congruent sides Sides of a figure that are equal in length.

convert To change one unit to another.

coordinate One of two numbers in an ordered pair.

The 1 is the number on the *x*-axis, the 5 is on the *y*-axis.

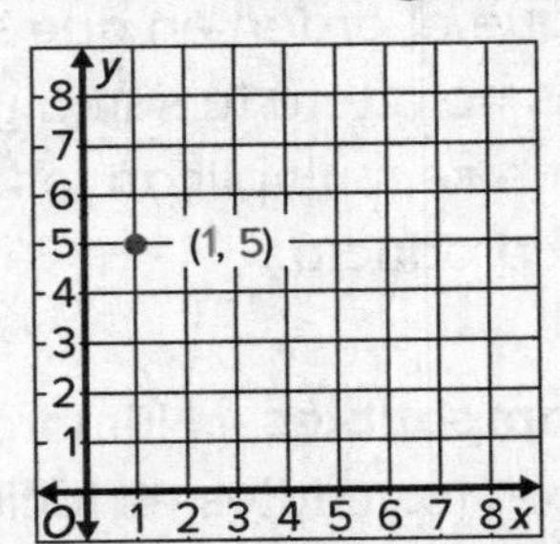

coordinate plane A plane that is formed when two number lines intersect.

número compuesto Número natural que tiene más de dos factores.

12 tiene a los factores 1, 2, 3, 4, 6 y 12.

congruentes Que tienen la misma medida.

ángulos congruentes Ángulos de una figura que tienen la misma medida.

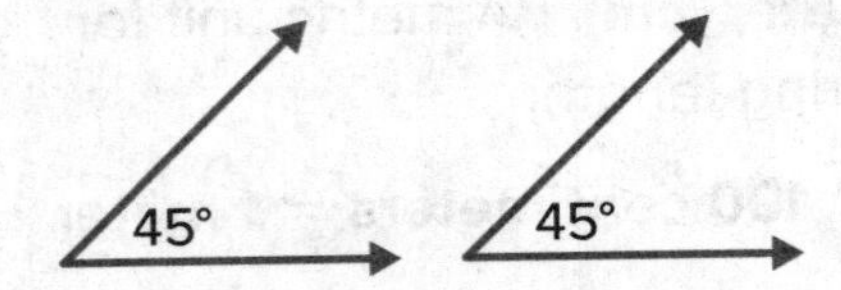

figuras congruentes Dos figuras que tienen el mismo tamaño y la misma forma.

lados congruentes Lados de una figura que tienen la misma longitud.

convertir Transformar una unidad en otra.

coordenada Cada uno de los números de un par ordenado.

El 1 es la coordenada *x* y el 5 es la coordenada *y*.

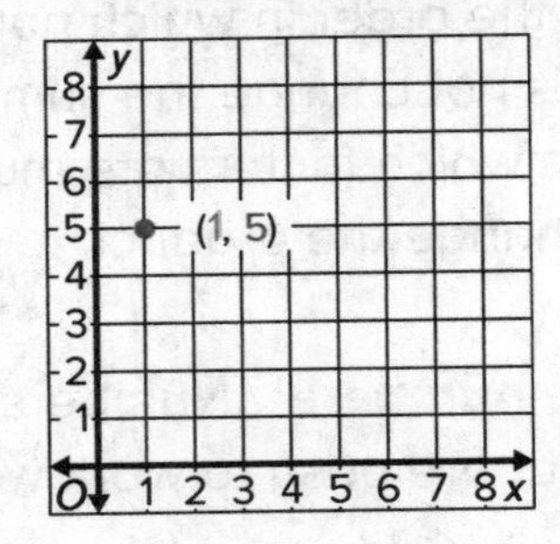

plano de coordenadas Plano que se forma cuando dos rectas numéricas se intersecan formando un ángulo recto.

cube A rectangular prism with six faces that are congruent squares.

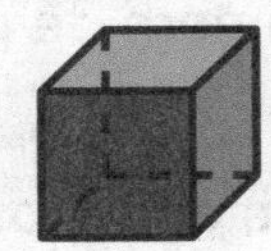

cubed A number raised to the third power; $10 \times 10 \times 10$, or 10^3.

cubic unit A unit for measuring volume, such as a cubic inch or a cubic centimeter.

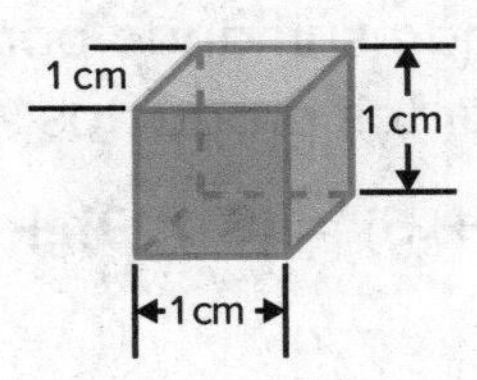

cup A customary unit of capacity equal to 8 fluid ounces.

customary system The units of measurement most often used in the United States. These include foot, pound, and quart.

cubo Prisma rectangular con seis caras que son cuadrados congruentes.

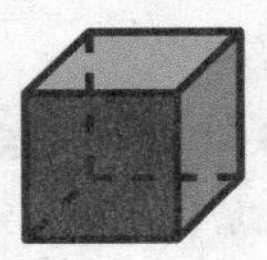

al cubo Número elevado a la tercera potencia; $10 \times 10 \times 10$ o 10^3.

unidad cúbica Unidad de volumen, como una pulgada cúbica o un centímetro cúbico.

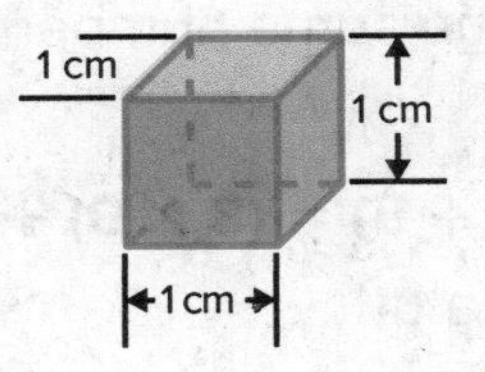

taza Unidad usual de capacidad que equivale a 8 onzas líquidas.

sistema usual Conjunto de unidades de medida de uso más frecuente en Estados Unidos. Incluyen el pie, la libra y el cuarto.

Dd

decimal A number that has a digit in the tenths place, hundredths place, and beyond.

decimal point A period separating the ones and the tenths in a decimal number.

0.8 or $3.77

degree (°) a. A unit of measure used to describe temperature. b. A unit for measuring angles.

decimal Número que tiene al menos un dígito en el lugar de las décimas, centésimas etcétera.

punto decimal Punto que separa las unidades y las décimas en un número decimal.

0.8 o $3.77

grado (°) a. Unidad de medida que se usa para describir la temperatura. b. Unidad que se usa para medir ángulos.

denominator The bottom number in a fraction. It represents the number of parts in the whole.

In $\frac{5}{6}$, 6 is the denominator.

digit A symbol used to write numbers. The ten digits are 0, 1, 2, 3, 4, 5, 6, 7, 8, and 9.

Distributive Property To multiply a sum by a number, you can multiply each addend by the same number and add the products.

$$8 \times (9 + 5) = (8 \times 9) + (8 \times 5)$$

divide (division) An operation on two numbers in which the first number is split into the same number of equal groups as the second number.

12 ÷ 3 means 12 is divided into 3 equal-size groups

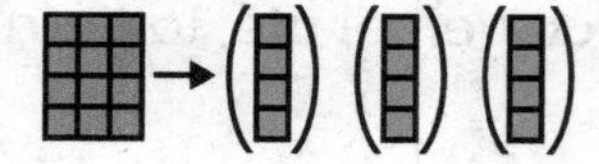

dividend A number that is being divided.

$3\overline{)19}$ ← 19 is the dividend.

divisible Describes a number that can be divided into equal parts and has no remainder.

39 is divisible by 3 with no remainder.

divisor The number that divides the dividend.

3 is the divisor. → $3\overline{)19}$

denominador Numero que se escribe debajo de la barra en una fracción. Representa el número de partes en que se divide un entero.

En $\frac{5}{6}$, 6 es el denominador.

dígito Símbolo que se usa para escribir los números. Los diez dígitos son 0, 1, 2, 3, 4, 5, 6, 7, 8 y 9.

propiedad distributiva Para multiplicar una suma por un número, puedes multiplicar cada sumando por ese número y luego sumar los productos.

$$8 \times (9 + 5) = (8 \times 9) + (8 \times 5)$$

dividir (división) Operación entre dos números en la cual el primer número se separa en tantos grupos iguales como indica el segundo número.

12 ÷ 3 significa que 12 se divide entre 3 grupos de igual tamaño.

dividendo Número que se divide.

$3\overline{)19}$ ← 19 es el dividendo.

divisible Describe un número que puede dividirse en partes iguales, sin residuo.

39 es divisible entre 3 sin residuo.

divisor Número entre el cual se divide el dividendo.

3 es el divisor. → $3\overline{)19}$

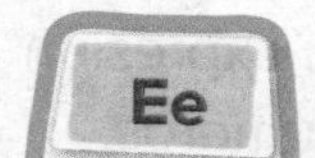

edge The line segment where two faces of a three-dimensional figure meet.

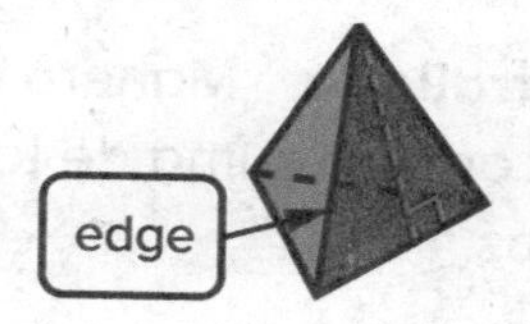

equation A number sentence that contains an equal sign, showing that two expressions are equal.

equilateral triangle A *triangle* with three *congruent* sides.

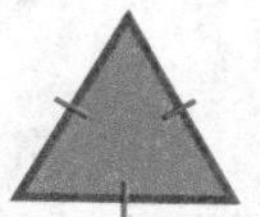

equivalent decimals Decimals that have the same value.

0.3 and 0.30

equivalent fractions Fractions that have the same value.

$$\frac{3}{4} = \frac{6}{8} = \frac{9}{12}$$

estimate A number close to an exact value. An estimate indicates about how much.

47 + 22 (round to 50 + 20)

The estimate is 70.

arista Segmento de recta donde se unen dos caras de una figura tridimensional.

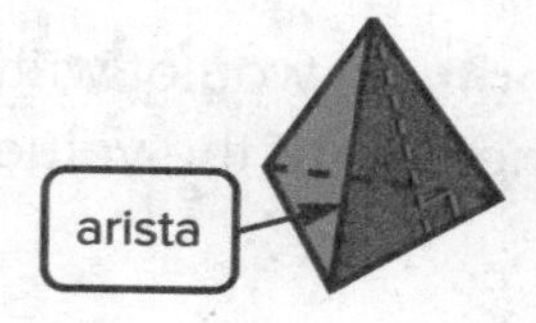

ecuación Expresión numérica que contiene un signo igual y que muestra que dos expresiones son iguales.

triángulo equilátero *Triángulo* con tres lados *congruentes*.

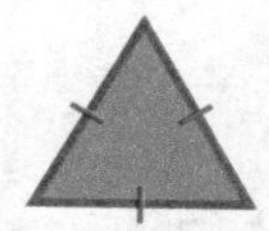

decimales equivalentes Decimales que tienen el mismo valor.

0.3 y 0.30

fracciones equivalentes Fracciones que tienen el mismo valor.

$$\frac{3}{4} = \frac{6}{8} = \frac{9}{12}$$

estimación Número cercano a un valor exacto. Una estimación indica una cantidad aproximada.

47 + 22 (se redondea a 50 + 20)

La estimación es 70.

Ee

evaluate To find the value of an expression by replacing variables with numbers.

even number A whole number that is divisible by 2.

expanded form A way of writing a number as the sum of the values of its digits.

exponent In a power, the number of times the base is used as a factor. In 5^3, the exponent is 3.

expression A combination of numbers, variables, and at least one operation.

evaluar Calcular el valor de una expresión reemplazando las variables por números.

número par Número natural divisible entre 2.

forma desarrollada Manera de escribir un número como la suma de los valores de sus dígitos.

exponente En una potencia, el número de veces que se usa la base como factor. En 5^3, el exponente es 3.

expresión Combinación de números, variables y por lo menos una operación.

Ff

face A flat surface.

A square is a face of a cube.

fact family A group of related facts using the same numbers.

factor A number that is multiplied by another number.

Fahrenheit (°F) A unit used to measure temperature.

fair share An amount divided equally.

fluid ounce A customary unit of capacity.

cara Superficia plana.

Cada cara de un cubo es un cuadrado.

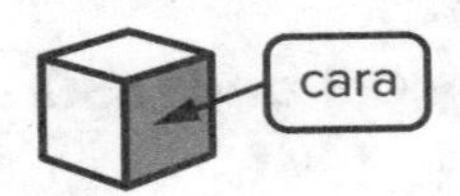

familia de operaciones Grupo de operaciones relacionadas que usan los mismos números.

factor Número que se multiplica por otro número.

Fahrenheit (°F) Unidad que se usa para medir la temperatura.

partes iguales Partes entre las que se divide equitativamente un entero.

onza líquida Unidad usual de capacidad.

foot (ft) A customary unit for measuring length. Plural is feet.

1 foot = 12 inches

pie (pie) Unidad usual de longitud.

1 pie = 12 pulgadas

fraction A number that represents part of a whole or part of a set.

$\frac{1}{2}, \frac{1}{3}, \frac{1}{4}, \frac{3}{4}$

fracción Número que representa una parte de un todo o una parte de un conjunto.

$\frac{1}{2}, \frac{1}{3}, \frac{1}{4}, \frac{3}{4}$

gallon (gal) A customary unit for measuring capacity for liquids.

1 gallon = 4 quarts

galón (gal) Unidad de medida usual de capacidad de líquidos.

1 galón = 4 cuartos

gram (g) A metric unit for measuring mass.

gramo (g) Unidad métrica para medir la masa.

graph To place a point named by an ordered pair on a coordinate plane.

graficar Colocar un punto nombrado por un par ordenado en un plano de coordenadas.

Greatest Common Factor (GCF) The greatest of the common factors of two or more numbers.

The greatest common factor of 12, 18, and 30 is 6.

máximo común divisor (M.C.D.) El mayor de los factores comunes de dos o más números.

El máximo común divisor de 12, 18 y 30 es 6.

height The shortest distance from the base of a parallelogram to its opposite side.

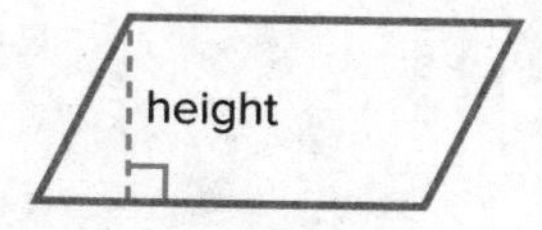

altura La distancia más corta desde la base de un paralelogramo hasta su lado opuesto.

Hh

hexagon A polygon with six sides and six angles.

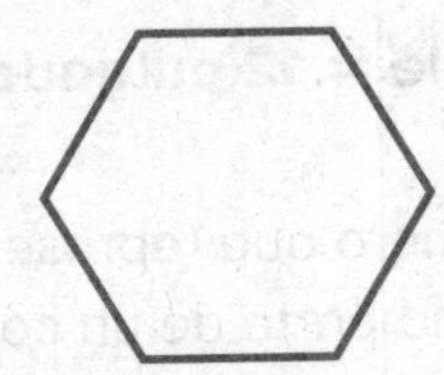

hexágono Polígono con seis lados y seis ángulos.

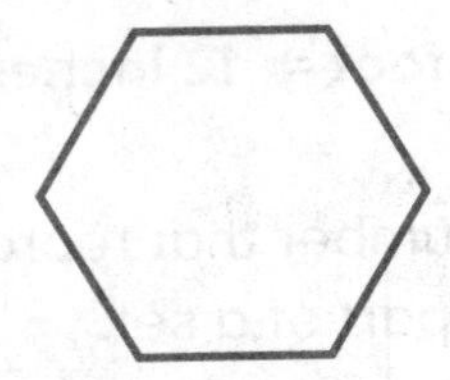

horizontal axis The axis in a coordinate plane that runs left and right (↔). Also known as the *x*-axis.

eje horizontal Eje en un plano de coordenadas que va de izquierda a derecha (↔). También conocido como eje *x*.

hundredth A place value position. One of one hundred equal parts. In the number 0.57, 7 is in the hundredths place.

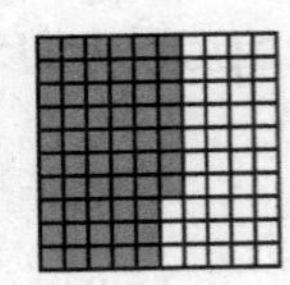

centésima Valor posícional. Una de cien partes iguales. En el número 0.57, 7 está en el lugar de las centésimas.

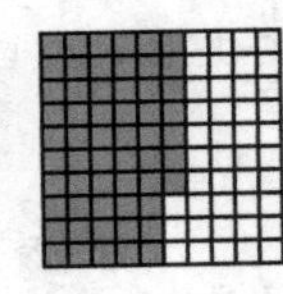

Ii

Identity Property Property that states that the sum of any number and 0 equals the number and that the product of any number and 1 equals the number.

propiedad de identidad Propiedad que establece que la suma de cualquier número y 0 es igual al número y que el producto de cualquier número y 1 es igual al número.

improper fraction A fraction with a numerator that is greater than or equal to the denominator.

$\frac{17}{3}$ or $\frac{5}{5}$

fracción impropia Fracción con un numerador mayor que él igual al denominador.

$\frac{17}{3}$ o $\frac{5}{5}$

inch (in.) A customary unit for measuring length. The plural is inches.

pulgada (pulg) Unidad usual de longitud.

inequality Two quantities that are not equal.

desigualdad Dos cantidades que no son iguales.

intersecting lines *Lines* that meet or cross at a common *point.*

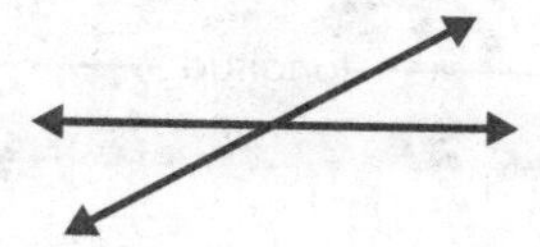

interval The distance between successive values on a scale.

inverse operations Operations that undo each other.

isosceles triangle A *triangle* with at least 2 *sides* of the same *length.*

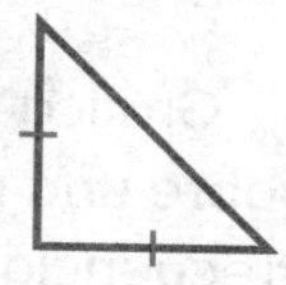

rectas secantes *Rectas* que se intersecan o se cruzan en un *punto* común.

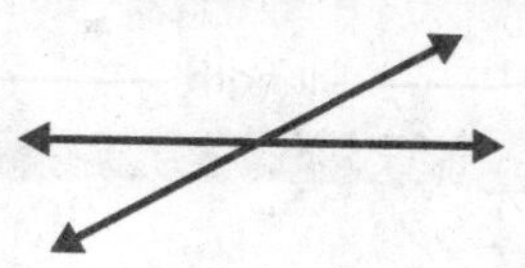

intervalo Distancia entre valores sucesivos en una escala.

operaciones inversas Operaciones que se cancelan entre sí.

triángulo isósceles *Triángulo* que tiene por lo menos 2 *lados* del mismo largo.

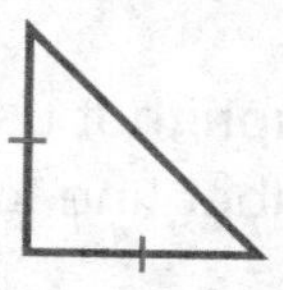

Kk

kilogram (kg) A metric unit for measuring mass.

kilometer (km) A metric unit for measuring length.

kilogramo (kg) Unidad métrica de masa.

kilómetro (km) Unidad métrica de longitud.

Ll

Least Common Denominator (LCD) The least common multiple of the denominators of two or more fractions.

$\frac{1}{12}, \frac{1}{6}, \frac{1}{8}$; **LCD is 24.**

Least Common Multiple (LCM) The smallest whole number greater than 0 that is a common multiple of each of two or more numbers.

The LCM of 2 and 3 is 6.

mínimo común denominador (m.c.d.) El mínimo común múltiplo de los denominadores de dos o más fracciones.

$\frac{1}{12}, \frac{1}{6}, \frac{1}{8}$; **el m.c.d. es 24.**

mínimo común múltiplo (m.c.m.) El menor número natural, mayor que 0, múltiplo común de dos o más números.

El m.c.m. de 2 y 3 es 6.

Ll

length Measurement of the distance between two points.

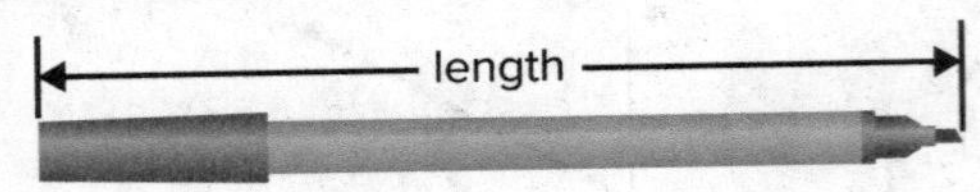

like fractions Fractions that have the same denominator.

$\frac{1}{5}$ **and** $\frac{2}{5}$

line A set of *points* that form a straight path that goes on forever in opposite directions.

line plot A graph that uses columns of Xs above a number line to show frequency of data.

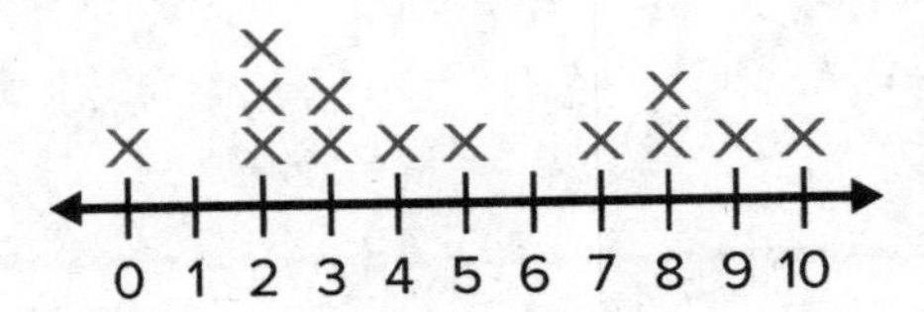

line segment A part of a *line* that connects two points.

liter (L) A metric unit for measuring volume or capacity.

1 liter = 1,000 milliliters

longitud Medida de la distancia entre dos puntos.

fracciones semejantes Fracciones que tienen el mismo denominador.

$\frac{1}{5}$ **y** $\frac{2}{5}$

recta Conjunto de *puntos* que forman una trayectoria recta sin fin en direcciones opuestas.

diagrama lineal Gráfica que usa columnas de X sobre una recta numérica para mostrar la frecuencia de los datos.

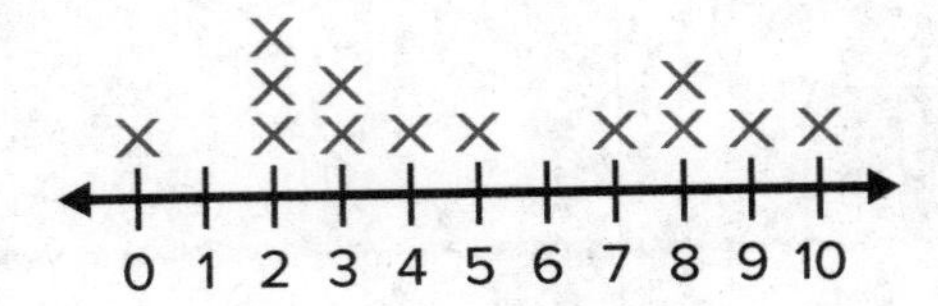

segmento de recta Parte de una *recta* que conecta dos puntos.

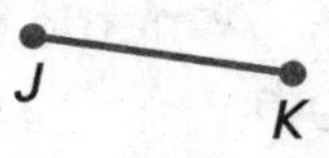

litro (L) Unidad métrica de volumen o capacidad.

1 litro = 1,000 mililitros

Mm

mass Measure of the amount of matter in an object.

meter (m) A metric unit used to measure length.

masa Medida de la cantidad de materia en un cuerpo.

metro (m) Unidad métrica que se usa para medir la longitud.

metric system (SI) The decimal system of measurement. Includes units such as meter, gram, and liter.

mile (mi) A customary unit of measure for length.

1 mile = 5,280 feet

milligram (mg) A metric unit used to measure mass.

1,000 milligrams = 1 gram

milliliter (mL) A metric unit used for measuring capacity.

1,000 milliliters = 1 liter

millimeter (mm) A metric unit used for measuring length.

1,000 millimeters = 1 meter

mixed number A number that has a whole number part and a fraction part. $3\frac{1}{2}$ is a mixed number.

multiple (multiples) A multiple of a number is the product of that number and any whole number.

15 is a multiple of 5 because $3 \times 5 = 15$.

multiplication An operation on two numbers to find their product. It can be thought of as repeated addition. 4×3 is another way to write the sum of four 3s, which is $3 + 3 + 3 + 3$ or 12.

sistema métrico (SI) Sistema decimal de medidas que incluye unidades como el metro, el gramo y el litro.

milla (mi) Unidad usual de longitud.

1 milla = 5,280 pies

miligramo (mg) Unidad métrica de masa.

1,000 miligramos = 1 gramo

mililitro (mL) Unidad métrica de capacidad.

1,000 mililitros = 1 litro

milímetro (mm) Unidad métrica de longitud.

1,000 milímetros = 1 metro

número mixto Número formado por un número natural y una parte fraccionaria. $3\frac{1}{2}$ es un número mixto.

múltiplo Un múltiplo de un número es el producto de ese número por cualquier otro número natural.

15 es múltiplo de 5 porque $3 \times 5 = 15$.

multiplicación Operación entre dos números para hallar su producto. También se puede interpretar como una suma repetida. 4×3 es otra forma de escribir la suma de cuatro veces 3, la cual es $3 + 3 + 3 + 3$ o 12.

Nn

net A two-dimensional pattern of a three-dimensional figure.

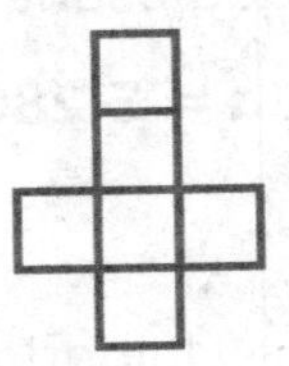

modelo plano Patrón bidimensional de una figura tridimensional.

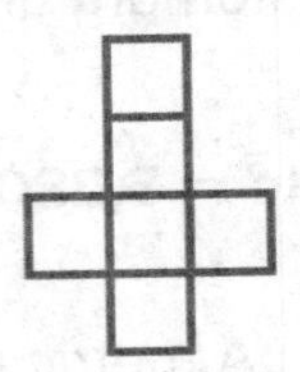

number line A line that represents numbers as points.

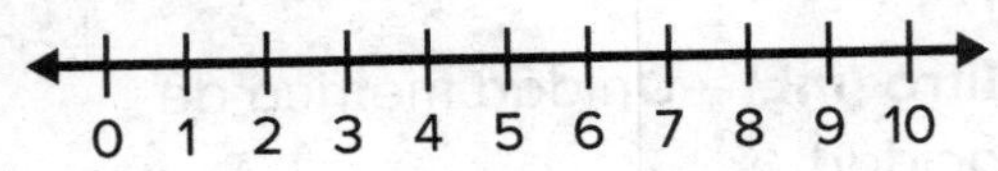

recta numérica Recta que representa números como puntos.

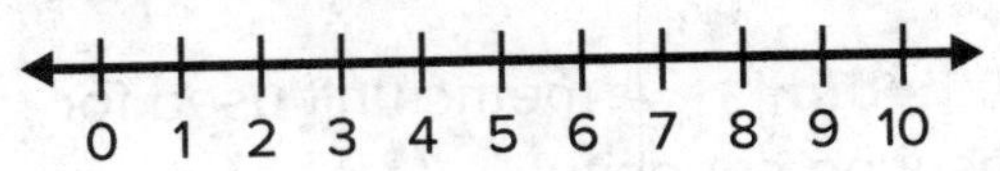

numerator The top number in a fraction; the part of the fraction that tells the number of parts you have.

numerador Número que se escribe sobre la barra de fracción; la parte de la fracción que indica el número de partes que hay.

numerical expression A combination of numbers and operations.

expresión numérica Combinación de números y operaciones.

Oo

obtuse angle An angle that measures between 90° and 180°.

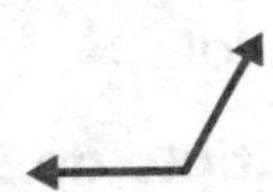

ángulo obtuso Ángulo que mide entre 90° y 180°.

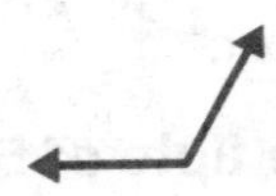

obtuse triangle A *triangle* with one *obtuse angle*.

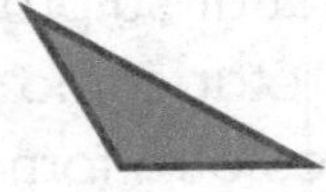

triángulo obtusángulo *Triángulo* con un *ángulo obtuso*.

octagon A polygon with eight sides.

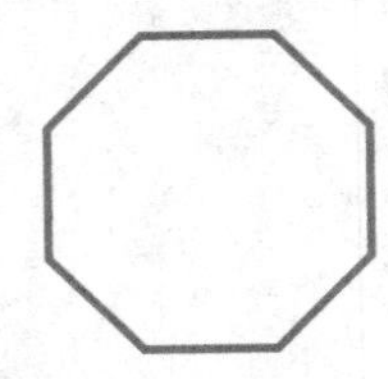

octágono Polígono de ocho lados.

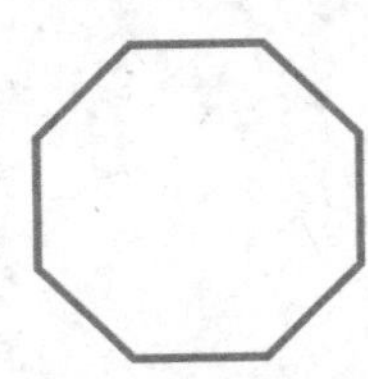

odd number A number that is not divisible by 2; such a number has 1, 3, 5, 7, or 9 in the ones place.

número impar Número que no es divisible entre 2. Los números impares tienen 1, 3, 5, 7, o 9 en el lugar de las unidades.

order of operations A set of rules to follow when more than one operation is used in an expression.

1. Perform operations in parentheses.
2. Find the value of exponents.
3. Multiply and divide in order from left to right.
4. Add and subtract in order from left to right.

orden de las operaciones Conjunto de reglas a seguir cuando se usa más de una operación en una expresión.

1. Realiza las operaciones dentro de los paréntesis.
2. Halla el valor de las potencias.
3. Multiplica y divide de izquierda a derecha.
4. Suma y resta de izquierda a derecha.

ordered pair A pair of numbers that is used to name a point on the coordinate plane.

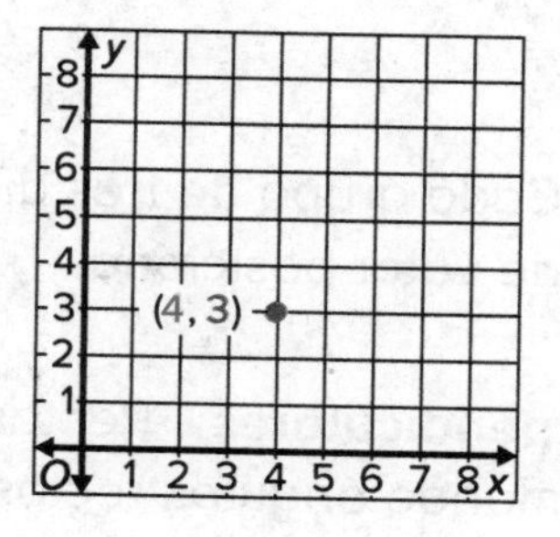

par ordenado Par de números que se usa para nombrar un punto en un plano de coordenadas.

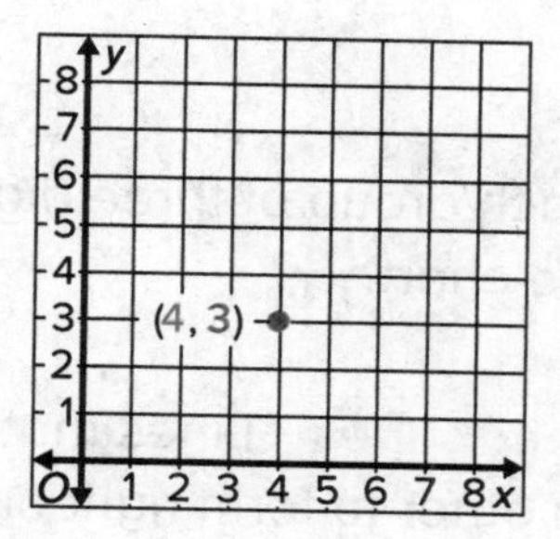

origin The point (0, 0) on a coordinate plane where the vertical axis meets the horizontal axis.

origen El punto (0, 0) en un plano de coordenadas donde el eje vertical interseca el eje horizontal.

ounce (oz) A customary unit for measuring weight or capacity.

onza (oz) Unidad usual de peso o capacidad.

Pp

parallel lines Lines that are the same distance apart. Parallel lines do not meet.

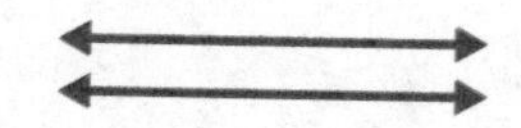

rectas paralelas Rectas separadas por la misma distancia en cualquier punto. Las rectas paralelas no se intersecan.

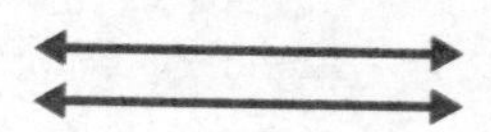

Pp

parallelogram A quadrilateral with four sides in which each pair of opposite sides are parallel and congruent.

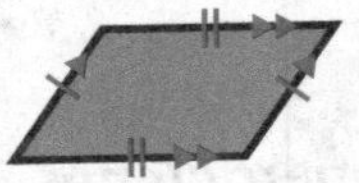

paralelogramo Cuadrilátero en el cual cada par de lados opuestos son paralelos y congruentes.

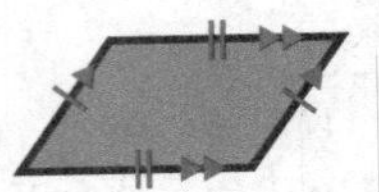

partial quotients A method of dividing where you break the dividend into sections that are easy to divide.

cocientes parciales Método de división por el cual se descompone el dividendo en secciones que son fáciles de dividir.

pentagon A polygon with five sides.

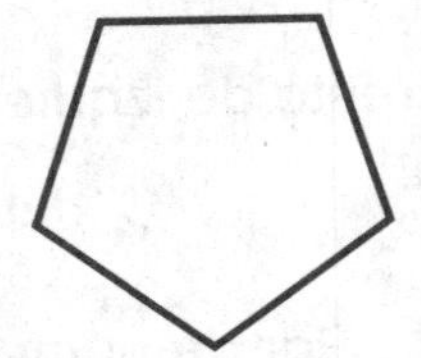

pentágono Polígono de cinco lados.

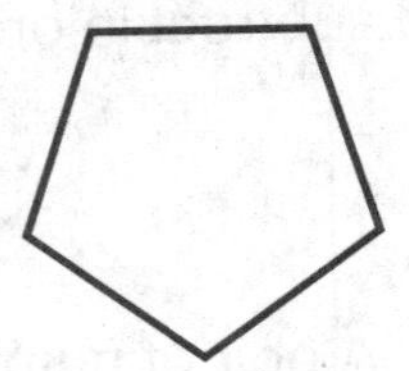

perimeter The *distance* around a polygon.

perímetro *Distancia* alrededor de un polígono.

period Each group of three digits on a place-value chart.

período Cada grupo de tres dígitos en una tabla de valor posicional.

perpendicular lines Lines that meet or cross each other to form right angles.

rectas perpendiculares Rectas que se cruzan formando ángulos rectos.

pint (pt) A customary unit for measuring capacity.

1 pint = 2 cups

pinta (pt) Unidad usual de capacidad.

1 pinta = 2 tazas

place The position of a digit in a number.

posición Lugar que ocupa un dígito en un número.

place value The value given to a digit by its position in a number.

valor posicional Valor dado a un dígito según su posición en el número.

place-value chart A chart that shows the value of the digits in a number.

tabla de valor posicional Tabla que muestra el valor de los dígitos en un número.

plane A flat surface that goes on forever in all directions.

point An exact location in space that is represented by a dot.

polygon A closed figure made up of line segments that do not cross each other.

positive number Number greater than zero.

pound (lb) A customary unit for measuring weight or mass.

1 pound = 16 ounces

power A number obtained by raising a base to an exponent.

$5^2 = 25$ 25 is a power of 5.

power of 10 A number like 10, 100, 1,000 and so on. It is the result of using only 10 as a factor.

prime factorization A way of expressing a composite number as a product of its prime factors.

prime number A whole number with exactly two factors, 1 and itself.

7, 13, and 19

prism A three-dimensional figure with two parallel, congruent faces, called bases. At least three faces are rectangles.

product The answer to a multiplication problem.

plano Superficie plana que se extiende infinitamente en todas direcciones.

punto Ubicación exacta en el espacio que se representa con una marca puntual.

polígono Figura cerrada compuesta por segmentos de recta que no se intersecan.

número positivo Número mayor que cero.

libra (lb) Unidad usual de peso.

1 libra = 16 onzas

potencia Número que se obtiene elevando una base a un exponente.

$5^2 = 25$ 25 es una potencia de 5.

potencia de 10 Número como 10, 100, 1,000, etc. Es el resultado de solo usar 10 como factor.

factorización prima Manera de escribir un número compuesto como el producto de sus factores primos.

número primo Número natural que tiene exactamente dos factores: 1 y sí mismo.

7, 13 y 19

prisma Figura tridimensional con dos caras congruentes y paralelas llamadas bases. Al menos tres caras son rectangulares.

producto Repuesta a un problema de multiplicación.

Pp

proper fraction A fraction in which the numerator is less than the denominator.

$\frac{1}{2}$

property A rule in mathematics that can be applied to all numbers.

protractor A tool used to measure and draw angles.

fracción propia Fracción en la que el numerador es menor que el denominador.

$\frac{1}{2}$

propiedad Regla de las matemáticas que puede aplicarse a todos los números.

transportador Instrumento que se usa para medir y trazar ángulos.

Qq

quadrilateral A polygon that has 4 sides and 4 angles.

quart (qt) A customary unit for measuring capacity.

1 quart = 4 cups

quotient The result of a division problem.

cuadrilátero Polígono con 4 lados y 4 ángulos.

cuarto (ct) Unidad usual de capacidad.

1 cuarto = 4 tazas

cociente Resultado de un problema de división.

Rr

ray A line that has one endpoint and goes on forever in only one direction.

rectangle A quadrilateral with four right angles; opposite sides are equal and parallel.

rectangular prism A prism that has six rectangular bases.

semirrecta Parta de una recta que tiene un extremo que se extiende infinitamente en una sola dirección.

rectángulo Cuadrilátero con cuatro ángulo rectos; los lados opuestos son iguales y paralelos.

prisma rectangular Prisma que tiene bases rectangulares.

regular polygon A polygon in which all sides are congruent and all angles are congruent.

remainder The number that is left after one whole number is divided by another.

rhombus A *parallelogram* with four *congruent sides.*

right angle An angle with a measure of 90°.

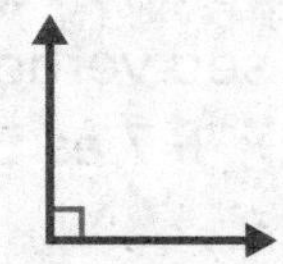

right triangle A *triangle* with one *right angle.*

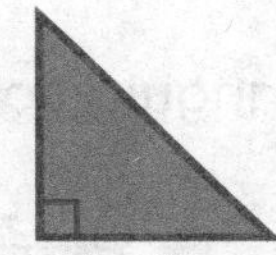

rounding To find the approximate value of a number.

6.38 rounded to the nearest tenth is 6.4.

polígono regular Polígono que tiene todos los lados congruentes y todos los ángulos congruentes.

residuo Número que queda después de dividir un número natural entre otro.

rombo *Paralelogramo* con cuatro *lados congruentes.*

ángulo recto Ángulo que mide 90°.

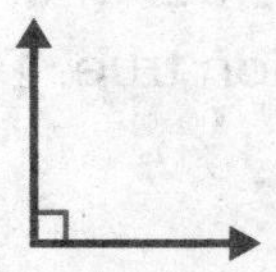

triángulo rectángulo *Triángulo* con un *ángulo recto.*

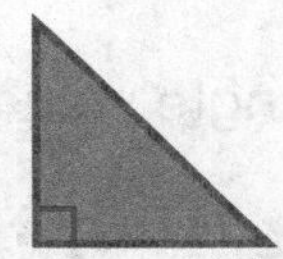

redondear Hallar el valor aproximado de un número.

6.38 redondeado a la décima más cercana es 6.4.

scale A set of numbers that includes the least and greatest values separated by equal intervals.

scalene triangle A *triangle* with no *congruent sides.*

escala Conjunto de números que incluye los valores menor y mayor separados por intervalos iguales.

triángulo escaleno *Triángulo* sin *lados congruentes.*

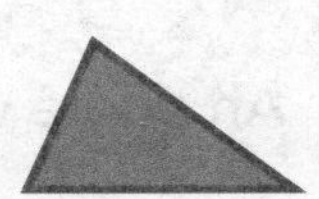

Ss

scaling The process of resizing a number when it is multiplied by a fraction that is greater than or less than 1.

simplificar Proceso de redimensionar un número cuando se multiplica por una fracción que es mayor que o menor que 1.

sequence A list of numbers that follow a specific pattern.

secuencia Lista de números que sigue un patrón específico.

simplest form A fraction in which the GCF of the numerator and the denominator is 1.

forma simplificada Fracción en la cual el M.C.D. del numerador y del denominador es 1.

solution The value of a variable that makes an equation true. The solution of $12 = x + 7$ is 5.

solución Valor de una variable que hace que la ecualción sea verdadera. La solución de $12 = x + 7$ es 5.

solve To replace a variable with a value that results in a true sentence.

resolver Remplazar una variable por un valor que hace que la expresión sea verdadera.

square A rectangle with four *congruent sides.*

cuadrado Rectángulo con cuatro *lados congruentes.*

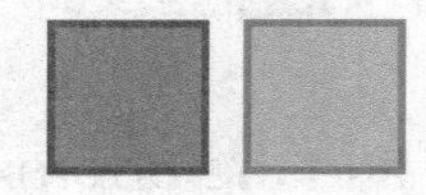

square number A number with two identical factors.

número al cuadrado Número con dos factores idénticos.

square unit A unit for measuring area, such as square inch or square centimeter.

unidad cuadrada Unidad de área, como una pulgada cuadrada o un centímetro cuadrado.

squared A number raised to the second power; 3×3, or 3^2.

al cuadrado Número elevado a la segunda potencia; 3×3 o 3^2.

standard form The usual or common way to write a number using digits.

forma estándar Manera usual o común de escribir un número usando dígitos.

straight angle An angle with a measure of 180°.

ángulo llano Ángulo que mide 180°.

sum The answer to an addition problem.

suma Respuesta que se obtiene al sumar.

Tt

tenth A place value in a decimal number or one of ten equal parts or $\frac{1}{10}$.

term A number in a pattern or sequence.

thousandth(s) One of a thousand equal parts or $\frac{1}{1000}$. Also refers to a place value in a decimal number. In the decimal 0.789, the 9 is in the thousandths place.

three-dimensional figure A figure that has length, width, and height.

ton (T) A customary unit for measuring weight. 1 ton = 2,000 pounds

trapezoid A quadrilateral with exactly one pair of parallel sides.

triangle A polygon with three sides and three angles.

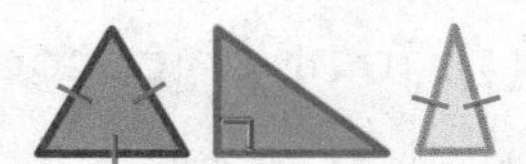

triangular prism A prism that has triangular bases.

décima Valor posicional en un número decimal o una de diez partes iguales o $\frac{1}{10}$.

término Cada número en un patrón o una secuencia.

milésima(s) Una de mil partes iguales o $\frac{1}{1000}$. También se refiere a un valor posicional en un número decimal. En el decimal 0.789, el 9 está en el lugar de las milésimas.

figura tridimensional Figura que tiene largo, ancho y alto.

tonelada (T) Unidad usual de peso. 1 tonelada = 2,000 libras

trapecio Cuadrilátero con exactamente un par de lados paralelos.

triángulo Polígono con tres lados y tres ángulos.

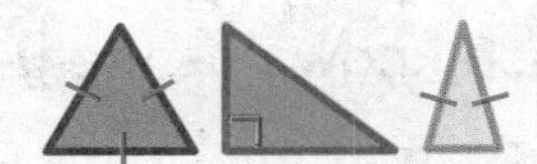

prisma triangular Prisma con bases triangulares.

unit cube A cube with a side length of one unit.

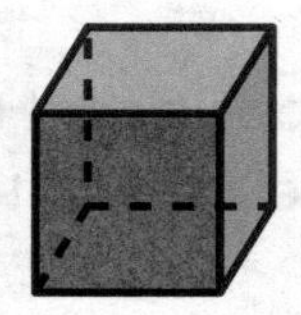

cubo unitario Cubo con lados de una unidad de longitud.

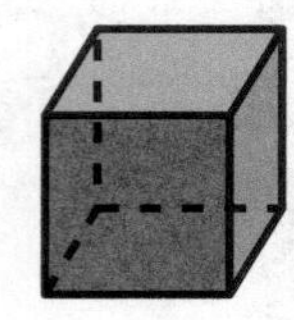

unit fraction A fraction with 1 as its numerator.

fracción unitaria Fracción que tiene 1 como su numerador.

unknown A missing value in a number sentence or equation.

incógnita Valor que falta en una oración numérica o una ecuación.

unlike fractions Fractions that have different denominators.

fracciones no semejantes Fracciones que tienen denominadores diferentes.

variable A letter or symbol used to represent an unknown quantity.

variable Letra o símbolo que se usa para representar una cantidad desconocida.

vertex The point where two rays meet in an angle or where three or more faces meet on a three-dimensional figure.

vértice a. Punto donde se unen los dos lados de un ángulo. b. Punto en una figura tridimensional donde se intersecan 3 o más aristas.

vertical axis A vertical number line on a graph (↕). Also known as the y-axis.

eje vertical Recta numérica vertical en una gráfica (↕). También conocido como eje y.

volume The amount of space inside a three-dimensional figure.

volumen Cantidad de espacio que contiene una figura tridimensional.

weight A measurement that tells how heavy an object is.

peso Medida que indica cuán pesado o liviano es un cuerpo.

***x*-axis** The horizontal axis (↔) in a coordinate plane.

eje *x* Eje horizontal (↔) en un plano de coordenadas.

***x*-coordinate** The first part of an ordered pair that indicates how far to the right of the *y*-axis the corresponding point is.

coordenada *x* Primera parte de un par ordenado; indica a qué distancia a la derecha del eje *y* está el punto correspondiente.

yard (yd) A customary unit of length equal to 3 feet or 36 inches.

yarda (yd) Unidad usual de longitud igual a 3 pies o 36 pulgadas.

***y*-axis** The vertical axis (↕) in a coordinate plane.

eje *y* Eje vertical (↕) en un plano de coordenadas.

***y*-coordinate** The second part of an ordered pair that indicates how far above the *x*-axis the corresponding point is.

coordenada *y* Segunda parte de un par ordenado; indica a qué distancia por encima del eje *x* está el punto correspondiente.

Name ______________________________

Work Mat 1: Number Lines

0 1 2 3 4 5 6 7 8 9 10

0 1 2 3 4 5 6 7 8 9 10 11 12 13 14 15 16 17 18 19 20

Work Mat 3: Centimeter Grid

Name ....

Work Mat 4: Place-Value Chart (Hundreds to Thousandths)

Ones			Decimals		
hundreds	tens	ones	tenths	hundredths	thousandths

Work Mat 5: Tenths and Hundredths Models

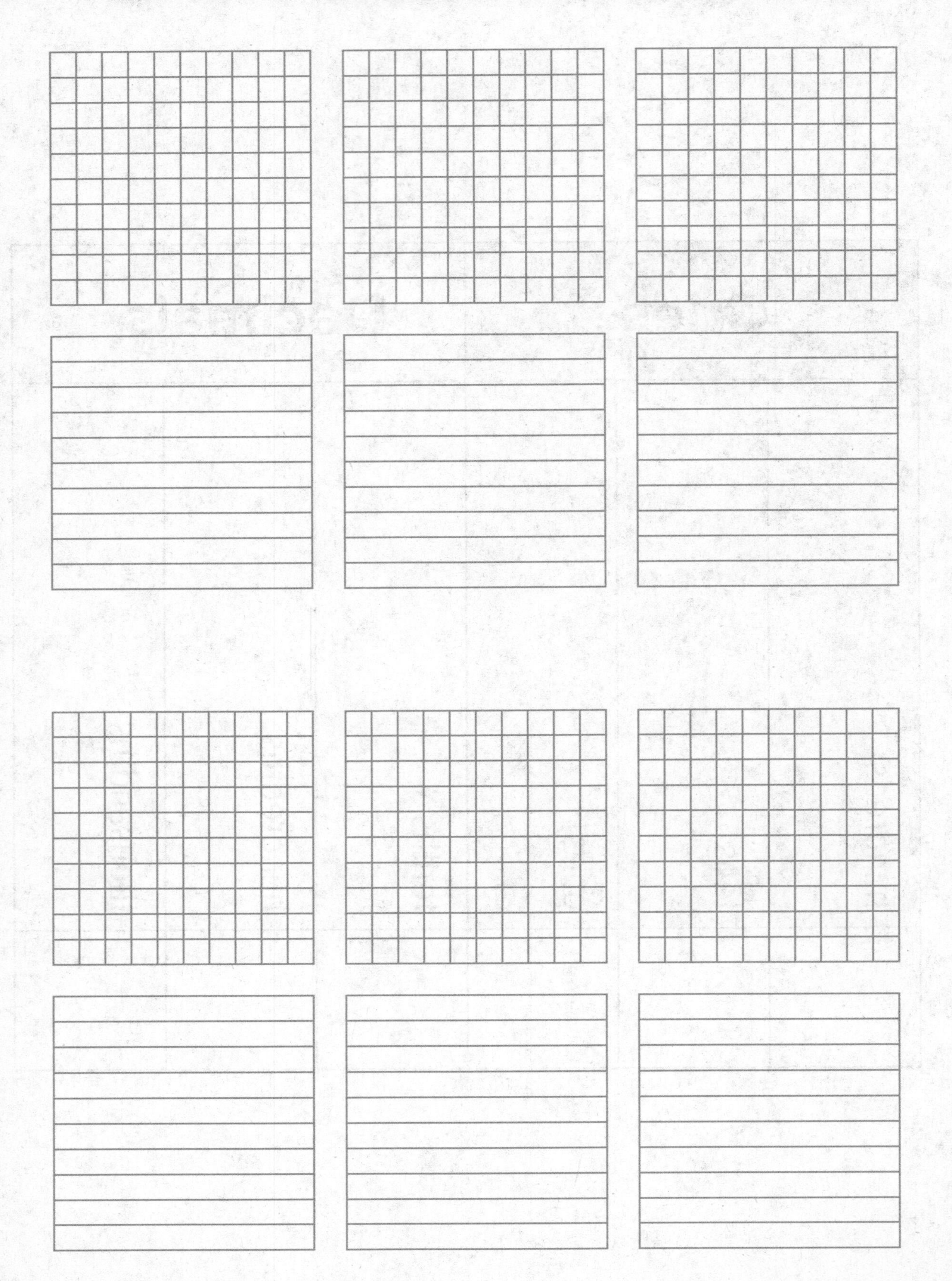

Name

Work Mat 6: Algebra Mat

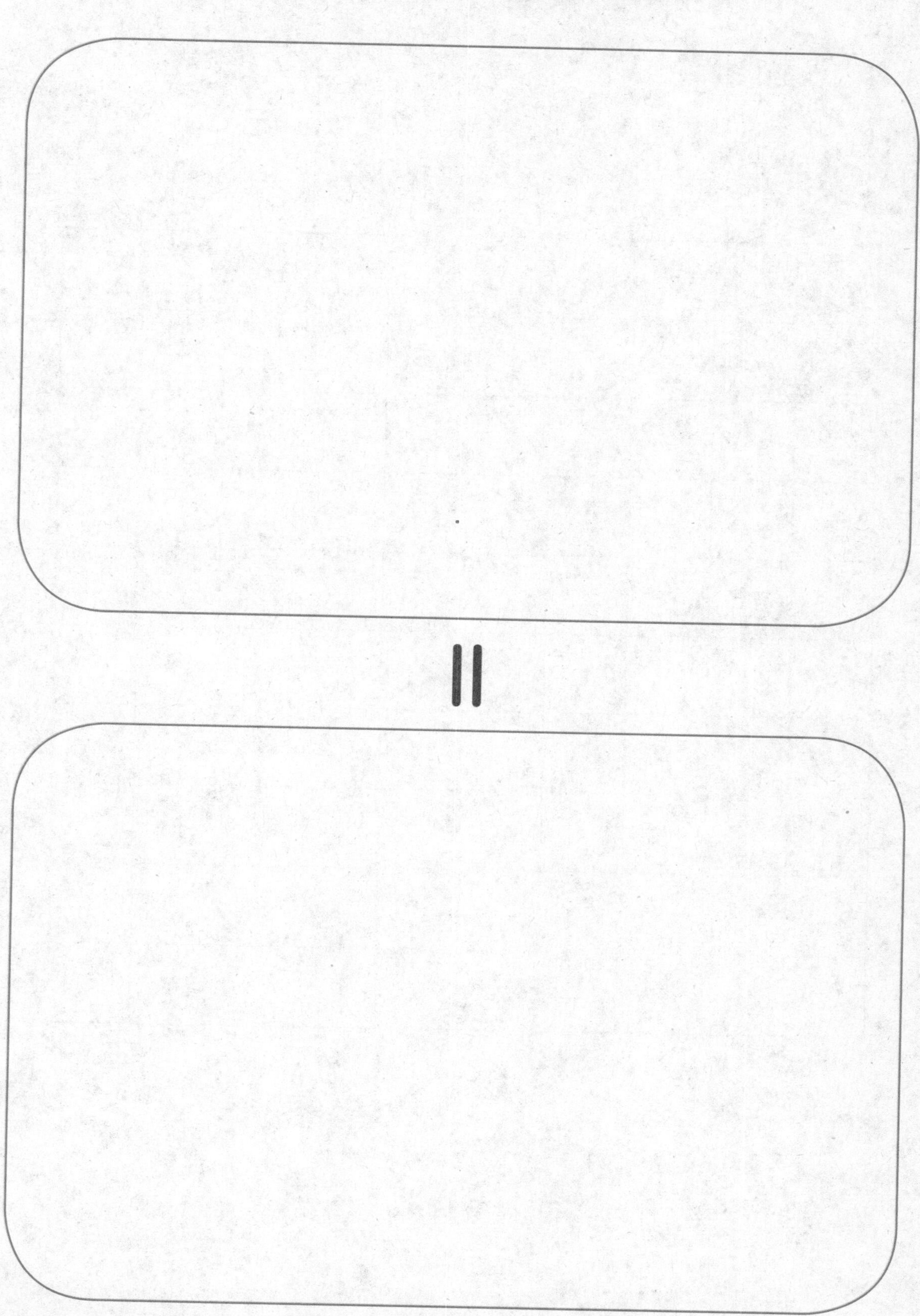

Work Mat 8: First-Quadrant Grid (blank)